Astrobiology

Current, Evolving and Emerging Perspectives

Edited by

André Antunes

Macau University of Science and Technology
Macau SAR
China

Caister Academic Press   www.caister.com

Copyright © 2020
Caister Academic Press, UK
www.caister.com

All rights reserved. No part of this publication may be reproduced, stored in a retrieval system, or transmitted, in any form or by any means, electronic, mechanical, photocopying, recording or otherwise, without the prior permission of the publisher. No claim to original government works.

ISBN: 978-1-912530-30-4 (paperback)
ISBN: 978-1-912530-31-1 (ebook)

DOI: https://doi.org/10.21775/9781912530304

Contents

1. Following the Astrobiology Roadmap: Origins, Habitability and Future Exploration 1
   Aubrie O'Rourke, Angela Zoumplis, Paul Wilburn, Michael D. Lee, Zhi Lee, Marissa Vecina and Kysha Mercader

2. Are we There Yet? Understanding Interplanetary Microbial Hitchhikers using Molecular Methods 33
   Alexander J. Probst and Parag Vaishampayan

3. Detection of Organic Matter and Biosignatures in Space Missions 53
   Zita Martins

4. Microbial Life in Impact Craters 75
   Charles S. Cockell, Gordon Osinski, Haley Sapers, Alexandra Pontefract and John Parnell

5. Impact of Simulated Martian Conditions on (Facultatively) Anaerobic Bacterial Strains from Different Mars Analogue Sites 103
   Kristina Beblo-Vranesevic, Maria Bohmeier, Sven Schleumer, Elke Rabbow, Alexandra K. Perras, Christine Moissl-Eichinger, Petra Schwendner, Charles S. Cockell, Pauline Vannier, Viggo T. Marteinsson, Euan P. Monaghan, Andreas Riedo, Pascale Ehrenfreund, Laura Garcia-Descalzo, Felipe Gómez, Moustafa Malki, Ricardo Amils, Frédéric Gaboyer, Keyron Hickman-Lewis, Frances Westall, Patricia Cabezas, Nicolas Walter and Petra Rettberg

6. Exploring Deep-Sea Brines as Potential Terrestrial Analogues of Oceans in the Icy Moons of the Outer Solar System 123
   André Antunes, Karen Olsson-Francis and Terry J. McGenity

7. Exploring Microbial Activity in Low-pressure Environments 163
   Petra Schwendner and Andrew C. Schuerger

8. Earth's Stratosphere and Microbial Life 197
   Priya DasSarma, André Antunes, Marta Filipa Simões and Shiladitya DasSarma

Current books of interest

- Microbial Biofilms: Current Research and Practical Implications   2020
- *Chlamydia* Biology: From Genome to Disease   2020
- Bats and Viruses: Current Research and Future Trends   2020
- SUMOylation and Ubiquitination: Current and Emerging Concepts   2019
- Avian Virology: Current Research and Future Trends   2019
- Microbial Exopolysaccharides: Current Research & Developments   2019
- Polymerase Chain Reaction: Theory and Technology   2019
- Pathogenic Streptococci: From Genomics to Systems Biology and Control   2019
- Insect Molecular Virology: Advances and Emerging Trends   2019
- Methylotrophs and Methylotroph Communities   2019
- Prions: Current Progress in Advanced Research (Second Edition)   2019
- Microbiota: Current Research and Emerging Trends   2019
- Microbial Ecology   2019
- Porcine Viruses: From Pathogenesis to Strategies for Control   2019
- *Lactobacillus* Genomics and Metabolic Engineering   2019
- Cyanobacteria: Signaling and Regulation Systems   2018
- Viruses of Microorganisms   2018
- Protozoan Parasitism: From Omics to Prevention and Control   2018
- Genes, Genetics and Transgenics for Virus Resistance in Plants   2018
- Plant-Microbe Interactions in the Rhizosphere   2018
- DNA Tumour Viruses: Virology, Pathogenesis and Vaccines   2018
- Pathogenic *Escherichia coli*: Evolution, Omics, Detection and Control   2018
- Postgraduate Handbook   2018
- Enteroviruses: Omics, Molecular Biology, and Control   2018
- Molecular Biology of Kinetoplastid Parasites   2018
- Bacterial Evasion of the Host Immune System   2017
- Illustrated Dictionary of Parasitology in the Post-Genomic Era   2017
- Next-generation Sequencing and Bioinformatics for Plant Science   2017
- Brewing Microbiology: Current Research, Omics and Microbial Ecology   2017
- Metagenomics: Current Advances and Emerging Concepts   2017
- The CRISPR/Cas System: Emerging Technology and Application   2017
- *Bacillus*: Cellular and Molecular Biology (Third edition)   2017
- Cyanobacteria: Omics and Manipulation   2017
- Foot-and-Mouth Disease Virus: Current Research and Emerging Trends   2017
- *Staphylococcus*: Genetics and Physiology   2016

www.caister.com

Preface

We live in an exciting era of scientific exploration. The upcoming wave of new space missions is strongly focused on the exploration of Mars (with planned sample-return) and the study of the oceans in the icy moons of Jupiter and Saturn. Since its dawn, Humankind has often wondered whether we are alone in the Universe. In the next few years, research in the booming field of Astrobiology will bring us closer than ever to finally getting an answer.

Astrobiology combines approaches from Biology, Geology, Chemistry, and Planetary Sciences to study the origin and development of Life in the universe, and search for extinct and currently existing organisms in other worlds. Research in Astrobiology is heavily anchored on the study of microbes from terrestrial analogue sites. Gaining new insights into how life copes with such extreme conditions or whether such extremophiles can survive under extraterrestrial conditions is vital for future missions.

The last decades have brought remarkable scientific advances and have shattered our long-standing misconceptions about Life's diversity and resilience. As a result of the introduction, and increased use and sophistication of molecular-based approaches, we have now realized that microbial life thrives under several of the most extreme conditions present on Earth, many of which previously thought to be anathema to Life. The physical-chemical similarities shared between some of these sites and other parts of our Solar System has brought new hope to the possible existence of Life on Mars or in the oceans of several icy moons of the outer Solar System (namely Europa and Enceladus). Microbes are now known to survive exposure to space travel, and even impact events.

The pervasiveness and new-found resilience of microbes raise some unexpected challenges. They can pose serious contamination risks associated with space missions, which might compromise results of experiments looking for the detection of organic matter and biosignatures- evidence of present or past life in other worlds. In a more extreme scenario, transport of microbial hitchhikers and contamination of other worlds could lead to the collapse of entire ecosystems before we even know that they exist.

This book combines the views of several leading experts across the globe and provides a current overview of this exciting cross-disciplinary research field. Its publication is rather timely, given the increased visibility and relevance of this area, and the upcoming wave of challenges and opportunities resulting from the new age of exploration of our Solar System.

I would like to take this opportunity to acknowledge Hugh Griffin, from Caister Academic Press, not only for challenging me to edit this book but also for his helpful support throughout the whole process. I thank all authors and reviewers for their availability to contribute to this process and for working together to producing such a remarkable book. Finally, I would also like to thank my family for their undying encouragement throughout the years, and my beloved wife Marta for her unwavering support and help in putting this project together.

André Antunes
Macau

Chapter 1

Following the Astrobiology Roadmap: Origins, Habitability and Future Exploration

Aubrie O'Rourke[1†*], Angela Zoumplis[1,2†], Paul Wilburn[3†], Michael D. Lee[3], Zhi Lee[4], Marissa Vecina[4] and Kysha Mercader[4]

[1]J. Craig Venter Institute, La Jolla, CA, USA
[2]Scripps Institution of Oceanography, La Jolla, CA, USA
[3]NASA Ames Research Center, Moffett Blvd, Mountain View, CA 94035, USA
[4]UCSD, School of Engineering, Senior Design Group, La Jolla, CA, USA

*aorourke@jcvi.org   †equal contribution

DOI: https://doi.org/10.21775/9781912530304.01

Abstract

Astrobiology asks three fundamental questions as outlined by the NASA Astrobiology Roadmap: 1. How did Life begin and evolve?; Is there Life elsewhere in the Universe?; and, What is the future of Life on Earth? As we gain perspective on how Life on Earth arose and adapted to its many niches, we too gain insight into how a planet achieves habitability. Here on Earth, microbial Life has evolved to exist in a wide range of habitats from aquatic systems to deserts, the human body, and the International Space Station (ISS). Landers, rovers, and orbiter missions support the search for signatures of Life beyond Earth, by generating data on surface and subsurface conditions of other worlds. These have provided evidence for water activity, supporting the potential for extinct or extant Life. To investigate the putative ecologies of these systems, we study extreme environments on Earth. Several locations on our planet provide analog settings to those we have detected or expect to find on neighboring and distant worlds. Whereas, the field of space biology uses the ISS and low gravity analogs to gain insight on how transplanted Earth-evolved organisms will respond to extraterrestrial environments. Modern genomics allows us to chronicle the genetic makeup of such organisms and provides an understanding of how Life adapts to various extreme environments.

Introduction

Several efforts, such as the Human Microbiome Project (HMP), the Earth Microbiome Project, and Microbes of the Built Environment (MoBE) research have contributed an understanding of the pervasiveness of the microbial world. Select microbes can survive the hyper-arid environment of Chile's Atacama Desert, the low temperatures of Antarctica, and the high radiation experienced on the exterior surfaces of the ISS. Such extreme environments found on Earth and in the outer limits of its atmosphere serve as analogs for the environments likely to be encountered in transit to and on other worlds. While we are in the early phases of hypothesizing on the nature of extraterrestrial Life, we must be particularly mindful of the potential for our Earth-evolved microbes to affect or inhabit extraterrestrial habitats (e.g. forward contamination), as we explore signatures of Life in the solar system. The goal of sending humans, deeper than ever before, into our solar system entails addressing additional questions and challenges.

The field of space biology uses phenotypic and genotypic characterization to better understand the effect the space environment has on Earth-evolved organisms. In the current -omics era, we have the capacity to better investigate the questions posed by both astrobiologists and space biologists alike. Next-generation sequencing (NGS) technologies have allowed us to understand the effects that environmental conditions, like those experienced on Earth in extreme environments or outside of our atmosphere, have upon the genomes, transcriptomes, and proteomes of prokaryotic and eukaryotic populations. In this chapter, we follow the Astrobiology Roadmap (Des Marais et al., 2008) by first describing the origin of Life on Earth, and then by discussing how life evolved to survive in extreme Earth environments (Figure 1). In the final section, we present the important role Planetary Protection plays in shaping the future of space exploration and space biology research.

Origins: how did Life begin and evolve?

There are two primary theories as to how Life came to be on Earth: I) the prebiotic theory and II) the theory of panspermia. The prebiotic or primordial soup theory, contends that chemical evolution involving the CNOP elements (carbon, nitrogen, oxygen, and phosphorus) occurred in response to physical and chemical selective pressures impressed by early Earth's reducing atmosphere (Farmer et al. 2018). The primordial soup was greatly affected by the energy inputs coming from meteor impacts, volcanic eruptions, lightning, and super tides. Utilizing these energy inputs, chemical elements formed monomers through the process known as abiogenesis (Emeline et al. 2003). These monomers would later polymerize into the four major macromolecules: carbohydrates, lipids, proteins, and nucleic acids – and eventually begin to self-replicate and allow for the formation of organic Life as we know it.

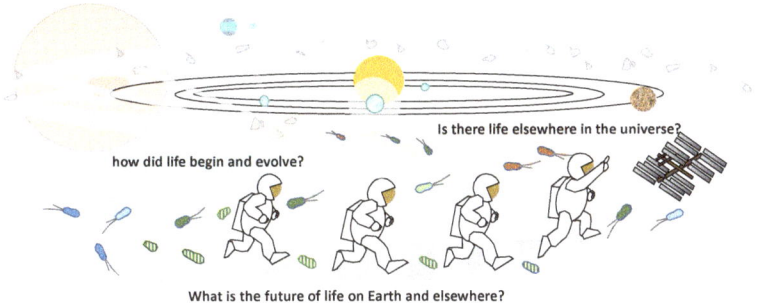

Figure 1. The Astrobiology Roadmap.

A proposed setting for the occurrence of abiogenesis is at the site of hydrothermal vents, where elements from the Earth's core could be shaped by ocean chemistry to form organic molecules (Martin et al., 2008). After millions of years of chemical evolution, circumstances may have arisen enabling a semipermeable membrane to encapsulate a self-replicating molecule, such as RNA, DNA, or possibly prion-like proteins. These progenitors of modern cells served to concentrate chemical reactions, which allowed cellular compartmentalization to continue to evolve. Then around 3.6 billion years ago, a successful combination of macromolecules within a membrane is believed to have led to, what is now known as, the Last Universal Common Ancestor (LUCA) - the cell from which all currently known biological cells arose (Koonin, 2003).

The alternative theory of panspermia contends that complex, organic macromolecules, or even possibly LUCA, was seeded to the Earth from another planet or celestial body. In our solar system, we have the four terrestrial or rocky planets closest to the sun – Mercury, Venus, Earth, and Mars; farther away are the four gas giants or Jovian planets: Jupiter, Saturn, Uranus, and Neptune. Each of these planets formed as it attracted matter into its gravitational pull through the process of accretion. The material that did not form planets such as rock, dust, and ice, instead formed the asteroids, comets, and dwarf planets. Asteroids are predominantly rocky, and mostly occupy the asteroid belt between the rocky inner planets and the outer Jovian planets. Whereas, comets are predominantly composed of ice (Lissauer,1993). In the 4.6 billion years since the accretion of our planet, Earth has experienced 5 mass extinctions punctuated by meteorite (asteroid and comets that have entered our atmosphere) impacts. These Earth- changing impacts define the geological time periods (Sleep et al. 1989). During the Hadean Period, 4.6 to 4 billion years ago (unit of time also known as Gya), the Earth and the rest of the inner terrestrial planets experienced a time known as the Late Heavy Bombardment – where there was an increased frequency of impact events from asteroids and other impactors (Fassett et al., 2013). Many of these, struck with enough force to cause fragments to be expelled from the planets' atmospheres. Our moon itself, formed as the result of a body estimated to be roughly the size of Mars, known as Theia, slamming into Earth (Herwartz et al., 2014).

Asteroids and comets are the two vehicles that may have helped enrich our early planet and Mars with water (Lunine et al., 2003; Raymond et al., 2004). Asteroids are also known to hold amino acids and other biomolecules (Kvenvolden et al., 1970). Meteorites have been argued to contain evidence for life by some, though others contend that their properties can arise entirely abiotically (Martel et al.,2012). Regardless of its origins, Life on Earth was able to take a foothold due to a reduction in the frequency of impacts, a sustained atmosphere and magnetosphere, and other properties of Earth, which allowed for a more stable environment for evolution. Mars, in contrast, lost its magnetic field and atmosphere (Dehant et al., 2007) relatively quickly. With the loss of this critical protection, it was no longer shielded from the Sun's UV light and radiation. In some respects, Mars could be viewed as an old Earth.

Search for past life on Mars
The exploration of Mars began with the Viking rover in 1976 then flybys by the Odyssey satellite in 2001. This continued with the lander missions which included the Opportunity rover in 2003 and later the Curiosity rover in 2011. Curiosity had the ability to drill shallow holes (5 centimeters deep), laser-blast rock surfaces, and perform mass spectrometry; yet, these efforts revealed no evidence of extant Life. However, as Curiosity surveyed the Gale Crater on Mars, it collected evidence to suggest that water once existed at the site (Martín-Torres et al., 2015). If we are to obey the Earth adage of "follow the water", as water fosters Life, we can hypothesize that at one point in the history of Mars, it may have been habitable. The layered silica sediments, as well as images of what the rocks look like, tell a story of a once wet Mars where Gale Crater was the site of an ancient spring-fed lake (Cabrol et al.,1999). The presence of silica in Gale Crater also suggested it would be an ideal location to find preserved biosignatures (Williams et al., 2013).

Additional evidence for organic Life on Mars includes the presence of carbonates in Martian dust; these carbonates are a sign of water and carbon dioxide, either past or present (Bridges et al., 2019). Carbonates can be generated by biotic processes as well as abiotic processes. On Earth, the bacterial carbonate deposits known as stromatolites and microbialites, serve as some of the oldest fossil evidence for biological Life (Nutman et al., 2016). In a recent discovery, NASA reported the presence of organic molecules in the rocks and methane in the atmosphere of Mars

(Ten Kate, 2018). The search for Life on Mars continues. Future projects such as MAVEN (Mars Atmosphere and Volatile Evolution mission) will provide a quantitative analysis of the gases of Mars (Jakosky et al., 2015), while next-generation rovers, such as the NASA's InSight lander and the Mars 2020 rover, as well as the European Space Agency's (ESA) ExoMars program will survey the Mars lithosphere using deeper coring methods (Vago et al., 2015). The geological results may reveal signatures reminiscent of the beginnings of Life on Earth.

Our capacity to evaluate extraterrestrial Life at the molecular and polymer level continues to build. Sequencing platforms are the workhorses of targeted amplicon, metagenomic and metatranscriptomic sequencing. Further development of NGS techniques allows for nucleic acid polymerization to be sequenced with fluorescence-based and the pore-based methods. In the fluorescence-based method, the sequence is read by tagging nucleotides with unique fluorophores. Alternatively, the pore-based method uses an electric field and a nanometer-sized pore to read the sequence of a polymer according to how the monomers alter the electric density of the current. This pore-based technology, known as Nanopore, may be able to characterize the building blocks of extraterrestrial Life, even of a different composition. Nanopore's small footprint, roughly the size of a human finger, makes it especially suitable for transport on limited capacity payloads, and it has been used to sequence DNA on the ISS (Castro-Wallace et al., 2017). These methods have been used extensively to characterize the diversity and abundance of microbes present in an environment, the constituents of their genomes, and the genes they express. Such approaches are applied to the extreme environments found on Earth as well, and help develop our understanding of how microbes survive in extraterrestrial analog settings.

Habitability: is there Life elsewhere in the universe?

Exoplanets
The astrobiology roadmap aims to guide the exploration of the nature and distribution of habitable environments within the universe. However, to begin, a distinction should be made between the commonly thought of "habitable zone" and the concept of "habitability". Herein, the "habitable zone" refers to the concept, theorized by astrophysicist Su-Shu Huang, in which a planet must orbit a star at a distance such that the energy from the star sustains liquid water on the planet's surface (Huang, 1960) – this is often referred to as the "Goldilocks zone". The term "habitability" herein, refers to being able to sustain Life from the point of view of our current information, meaning requiring liquid water, an energy source, and conditions suitable for complex chemistry (but no stellar requirements, distance-based or otherwise). The Kepler space telescope has been a key player in identifying exoplanets (extrasolar planets) which fall in the habitable zone (Batalha et al., 2014).

One of the most common methods for detecting exoplanets is known as the transit method, wherein an exoplanet may be detected as it passes in front of its parent star from our point of view. It is then sometimes possible to use transmission spectroscopy to gain insight into the atmosphere of the prospective exoplanet to look for compounds such as water vapor (Sing et al., 2011). There are an estimated 40 billion exoplanets within the habitable zone (Farmer, 2018). To date, the Kepler mission has identified greater than 3,000 and confirmed more than 100 in the habitable zone (Batalha, 2014). A number of these are in tidal lock and therefore experience temperature extremes on either side of the planet. It remains to be seen if Earth's life-generating features, such as liquid water, a Life-sustaining atmosphere, a radiation-shielding magnetosphere, and a UV-blocking ozone have been reproduced on any of these exoplanets.

Analog environments: Adaptive strategies in cold, dry environments
Icy moons and tidally locked exoplanets may hold microbial Life which can survive the cold. Metabolism and energy production are typically repressed in psychrotrophs (organisms that can survive at low temperatures but have optimal growth temperatures >20°C) in colder conditions, while psychrophiles (organisms with optimal growth temperatures <10°C) tend to increase nutrient acquisition, photosynthetic efficiency, and upregulate genes involved in metabolic pathways (Raymond-Bouchard et al., 2017). Low temperatures have been shown to cause increased expression in the following for both psychrotrophs and psychrophiles: lipid biosynthesis, osmoprotectants, cold shock, and antifreeze proteins (Raymond-Bouchard et al., 2017). Fatty acid desaturases help limit the increased rigidity of the cell membrane upon freeze events (Králová, 2017). The biosynthesis and transport of compatible solutes including glycine betaine, ectoine, and trehalose help to prevent water loss from cells as ice forms.

At the transcription and translation levels, a suite of molecular adaptations can help deal with the formation of secondary structures in mRNA. Many of these adaptations come with trade-offs for the cell, either as energy costs or by compromising performance in mesophilic environments. For example, the so-called DEAD-box RNA helicases are experimentally shown to induce ATP-dependent unwinding of RNA secondary structures formed during cold stress in *E.coli* (Cartier et al., 2010). A broad class of *csp* (e.g. *cspA*) cold-shock RNA chaperones has been detected in many Bacteria (Wouters et al. 2001) , and *ctr* (cold-responsive TRAM domain) are the small RNA binding proteins with a TRAM domain that have been proposed as putative cold-induced chaperones in some Archaea (Zhang et al., 2017). Small RNA-binding proteins (Rbps) play important roles in transcription termination, especially in cyanobacteria, and have been reported to accumulate in the wake of cold stress (Mori et al., 2003). RNA degradosomes (protein complexes built on top of RNases) have also been reported to show increased expression in the cold, presumably to recycle ribonucleotides from putatively secondarily structured RNAs. Chaperone and other accessory proteins also come at a cost, as protein synthesis is the most energetically costly process in the cell. Additionally, the non-planar structure of dihydrouridine can enhance tRNA flexibility and is elevated in some psychrophilic Bacteria and Archaea (Dalluge et al., 1997).

At the protein level, ClpB and the oxidation-activated Hsp33 are chaperones that stabilize other proteins, and have also been reported as upregulated in psychrophiles (Bakermans et al. 2009; Campanaro et al., 2011; Ponder et al., 2005). In addition to using chaperones, psychrophilic enzyme homologs increase flexibility with residue substitutions that disrupt protein stability in the surrounding aqueous solution. Proteins exist in a so-called aqueous cage, created by water molecules hydrogen bonding with one another and the solutes they dissolve. Part of what stabilizes protein tertiary structures is their folding in a way that positions hydrophilic residues on the outside, allowing them to interact with the aqueous cage, while hydrophobic residues are tucked away in the core of the protein structure. Residue substitutions that disrupt this arrangement, by decreasing hydrophobicity of the core and increasing hydrophobicity of the protein outside, results in greater flexibility of the destabilized protein. Cold-adapted proteins may also decrease cysteine content to avoid formation of disulfide bridges (Ásgeirsson et al., 2003). Psychrophilic genomes tend to have fewer charged and polar amino acids that can interact to form salt bridges or hydrogen bonds (Margesin et al., 2008). The amino acids in lower abundance include acidic residues and the basic arginine (also positively charged), lysine and histidine. Glycine, on the other hand, is small and non-polar. Although whole psychrophilic genomes have shown weak support for glycine enrichment, one study showed that glycine enrichment around domain active sites can increase protein flexibility without affecting residue ratios (Feller, 2013).

At a cellular level, two hallmark stories for cold adaptation deal with maintenance of membrane fluidity and freezing prevention. Lower temperatures are usually associated with increased unsaturated fatty acid (UFA) content and decreased saturated fatty acid (SFA) to UFA ratio (Mock et al., 2002; Morgan-Kiss et al., 2006; Valledor et al., 2013). This maintains membrane fluidity and provides an electron sink (in the form of UFA) in cold, photoinhibitive conditions. Interestingly, in most organisms UFA are synthesized by reversible desaturation of SFA, which must be made first. The advantage is that the SFA:UFA ratio is easily adjustable. In contrast, a few cold specialists have evolved *de novo* synthesis of UFA (Morgan-Kiss et al., 2006). By avoiding mandatory SFA synthesis, the *de novo* pathway may be more efficient at producing UFA at the cost of having a responsive SFA:UFA system. These represent alternate metabolic strategies for thermal adaptation, such that species adapted to constant cold would exhibit the efficient *de novo* pathway, while species exposed to variable temperatures would benefit from the less efficient but reversible standard pathways.

Evaluating analog environments
Locations on our planet which approach the limits of currently understood Life, serve as analogs to the early Earth, Mars, other icy worlds in our solar system, and plausible exoplanets. Quantifying the biological potential or habitability of candidate planets, moons, and asteroids is the first among equal goals of space exploration (Hubbard et al., 2002) central to NASA objectives (Des Marais et al., 2008). Characterization of extreme environments on Earth has been giving us many abiotic (environmental), biotic (genetic traits) and bioenergetic (reaction rates) insights. Each insight helps to parameterize a working definition of habitability, mostly aiming to constrain the search to key components – exemplified by "follow the water" and "follow the energy" narratives. Those approaches are rightfully useful, especially in scenarios where information available for distant exoplanets is restricted to chemical and perhaps isotopic composition.

Water is required for all known Life, and presence of liquid water on a planet's surface was classically recognized as a prerequisite for Life (Kasting et al., 1993) helping to usher in the "follow the water" guideline. Presence of liquid water has been the principal motivation for an interest in Europa and Enceladus, moons of Jupiter and Saturn respectively, and the recent discovery of water on the Moon is fueling a renewed interest in establishment of a lunar base (Milliken et al., 2017). Subsequent research included other abiotic factors, such as raw materials, a livable pH and temperature, and availability of light or chemical energy to drive biochemical processes – but water remains a necessary condition (Hoehler, 2007; Jones et al., 2010; Knoll et al., 2006; Tosca et al., 2008). On Mars, water exists as ice at the polar regions and in the subsurface, with the rest of the planet extremely dry and devoid of Life. In the following section, we will explore two examples of Mars analogs found here on Earth. One is the Atacama Desert which draws parallels to the arid Martian environment, and the second is the Dry Valleys of Antarctica from which we hope to learn about the ancient fluvial systems on Mars.

Analog environments: Atacama Desert, Chile
The Atacama Desert offers multiple analog conditions for Mars, as well as our Earth's Moon. It is the most ancient (90 My) and driest desert (Hartley et al, 2002; Houston et al., 2003). The ancient age of its constant environment, together with local geology, resulted in the accumulation of compounds that give Atacama's soils Mars-like properties. The hyper-arid interior of the Atacama lies at the end of a long aridity gradient, with little variation in temperature. At the dry end of that gradient, life persists in patchy endolithic communities, where photosynthesis is limited by the minimal persistence of water into daylight hours (Davilae t al., 2016; Warren-Rhodes et al., 2006). These outposts for life in the Atacama interior

are characterized by hygroscopic, or deliquescent, materials – materials that can absorb moisture from air. Deliquescent salts, including ignimbrite (Wierzchos et al., 2013), calcite or calcium carbonate (Crits-Christoph et al., 2016a) sodium chloride (Davila et al., 2010), and gypsum (Robinson et al., 2015), or calcium sulfate (Cámara et al., 2016), have been implicated in supporting habitats for endolithic cyanobacteria and a cohort of heterotrophic prokaryotes (Catling et al., 2010; Davila et al., 2008, 2010; Warren-Rhodes et al., 2007; Wierzchos et al., 2013, 2015; Wierzchos et al., 2006; Wierzchos et al., 2012). However, the Atacama is famous for its perchlorate deposits. Similarly, Mars soils are also rich in perchlorates. Catling *et al* used *in silico* modeling, which suggests that in both places perchlorates could be a result of photochemistry and ozone interacting with soil salts (Catling et al., 2010). While perchlorates can also be hygroscopic, their high oxidative reactivity provides additional challenges to survival.

One way of looking at habitability is through a biological fitness landscape. Organisms exist in fitness landscapes, dependent upon their traits, which process the environment in ways that make the environment habitable. These organismal traits are, in simplest form, often inferred descriptively through characterizing organisms or observing associations between traits and habitats. With the advent of high-throughput DNA sequencing, the associations are frequently gene-centric. Indeed, continuing discoveries of extremophiles have demonstrated the extent of habitability, and investigation of the metabolic diversity enabling this habitability has followed (Seckbach et al.,2010). Sampling of hyper-arid environments, including the Atacama, has revealed a few hallmark organisms, their community dynamics, and their metabolic potential. Members of the genus *Chroococcidiopsis* were found several millimeters below the crust surface (Wierzchos et al., 2006) and inside hygroscopic rocks (Warren-Rhodes et al., 2006). These primary producers support a cohort of heterotrophs in the Archae and Bacteria domains (Warren-Rhodes et al., 2006, 2007; Wierzchos et al., 2006).

Warren-Rhodes *et al* used molecular methods to investigate intra-genus diversity of the *Chroococcidiopsis* morphotype, with phylogenetic clades showing sampled-site specificity (Warren-Rhodes et al., 2007). Subsequent research used amplicons and shotgun metagenomics to reveal community dynamics, and assemble genes and complete genomes from organisms found in the Atacama (Crits-Christoph et al., 2016b; Crits-Christoph et al., 2016a; Wierzchos et al., 2015, 2012). Characterized organisms were shown to have a metabolic repertoire with putative features to prevent desiccation, carry out vital processes with limited water, resist UV damage, enter dormancy, and efficiently re-activate metabolism during short periods of water availability (Crits-Christoph et al., 2016b; Crits-Christoph et al., 2016a; Harel, 2004). In another study, Robinson *et al* used multivariate analyses of 16s ribosomal RNA amplicons to show trends in community composition between arid and hyper-arid sites, and identified phylotypes most responsible for microbial community differences and implicated air humidity as the strongest environmental (Robinson et al., 2015). Recent publications advanced a few more steps to infer adaptive strategies of sampled organisms. Crits-Christoph *et al* sequenced halite endoliths to reveal a simple community dominated by haloarchaea, all putatively supported by a single phylotype of cyanobacteria (Crits-Christoph et al., 2016b) (Figure 2).

In a separate publication, Crits-Christoph *et al* presented more near-complete genomes from two distinct geological sites in the Atacama, providing additional genomic information for representatives of the few dominant genera at the sampled sites – including *Gloeocapsa*, *Thermomicrobia*, *Conexibacter*, *Microlunatus*, *Frankineae*, and *Chroociccidiopsis* (Crits-Christoph et al., 2016a). For example, all organisms were enriched

Figure 2. Halites (salt rocks) are the last habitats for life in the hyperarid Atacama Desert, Chile. Three sites illustrate the aridity gradient, refleted in halite nodule appearance. An example nodule (from SG) is shown. Note the porosity, which allows light and gas exchange. Colonized areas (partly visible green areas) are in the salt portion of the rocks. Colonization is minimal in the sparse nodules from the driest Yungay site (not shown).

with genes involved in secondary metabolite production, which can serve as building blocks for conjugated hydrocarbons, isoprenoids, and other photoprotective compounds. These results were supported by Wierzchos et al's earlier detection of photoprotective secondary metabolites using chemical assays (Wierzchos et al., 2015), and qualitatively further described hyper-arid lifestyle potential.

These examples characterize the traits that are exhibited by organisms which are able to withstand life in the Atacama Desert and suggests that habitability is tied with unifying abiotic and biotic components. Understanding these traits alone offers a limited mechanistic understanding of organismal capabilities. It remains unclear which of the thousands of organismal traits contribute to their host's fitness along environmental gradients, including the extremes. Organisms exist in a fitness landscape, which is produced by those traits that process the environment in ways that ultimately allow that environment to be "habitable" for a given organism. Models of fitness that explicitly discover traits with environmental conditions need to be developed. This brings into focus the traits with significant impacts on success of their extremophilic hosts, and providing mechanistic insights into the minimal gene set and the limits of habitability.

Analog environments: McMurdo Dry Valleys, Antarctica
The McMurdo Dry Valleys (MDV) are considered a Martian analog environment due to extremes in aridity, high-speed winds, and low temperatures (Stanish, Nemergut, and McKnight, 2011). The use of omics tools in this environment has also led to new insights into the previously unknown abundance, diversity, and functional strategies of microbial life persisting in MDV ecosystems (Adams et al., 2014; Buelow et al., 2016; Kohler et al., 2015; Stanish et al., 2013; Van Horn et al., 2016). This cold, dry landscape consists of dynamic aquatic systems that provide unique niches for microbial communities. Although there are distinct community signatures between habitats, there is a significant degree of connectivity among glaciers, streams, lakes, and soils (Gooseff et al., 2011; Wlostowski et al., 2016). In this section we discuss MDV ecosystems in terms of their potential as Martian analogs, geochemical features, challenges to microbial life and diversity, and the functionality of their microbial communities.

McMurdo Dry Valley Lakes
Evidence that stable liquid water exists on Mars today, is supported by radar profiles of a subglacial brine pool below Mars' southern ice cap from the Mars Express spacecraft (Orosei et al., 2018). The Mars Advanced Radar for Subsurface and Ionosphere Sounding (MARSIS) instrument, on the Mars Express spacecraft, is an instrument designed to scan the surface of Mars looking for evidence of liquid water. Radar profiles collected by MARSIS, between 2012-2015, displayed strong basal echoes from a 20 km area in Mars' southern polar ice cap, known as South Polar Layered Deposits (SPLD). The radar profiles display bright reflections that are

indicative of an interface between ice and liquid. Further analysis of the radar data revealed a high relative dielectric permittivity (>15) of the subsurface reflection which is inferred as liquid water. The liquid water is predicted to be around 1.5km below the ice (Orosei et al., 2018). Data gathered yields support for the first known stable body of liquid water on Mars.

Decades of orbital imagery studies provide support for Martian hydrological activity with the identification of river channels, valley networks, and over 400 paleolake basins that potentially maintained bodies of water with unknown residence times (Cabrol and Grin, 1999; Goudge et al., 2016; Williams et al., 2001). Surface morphology analyses detect a variety of fluvial features that expose the diversity of these lakes in terms of formation (Cabrol et al., 1999). Lake basins can be closed, open, or exist in chains. Some possess outflow channels indicative of hypothesized flood events. Others include signs of erosion and deposition (Carr, 2012; Carr et al., 2010; Goudge et al., 2016). The geomorphic characterizations of these ancient lake basins, along with supporting Earth analog studies recreate possible geochemical features and physical processes that may have once controlled putative life on Mars.

The perennial ice-covered lakes of the MDV provide the closest Earth analog to recent subglacial pool discoveries, as well as the evolution of ancient paleolakes on Mars. Microbes that exist in the uniquely stratified MDV lakes survive in low-light levels, freezing temperatures, sharp chemical gradients, oxic and anoxic conditions, and chaotropic salts, which interfere with hydrogen bonding. Several lakes in the MDV are heavily studied for analogous characteristics to various freezing stages of Martian lakes (Mikucki et al., 2006).

Lake Vida is the largest lake in the MDV. The lake is in Victoria Valley which is northward of Wright Valley. Victoria Valley generally maintains lower temperatures than the southern valleys. During the austral summer, the area warms enough to produce short-lived meltwater streamflow. In early studies, Lake Vida was believed to have been completely frozen with no traces of water; however, in a later study by Doran et al. ,advancements in ground penetrating radar revealed a brine pocket 19 m under the ice (Doran et al., 2003). Isolated for around 2800 years, the aphotic, anoxic brine here has a temperature of -13°C and a salinity of 200 ppt (Murray et al., 2012). Further investigations of this brine system identified a bacterial-dominated microbial assemblage in the isolated cryo-environment. The obstacles posed to Life in this system would be like those experienced in the subglacial systems beyond Earth. Therefore, this discovery fundamentally informs perceptions of the ability of Life to persist on icy worlds.

McMurdo Dry Valley Meltwater streams
The meltwater streams of the MDV are an important analog to the fluvial systems that may have existed on ancient Mars (Murray et al., 2012). The valley networks in both Antarctica and Martian Noachian terrain (4.1-3.7 Gya) are comparable in surface features (Carr, 2012). Studies conducted on Martian clay mineral formation and paleolake basin stratigraphy indicate intermittent surface waters that carved valleys, transported sediments, and sustained lakes throughout the late Noachian/early Hesperian (Ehlmann et al., 2011; Goudge et al., 2016). Despite differences in nutrient availability, Martian and Dry Valley landscapes are comprised of similar basalt substrates, limited organic material, and experience long durations of dryness and freezing temperatures due to transient flow (Doran et al., 2010).

During the austral summer, rising temperatures cause glacial melting and transient freshwater streams begin to flow, downward through the valleys, carrying alluvium while rehydrating stream beds. Temperature fluctuations of the Dry Valley meltwater streams reach anywhere from 0.1 °C to a maximum of 15 °C with daily shifts between 6°C-9°C. Changes in temperature, flow duration, and speed are associated with rapid climate shifts making adaptations to a highly variable environment an important quality among inhabitants (Stanish et al., 2011). Life in the stream channels survives an absence of water for around 9 out of 12 months and is primarily comprised of cyanobacterial mats that harbor bacteria, eukaryotic algae, and micro-invertebrates (Adams et al., 2014; Esposito et al., 2008; McKnight et al., 1999). Throughout early and late summer, the cyanobacterial mats undergo numerous freeze-thaw cycles (Davey et al., 1992). With the onset of winter, these mats persist in a freeze-dried state until the following season (Figure 3).

McMurdo Dry Valley Don Juan Pond
Observations of dark flows on steep slopes that extend downhill recur each Mars year known as recurring slope lineae (RSL), as well as the presence of hydrated salts, are potential signs of a fleeting source of liquid water. The RSL have been linked to warm temperatures (Grimm et al., 2014). In addition, spectral data analyzed from the Compact Reconnaissance Imaging Spectrometer for Mars (CRISM) on the Mars Reconnaissance Orbiter (MRO), have indicated the presence of magnesium perchlorate, magnesium chlorate, and sodium perchlorate in four RSL locations (Ojha et al., 2015).

The atmospheric pressure on Mars confines the potential for any biologically available liquid water to low elevations where pressure may exceed 6.1mbars (Doran et al., 2010). Any water formed exists for a short duration before evaporating or freezing (Grimm et al., 2014). However, an incorporation of salts lowers the freezing point and evaporation rate which may increase the stability of water (Massé et al., 2016).

Figure 3. A gauge box set up by the Long-Term Ecological Research team to monitor water stream chemistry and hydrology of Von Guerard stream in the McMurdo Dry Valleys. Image zoom is of the cyanobacteria mats that are found in the meltwater streambeds.

Don Juan Pond (DJP) is an ice-free hypersaline pond located in the Wright Valley of the MDV. It is one of the saltiest bodies of water on Earth, comprised of 40% salt by weight. Salt in DJP is primarily (>90%) $CaCl_2$ (Toner, Catling, and Sletten, 2017). The high salinity ensures that the pond remains unfrozen, during the austral winters, at temperatures below -50°C (Marion, 1997). The input sources for the pond are widely debated in the literature. Arguments have been made for the upwelling of a deep groundwater flow system that replenishes the brines, and also for the deliquescence of salts in the shallow subsurface (Dickson et al., 2013; Nuding et al., 2014; Toner et al., 2017). Deliquescence is the process in which salts absorb atmospheric moisture at a critical relative humidity. The salts then dissolve in the absorbed water resulting in the formation of an aqueous brine (Nuding et al., 2014). Many studies on RSL suggest deliquescence as the formation mechanism for the darkened slope streaks (Dickson et al., 2013; Gough et al., 2017). Laboratory studies on DJP salts suggest brines on Mars could be rich in $CaCl_2$ given the melting temperature of the mixture. The unique geochemistry of this system is an important analog to the Martian subsurface and may provide clues in the ongoing investigations of RSL (Dickson et al., 2013; Mikucki et al., 2006).

DJP is an important site in the search for extremophiles and for Life that may be able to exist in the ephemeral brines of Mars. Water activity (a_w) thresholds for life on Earth appear to be limited to an $a_w \geq 0.605$ (Grant,

2004; Stevenson et al., 2015). It is uncertain whether any actively metabolizing Life exists in the DJP environment due to extremes in salt content, low fluctuating water activity $a_w = 0.3 - 0.6$, and subzero temperatures. Early studies suspected the presence of microbial mats in DJP brine (Horowitz et al., 1972; Siegel et al., 1979). Several recent studies attempting to detect active life have yielded contrary results (Peters et al., 2014; Samarkin et al., 2010). A primary limitation on the existence of life in DJP is the chaotropic nature of the brine salts, which interferes with hydrogen bonding, weakens the hydrophobic effect and lends to the instability of macromolecules.

We have presented two extreme Earth environments that serve as analogs to better understand the potential for habitability in some extraterrestrial environments. However, as we look to the future of Life in other locations, we also learn a lot about the capacity of Earth-evolved organisms to continually adapt to their environments over time.

Space Exploration: what is the future of life on Earth and elsewhere?

In previous sections we were introduced to the (possibly) low-frequency event that is the emergence of Life – be it through chemical evolution on Earth or elsewhere in our galaxy or through the chance event that an asteroid with the right elements struck our planet at the right time. We have also been introduced to two extreme environments on Earth which demonstrate how the microbial world under nearly any circumstance will abound and rebound. If we consider these concepts in the context of human space travel, we begin to understand the future role humans will play in trafficking microbes. In this section we discuss Planetary Protection, which serves both as a policy and motivation for scientific investigation. Planetary Protection places humans as the custodians of our solar system and uses space biology to understand how Earth-evolved organisms will be affected by or will affect the extraterrestrial environment.

Planetary Protection

NASA complies with the Planetary Protection provisions set forth by the 1967 Outer Space Treaty. The Outer Space Treaty, which was signed by all space-faring nations in 1967, stipulates that every effort must be made to protect other worlds from biological contamination from Earth. In addition, the Earth must be protected from the potential hazards posed by extraterrestrial matter carried by a spacecraft returning from another planet or other extraterrestrial sources. Therefore, NASA has gone to great lengths to control any potential organic and biological contamination carried by spacecraft. The governing body on the regulation of interplanetary microbial contamination is NASA's Office of Planetary Protection. Now more than ever, as we travel into space with the intent to land on the surface of these other worlds, we need to consider the risk of contaminating the destination environment. The Committee on Space Research (COSPAR) has a guideline that any mission, designed to look for life on other worlds, must not have a probability greater than 1-in-10,000 that a single microbe carried on board will contaminate potential extraterrestrial habitats.

As a precautionary measure, space-bound instrumentation has been routinely swabbed in order to categorize and characterize any microbial hitchhikers (Crawford, 2005). Researchers have catalogued strains of bacteria that were able to survive sterilization processes in European Space Agency (ESA) clean rooms (Moissl-Eichinger, 2017; Moissl et al., 2007); similarly, microbial survivors post-sterilization have been observed in the clean rooms utilized by NASA to assemble spacecraft (Mahnert et al., 2015). The discovery of resistant microbes shows that previous methods of cleaning, such as radiation and desiccation, might not be sufficient to sterilize spacecraft. For example, *Acinetobacter* species are dominant members of the clean room community and have been found to be able to metabolize or biodegrade the utilized cleaning reagents such as ethanol, 2-propanol, and Kleenol 30 (alkaline floor detergent). This suggests that the organisms can metabolize disinfectants, metabolizing them as a nutrient source. This may also prime the bacteria to be tolerant to the oxidative stresses associated with the low-humidity environment of the clean rooms (Mogul et al., 2018). In addition, the external surfaces of spacecrafts are found to harbor highly resilient bacterial species. It was through the EXPOSE-E mission of the European Space Agency (ESA) that researchers found that endospores mounted on the ISS could still survive by escaping exposure to radiation (hiding in cracks or being shielded by the spacecraft) (Horneck et al., 2012). These observations will be factored into design considerations in order to uphold Planetary Protection policies.

Space biology and human-hosted microbes aboard the ISS
Given that humans host millions of microorganisms, and that microbes will be an essential component of any self-sustaining food-production system, human space travel is inextricably linked to microbial communities. The field of space biology actively investigates the effects of radiation and microgravity upon microbes, fungi, plants, and animals to understand how space travel may cause them to adapt, develop differently, or change at the cellular and molecular levels. In the effort to send humans deeper into space, exhaustive research into human health in the space environment is underway in both Earth-based analogs and on board the ISS. NASA's Human Exploration Research Analog (HERA), NASA's Extreme Environment Mission Operations (NEEMO), High Seas, and other Earth-based analogs work to understand the dynamics of long duration confinement and how the body responds to stressors. The Center for the Advancement of Science in Space (CASIS) is the manager of the ISS U.S. National Laboratory, where years of space biology research has taken place and been chronicled in NASA's GeneLab Project.

In the light of Planetary Protection, human-hosted microbes are a key element of deep-space travel that remain to be better understood. In a recently published astronaut twin study a number of genetic, immune function, metabolic and microbial differences were observed between the individual which spent a year in space and the ground control (Garrett-Bakelman et al., 2019). Their findings on the changes experienced in the gut microbial assemblage of the one astronaut, agreed with results from The Astronaut Microbiome Project (AMP). The AMP has characterized the gut, oral, skin, and nasal microbiomes from a cohort of astronauts before, during, and after spaceflight. This research sought to understand how these populations would change over the course of a space mission as a result of stress and altered immune function, and how the resident microbes on board the ISS might affect the astronaut microbiome and vice versa. Preliminary findings concluded that the gut microbiome does indeed change during spaceflight and in some astronauts returns to its preflight status within 60 days after returning to Earth, whereas in others it did not. It was additionally concluded that the microbiome of the ISS surfaces changes over time and this is influenced by the skin microbiota of the astronauts residing in the ISS at the associated point in time (Voorhies et al., 2016). A recent publication that details the total and viable bacterial and fungal communities associated with the ISS, provides further evidence that the microbiome of the ISS changes over time, while the dominant taxa are associated with the human microbiome and include potentially opportunistic species (Checinka Sielaff et al., 2019).

It remains to be understood how the resident microbes on board the ISS affect the astronaut microbiome, but a first step is to understand how the space environment affects the microorganisms themselves. In-flight experiments and ground-based, low-shear, modeled microgravity experiments on *Escherichia coli* (Ciferri et al., 1986; Kacena et al., 1999a; Kacena et al.,1997; Kacena et al., 1999b; Klaus et al., 1997; Volkmann, 1988), *Bacillus subtilis* (Kacena et al., 1999a; Kacena et al., 1997; Kacena et al., 1999c), *Salmonella enterica* (Foster and Spector, 1995), *Staphylococcus aureus* (Castro-Wallace et al., 2011; Castro-Wallace, 2012; Rosado et al., 2010), *Pseudomonas aeruginosa* (McLean et al., 2001) and *Ralstonia pickettii* (Mijnendonckx et al., 2013), it has been shown that bacterial morphology, growth, metabolism, gene transfer, antibiotic resistance, and infectious nature can be altered in response to the space environment (Lapchine et al., 1986; Nickerson et al., 2004; C. a Nickerson et al., 2000; Tixador et al., 1985). So far, any changes detected have primarily been ascribed to two characteristics of the microgravity environment – the reduction in hydrostatic pressure and the lower fluid-shear forces (K. J. Dickson, 1991).

NASA has routinely sampled the air, surface, and water on board the ISS. This has resulted in a collection of 424 bacterial isolates, collected over 12 years and 22 missions. While not experimentally designed per se (as it is impossible to know how long any individual isolate was on the ISS before it was isolated, regardless of the date of isolation), this library provides one window into the effect of the microgravity and enhanced radiation experienced by resident microbes. In the air and surface samples, *Staphylococcus* species account for 80% and 70% of the isolates recovered in this collection, respectively. These *Staphylococcus* species include isolates of *S. aureus, S. auricularis, S. capitis, S. cohnii, S. epidermidis, S. haemolyticus, S. hominis, S. lugdunensis, S. saprophyticus, and S. warneri* – all human commensals. Species of the closely related *Burkholderia* and *Ralstonia* genera comprise 60% of the isolates collected from the water system over the 12 years. Recovered *Burkholderia* species include *B. contaminans* and *B. cepacia*. In general, *Burkholderia* species exhibit a high degree of phenotypic plasticity as they can live in soil, water (Miller et al., 2002), and the human body (Valvano, Keith, and Cardona, 2005), known to be opportunistic pathogens and are equipped for life in extreme environments (Lee et al., 2014; Lester et al., 2007) (Figure 4). Select isolates from our culture collection display the ability to inhibit *Aspergillus* species (Figure 4B) and exhibit the ability to lyse red blood cells (Figure 4C).

In some bacteria, the microgravity environment can affect bacterial virulence, resulting in altered pathogenicity (C. A. Nickerson et al., 2004). For instance, members of *Salmonella enterica* have demonstrated increased virulence when grown under modeled microgravity simulated using a high-aspect rotating vessel (C. a Nickerson et al., 2000). Whereas, *S. aureus* grown under similar modeled microgravity conditions show an increase in the transcript levels of genes involved in biofilm formation but

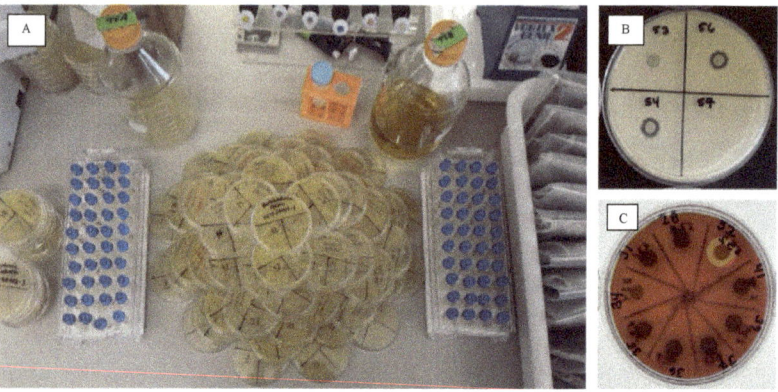

Figure 4. A. Isolates of Burkholderia and Ralstonia species cultured from the ISS potable water system. B. Select isolates display the ability to inhibit Aspergillus species (Burkholderia demonstrate zones of inhibition when spotted on agar also growing Aspergillus sp.). C. Select solates also exhibit the ability to lyse red blood cells as demonstrated by spotting the cultures on sheep's blood agar plates.

an overall decrease in virulence-factor transcripts (Castro et al., 2011; Rosado et al., 2006). Similarly, a terrestrially-derived strain of S. aureus exposed to simulated microgravity for 24 hours exhibited a reduction in the amount of secreted alpha-toxin and hemolysin (Rosado et al., 2006). Organisms that demonstrate decreased virulence, if maintained over time, would naturally appeal to us as more suitable for a deep-space exploration scenario where a human resides in a closed-environment for months or years at a time. Space Biology is inherently a systems-biology approach where the demand for robust life-support systems will fuel the exhaustive characterization of microbes and their effect on both the crew and engineered systems. As the distance to the target increases, the odds of success decrease, and out of billions of microbial hitchhikers, some may be more appropriate for the journey than others.

Conclusions

Here we have presented only a portion of the expansive territory of Astrobiology using the Astrobiology Roadmap as our guide. We introduced the contrasting theories as to how life began on our planet and contend that perhaps it is a question best informed by observing the formation of life on another planet. However, we can still observe how life adapts as microorganisms with fast generation times reveal the mechanisms used to survive in even the harshest of environmental conditions. Understanding of microbial activity and ecology at analog sites expands our toolkit of biological signatures and expands the boundaries of which extraterrestrial environments should be targets in the search for Life in the solar system. Microbial life is able to employ a variety of mechanisms in order to satisfy their metabolic requirements, and as we set our sights on exploring other habitable worlds, we must be cognizant of the billions of microbes we will bring with us, and the potential for these microbes to survive in the extreme conditions that will be encountered. Planetary Protection aims to prevent the non-deliberate delivery of space-derived microbes to our planet (back-contamination), and the delivery of Earth-derived microbes as an invasive species to another world (forward-contamination). Due to our limited understanding of the effect that the space environment has on Earth-evolved organisms, research in the field of space biology has gained traction and will continue to inform policy decisions and engineering guidelines for the future of space travel. A human presence in space is in our near future, and with it will come groundbreaking technologies and discoveries capable of improving life on Earth as well as provide us with answers as to Earth's biotic beginnings. Indeed, it seems almost science fiction when we think about the potential for Earth-evolved organisms to affect the trajectory of life on another planet, just as an intercepted asteroid may have changed the trajectory of our own.

Future trends

As genomic sequencing tools evolve and are combined with state-of-the-art analytical chemistry techniques, we will continue to improve our understanding of both abiotic and biotic components and refine our definitions of the requirements for habitability. Models of fitness that explicitly correlate genotypic traits with survival under specific environmental conditions need to be further developed, along with more comprehensive understandings of cellular activity and regulation. These will help to identify which particular traits have significant impacts (positive or negative) on the success of an organism and provide greater insight into the limits of habitability. Furthermore, once we can control each of the environmental and genomic factors, which influence a microbial phenotype, we can successfully plan for microbial containment. This extends to how to fulfill Planetary Protection guidelines, as well as how to keep a crew healthy and a spacecraft properly functioning, as we travel further out into our solar system.

References

Adams, B. J., Wall, D. H., Virginia, R. A., Broos, E., and Knox, M. A. (2014). Ecological biogeography of the terrestrial Nematodes of Victoria Land, Antarctica. *ZooKeys*. https://doi.org/10.3897/zookeys.419.7180

Alexander Voorhies, Manolito Torralba, K. M. and H. L. (n.d.). Study of the impact of longterm space travel on the Astronauts' microbiome.

Ásgeirsson, B., Nielsen, B. N., and Højrup, P. (2003). Amino acid sequence of the cold-active alkaline phosphatase from Atlantic cod (Gadus morhua). *Comparative Biochemistry and Physiology - B Biochemistry and Molecular Biology*. https://doi.org/10.1016/S1096-4959(03)00167-2

Bakermans, C., Bergholz, P. W., Ayala-del-Río, H., and Tiedje, J. (2009). Genomic Insights into Cold Adaptation of Permafrost Bacteria. In *Permafrost Soils, Soil Biology 16*. https://doi.org/10.1007/978-3-540-69371-0_11

Batalha, N. M. (2014). Exploring exoplanet populations with NASA's Kepler Mission. *Proceedings of the National Academy of Sciences*. https://doi.org/10.1073/pnas.1304196111

Bridges, J. C., Hicks, L. J., and Treiman, A. H. (2019). Chapter 5 - Carbonates on Mars. In J. Filiberto and S. P. B. T.-V. in the M. C. Schwenzer (Eds.) (pp. 89–118). Elsevier. https://doi.org/https://doi.org/10.1016/B978-0-12-804191-8.00005-2

Buelow, H. N., Winter, A. S., Van Horn, D. J., Barrett, J. E., Gooseff, M. N., Schwartz, E., and Takacs-Vesbach, C. D. (2016). Microbial community responses to increased water and organic matter in the arid soils of the mcmurdo dry valleys, antarctica. *Frontiers in Microbiology*. https://doi.org/10.3389/fmicb.2016.01040

Cabrol, N. A., and Grin, E. A. (1999). Distribution, Classification, and Ages of Martian Impact Crater Lakes. *Icarus*. https://doi.org/10.1006/icar.1999.6191

Cabrol, N. A., Grin, E. A., Newsom, H. E., Landheim, R., and McKay, C. P. (1999). Hydrogeologic Evolution of Gale Crater and Its Relevance to the Exobiological Exploration of Mars. *Icarus*. https://doi.org/10.1006/icar.1999.6099

Cámara, B., Souza-Egipsy, V., Ascaso, C., Artieda, O., De Los Ríos, A., and Wierzchos, J. (2016). Biosignatures and microbial fossils in endolithic microbial communities colonizing Ca-sulfate crusts in the Atacama Desert. *Chemical Geology*. https://doi.org/10.1016/j.chemgeo. 2016.09.019

Campanaro, S., Williams, T. J., Burg, D. W., De Francisci, D., Treu, L., Lauro, F. M., and Cavicchioli, R. (2011). Temperature-dependent global gene expression in the Antarctic archaeon Methanococcoides burtonii. *Environmental Microbiology*, *13*(8), 2018–2038. https://doi.org/10.1111/j. 1462-2920.2010.02367.x

Carr, M. H. (2012). The fluvial history of Mars. *Philosophical Transactions of the Royal Society A: Mathematical, Physical and Engineering Sciences*. https://doi.org/10.1098/rsta.2011.0500

Carr, M. H., and Head, J. W. (2010). Geologic history of Mars. *Earth and Planetary Science Letters*. https://doi.org/10.1016/j.epsl.2009.06.042

Cartier, G., Lorieux, F., Allemand, F., Dreyfus, M., and Bizebard, T. (2010). Cold Adaptation in DEAD-Box Proteins. *Biochemistry*, *49*(12), 2636–2646. https://doi.org/10.1021/bi902082d

Castro-Wallace, S. L., Chiu, C. Y., John, K. K., Stahl, S. E., Rubins, H., Mcintyre, A. B. R., Alexander, N. (2017). Nanopore DNA Sequencing and Genome Assembly on the International Space Station. *Scientific Reports*, (December), 1–12. https://doi.org/10.1038/s41598-017-18364-0

Castro-Wallace, S. L. (2012). The Response of *Staphylococcus aureus* to Culture in a Low-Fluid-Shear Environment.

Castro-Wallace, S. L., Nelman-Gonzalez, M., Nickerson, C. A., and Ott, C. M. (2011). Induction of attachment-independent biofilm formation and repression of hfq expression by low-fluid-shear culture of *Staphylococcus aureus*. *Applied and Environmental Microbiology*, *77*(18), 6368–6378. https://doi.org/10.1128/AEM.00175-11

Catling, D. C., Claire, M. W., Zahnle, K. J., Quinn, R. C., Clark, B. C., Hecht, M. H., and Kounaves, S. (2010). Atmospheric origins of perchlorate on mars and in the atacama. *Journal of Geophysical Research E: Planets*. https://doi.org/10.1029/2009JE003425

Checinska Sielaff, A., Urbaniak, C., Mohan, G. B. M., Stepanov, V. G., Tran, Q., Wood, J. M., Venkateswaran, K. (2019). Characterization of the total and viable bacterial and fungal communities associated with the International Space Station surfaces. *Microbiome*, 7(1), 50. https://doi.org/10.1186/s40168-019-0666-x

Ciferri, O., Tiboni, O., Pasquale, G., Orlandoni, A. M., and Marchesi, M. L. (1986). Effects of microgravity on genetic recombination in *Escherichia coli*; *Naturwissenschaften*, 73(7), 418–421. https://doi.org/10.1007/BF00367284

Crawford, R. L. (2005). Microbial diversity and its relationship to planetary protection. *Applied and Environmental Microbiology*. https://doi.org/10.1128/AEM.71.8.4163-4168.2005

Crits-Christoph, A., Gelsinger, D. R., Ma, B., Wierzchos, J., Ravel, J., Davila, A., DiRuggiero, J. (2016)a. Functional interactions of archaea, bacteria and viruses in a hypersaline endolithic community. *Environmental Microbiology.* https://doi.org/10.1111/1462-2920.13259

Crits-Christoph, A., Robinson, C. K., Ma, B., Ravel, J., Wierzchos, J., Ascaso, C., ... DiRuggiero, J. (2016)b. Phylogenetic and functional substrate specificity for endolithic microbial communities in hyper-arid environments. *Frontiers in Microbiology.* https://doi.org/10.3389/fmicb.2016.00301

Dalluge, J. J., Hamamoto, T., Horikoshi, K., Morita, R. Y., Stetter, K. O., and McCloskey, J. A. (1997). Posttranscriptional modification of tRNA in psychrophilic bacteria. *Journal of Bacteriology*, 179(6), 1918–1923. https://doi.org/10.1128/jb.179.6.1918-1923.1997

Davey, M. C., and Clarke, K. J. (1992).Fine structure of a terrestrial cyanobacterial mat from Antarctica. *Journal of Phycology.* https://doi.org/10.1111/j.0022-3646.1992.00199.x

Davila, A. F., Duport, L. G., Melchiorri, R., Jänchen, J., Valea, S., de los Rios, A., Wierzchos, J. (2010). Hygroscopic Salts and the Potential for Life on Mars. *Astrobiology.* https://doi.org/10.1089/ast.2009.0421

Davila, A. F., Gómez-Silva, B., de los Rios, A., Ascaso, C., Olivares, H., McKay, C. P., and Wierzchos, J. (2008). Facilitation of endolithic microbial survival in the hyperarid core of the Atacam Desert by mineral deliquescence. *Journal of Geophysical Research: Biogeosciences.* https://doi.org/10.1029/2007JG000561

Davila, A. F., and Schulze-Makuch, D. (2016). The Last Possible Outposts for Life on Mars. *Astrobiology.* https://doi.org/10.1089/ast.2015.1380

Dehant, V., Lammer, H., Kulikov, Y. N., Grießmeier, J. M., Breuer, D., Verhoeven, O., ... Lognonné, P. (2007). Planetary magnetic dynamo effect on atmospheric protection of early earth and mars. *Space Science Reviews.* https://doi.org/10.1007/s11214-007-9163-9

Des Marais, D. J., Nuth, J. A., Allamandola, L. J., Boss, A. P., Farmer, J. D., Hoehler, T. M., ... Spormann, A. M. (2008). The NASA Astrobiology Roadmap. *Astrobiology.* https://doi.org/10.1089/ast.2008.0819

Dickson, J. L., Head, J. W., Levy, J. S., and Marchant, D. R. (2013). Don Juan Pond, Antarctica: Near-surface CaCl 2-brine feeding Earth's most saline lake and implications for Mars. *Scientific Reports.* https://doi.org/10.1038/srep01166

Dickson, K. J. (1991). Summary of biological spaceflight experiments with cells. *ASGSB Bulletin : Publication of the American Society for Gravitational and Space Biology*, 4(2), 151–260. Retrieved from http://www.ncbi.nlm.nih.gov/pubmed/11537177

Doran, P. T., Fritsen, C. H., McKay, C. P., Priscu, J. C., and Adams, E. E. (2003). Formation and character of an ancient 19-m ice cover and underlying trapped brine in an "ice-sealed" east Antarctic lake. *Proceedings of the National Academy of Sciences.* https://doi.org/10.1073/pnas.222680999

Doran, P. T., Lyons, W. B., and McKnight, D. M. (2010). Life in Antarctic Deserts and other Cold Dry Environments: Astrobiological analogs. *Cambridge Astrobiology.* https://doi.org/10.1017/CBO9780511712258

Ehlmann, B. L., Mustard, J. F., Murchie, S. L., Bibring, J.-P., Meunier, A., Fraeman, A. A., and Langevin, Y. (2011). Subsurface water and clay mineral formation during the early history of Mars. *Nature*. https://doi.org/10.1038/nature10582

Emeline, A. V., Otroshchenko, V. A., Ryabchuk, V. K., and Serpone, N. (2003). Abiogenesis and photostimulated heterogeneous reactions in the interstellar medium and on primitive earth: Relevance to the genesis of life. *Journal of Photochemistry and Photobiology C: Photochemistry Reviews*. https://doi.org/10.1016/S1389-5567(02)00039-4

Esposito, R. M. M., Spaulding, S. A., McKnight, D. M., Van de Vijver, B., Kopalová, K., Lubinski, D., Whittaker, T. (2008). Inland diatoms from the McMurdo Dry Valleys and James Ross Island, Antarctica. *Botany*. https://doi.org/10.1139/B08-100

Farmer, J. D. (2018). Chapter 1 - Habitability as a Tool in Astrobiological Exploration. *From Habitability to Life on Mars*. (E. Grin and N. A. Cabrol, Eds.). Elsevier.

Fassett, C. I., and Minton, D. A. (2013). Impact bombardment of the terrestrial planets and the early history of the Solar System. *Nature Geoscience*. https://doi.org/10.1038/ngeo1841

Feller, G. (2013). Psychrophilic Enzymes: From Folding to Function and Biotechnology. *Scientifica*, *2013*, 1–28. https://doi.org/10.1155/2013/512840

Foster, J. W., and Spector, M. P. (1995). How *Salmonella* Survive Against the Odds. *Annual Review of Microbiology*. https://doi.org/10.1146/annurev.mi.49.100195.001045

Garrett-Bakelman, F. E., Darshi, M., Green, S. J., Gur, R. C., Lin, L., Macias, B. R., ... Turek, F. W. (2019). The NASA Twins Study: A multidimensional analysis of a year-long human spaceflight. *Science (New York, N.Y.)*, *364*(6436). https://doi.org/10.1126/science.aau8650

Gooseff, M. N., McKnight, D. M., Doran, P., Fountain, A. G., and Lyons, W. B. (2011). Hydrological connectivity of the landscape of the McMurdo Dry Valleys, Antarctica. *Geography Compass*. https://doi.org/10.1111/j.1749-8198.2011.00445.x

Goudge, T. A., Fassett, C. I., Head, J. W., Mustard, J. F., and Aureli, K. L. (2016). Insights into surface runoff on early Mars from paleolake basin morphology and stratigraphy. *Geology*. https://doi.org/10.1130/G37734.1

Gough, R. V., Wong, J., Dickson, J. L., Levy, J. S., Head, J. W., Marchant, D. R., and Tolbert, M. A. (2017). Brine formation via deliquescence by salts found near Don Juan Pond, Antarctica: Laboratory experiments and field observational results. *Earth and Planetary Science Letters*. https://doi.org/10.1016/j.epsl.2017.08.003

Grant, W. D. (2004). Life at low water activity. *Philosophical Transactions of the Royal Society B: Biological Sciences*. https://doi.org/10.1098/rstb.2004.1502

Grimm, R. E., Harrison, K. P., and Stillman, D. E. (2014). Water budgets of martian recurring slope lineae. *Icarus*. https://doi.org/10.1016/j.icarus.2013.11.013

Harel, Y. (2004). Activation of Photosynthesis and Resistance to Photoinhibition in Cyanobacteria within Biological Desert Crust. *Plant Physiology*. https://doi.org/10.1104/pp.104.047712

Hartley, A. J., and Chong, G. (2002). Late Pliocene age for the Atacama Desert: Implications for the desertification of western South America. *Geology*. https://doi.org/10.1130/0091-7613(2002)030<0043:LPAFTA>2.0.CO;2

Herwartz, D., Pack, A., Friedrichs, B., and Bischoff, A. (2014). Identification of the giant impactor Theia in lunar rocks. *Science*. https://doi.org/10.1126/science.1251117

Hoehler, T. M. (2007). An Energy Balance Concept for Habitability. *Astrobiology*. https://doi.org/10.1089/ast.2006.0095

Horneck, G., Moeller, R., Cadet, J., Douki, T., Mancinelli, R. L., Nicholson, W. L., ... Venkateswaran, K. J. (2012). Resistance of Bacterial Endospores to Outer Space for Planetary Protection Purposes—Experiment PROTECT of the EXPOSE-E Mission. *Astrobiology*. https://doi.org/10.1089/ast.2011.0737

Horowitz, N. H., Cameron, R. E., and Hubbard, J. S. (1972). Microbiology of the dry valleys of antarctica. *Science (New York, N.Y.)*. https://doi.org/10.1126/science.176.4032.242

Houston, J., and Hartley, A. J. (2003). The central andean west-slope rainshadow and its potential contribution to the origin of hyper-aridity in the Atacama Desert. *International Journal of Climatology*. https://doi.org/10.1002/joc.938

Huang, S.-S. (1960). Life-Supporting Regions in the Vicinity of Binary Systems. *Publications of the Astronomical Society of the Pacific*. https://doi.org/10.1086/127489

Hubbard, G. S., Naderi, F. M., and Garvin, J. B. (2002). Following the water, the new program for Mars exploration. *Acta Astronautica*. https://doi.org/10.1016/S0094-5765(02)00067-X

Jakosky, B. M., Lin, R. P., Grebowsky, J. M., Luhmann, J. G., Mitchell, D. F., Beutelschies, G., Zurek, R. (2015). The Mars Atmosphere and Volatile Evolution (MAVEN) Mission. *Space Science Reviews*. https://doi.org/10.1007/s11214-015-0139-x

Jones, E. G., and Lineweaver, C. H. (2010). To What Extent Does Terrestrial Life "Follow The Water"? *Astrobiology*. https://doi.org/10.1089/ast.2009.0428

Kacena, M. A., Leonard, P. E., Todd, P., and Luttges, M. W. (1997). Low gravity and inertial effects on the growth of *E. coli* and *B. subtilis* in semi-solid media. *Aviation, Space, and Environmental Medicine*.

Kacena, M. A., Merrell, G. A., Manfredi, B., Smith, E. E., Klaus, D. M., and Todd, P. (1999)a. Bacterial growth in space flight: Logistic growth curve parameters for *Escherichia coli* and *Bacillus subtilis*. *Applied Microbiology and Biotechnology*. https://doi.org/10.1007/s002530051386

Kacena, M. A., Smith, E. E., and Todd, P. (1999)b. Autolysis of *Escherichia coli* and *Bacillus subtilis* cells in low gravity. *Applied Microbiology and Biotechnology*. https://doi.org/10.1007/s002530051543

Kacena, M. A., and Todd, P. (1999)c. Gentamicin: effect on *E. coli* in space. *Microgravity Science and Technology*.

Kasting, J. F., Whitmire, D. P., and Reynolds, R. T. (1993). Habitable Zones around Main Sequence Stars. *Icarus*. https://doi.org/10.1006/icar.1993.1010

Klaus, D., Simske, S., Todd, P., and Stodieck, L. (1997). Investigation of space flight effects on *Escherichia coli* and a proposed model of underlying physical mechanisms. *Microbiology*. https://doi.org/10.1099/00221287-143-2-449

Knoll, A. H., and Grotzinger, J. (2006). Water on Mars and the prospect of martian life. *Elements*. https://doi.org/10.2113/gselements.2.3.169

Kohler, T. J., Stanish, L. F., Crisp, S. W., Koch, J. C., Liptzin, D., Baeseman, J. L., and McKnight, D. M. (2015). Life in the Main Channel: Long-Term Hydrologic Control of Microbial Mat Abundance in McMurdo Dry Valley Streams, Antarctica. *Ecosystems*. https://doi.org/10.1007/s10021-014-9829-6

Koonin, E. V. (2003). Comparative genomics, minimal gene-sets and the last universal common ancestor. *Nature Reviews Microbiology*. https://doi.org/10.1038/nrmicro751

Králová, S. (2017). Role of fatty acids in cold adaptation of Antarctic psychrophilic Flavobacterium spp. *Systematic and Applied Microbiology*. https://doi.org/10.1016/j.syapm.2017.06.001

Kvenvolden, K., Lawless, J., Pering, K., Peterson, E., Flores, J., Ponnamperuma, C., ... Moore, C. (1970). Evidence for Extraterrestrial Amino-acids and Hydrocarbons in the Murchison Meteorite. *Nature*, 228, 923. Retrieved from http://dx.doi.org/10.1038/228923a0

Lapchine, L., Moatti, N., Gasset, G., Richoilley, G., Templier, J., and Tixador, R. (1986). Antibiotic activity in space. *Drugs under Experimental and Clinical Research*, 12(12), 933–938. Retrieved from http://www.ncbi.nlm.nih.gov/pubmed/3569006

Lee, Y. M., Kim, E. H., Lee, H. K., and Hong, S. G. (2014). Biodiversity and physiological characteristics of Antarctic and Arctic lichens-associated bacteria. *World Journal of Microbiology and Biotechnology*. https://doi.org/10.1007/s11274-014-1695-z

Lester, E. D., Satomi, M., and Ponce, A. (2007). Microflora of extreme arid Atacama Desert soils. *Soil Biology and Biochemistry*. https://doi.org/10.1016/j.soilbio.2006.09.020

Lissauer, J. (1993). Planet Formation. *Araand a*, 31, 129.

Lunine, J. I., Chambers, J., Morbidelli, A., and Leshin, L. A. (2003). The Origin of water on Mars. *Icarus*. https://doi.org/10.1016/S0019-1035(03)00172-6

Mahnert, A., Moissl-Eichinger, C., Berg, G., Vaishampayan, P. A., Probst, A. J., Auerbach, A. K., Miller, W. (2015). Molecular bacterial community analysis of clean rooms where spacecraft are assembled. *Astrobiology*. https://doi.org/10.1126/science.aad1329

Margesin, R., Schinner, F., Marx, J. C., and Gerday, C. (2008). *Psychrophiles: From biodiversity to biotechnology. Psychrophiles: From Biodiversity to Biotechnology.* https://doi.org/10.1007/978-3-540-74335-4

Marion, G. M. M. (1997). A theoretical evaluation of mineral stability in Don Juan Pond, Wright Valley, Victoria Land. *Antarctic Science.* https://doi.org/10.1017/S0954102097000114

Martel, J., Young, D., Peng, H.-H., Wu, C.-Y., and Young, J. D. (2012). Biomimetic Properties of Minerals and the Search for Life in the Martian Meteorite ALH84001. *Annual Review of Earth and Planetary Sciences.* https://doi.org/10.1146/annurev-earth-042711-105401

Martín-Torres, F. J., Zorzano, M.-P., Valentín-Serrano, P., Harri, A.-M., Genzer, M., Kemppinen, O., Vaniman, D. (2015). Transient liquid water and water activity at Gale crater on Mars. *Nature Geoscience, 8*, 357. Retrieved from https://doi.org/10.1038/ngeo2412

Martin, W., Baross, J., Kelley, D., and Russell, M. J. (2008). Hydrothermal vents and the origin of life. *Nature Reviews Microbiology.* https://doi.org/10.1038/nrmicro1991

Massé, M., Conway, S. J., Gargani, J., Patel, M. R., Pasquon, K., McEwen, A., Jouannic, G. (2016). Transport processes induced by metastable boiling water under Martian surface conditions. *Nature Geoscience.* https://doi.org/10.1038/ngeo2706

McKnight, D. M., Niyogi, D. K., Alger, A. S., Bomblies, A., Conovitz, P. A., and Tate, C. M. (1999). Dry Valley Streams in Antarctica: Ecosystems Waiting for Water. *BioScience.* https://doi.org/10.2307/1313732

McLean, R. J., Cassanto, J. M., Barnes, M. B., and Koo, J. H. (2001). Bacterial biofilm formation under microgravity conditions. *FEMS Microbiology Letters.* https://doi.org/S0378-1097(00)00549-8 [pii]

Mijnendonckx, K., Provoost, A., Ott, C. M., Venkateswaran, K., Mahillon, J., Leys, N., and van Houdt, R. (2013). Characterization of the Survival Ability of *Cupriavidus metallidurans* and *Ralstonia pickettii* from Space-Related Environments. *Microbial Ecology.* https://doi.org/10.1007/s00248-012-0139-2

Mikucki, J. A., Priscu, J. C., Lyons, W. B., Welch, K. A., Tranter, M., and Pearson, A. (2006). Geomicrobiology of an Antarctic subglacial brine: A plausible Martian ecosystem. *Geochimica et Cosmochimica Acta.* https://doi.org/10.1016/j.gca.2006.06.844

Miller, S. C. M., LiPuma, J. J., and Parke, J. L. (2002). Culture-based and non-growth-dependent detection of the *Burkholderia cepacia* complex in soil environments. *Applied and Environmental Microbiology, 68*(8), 3750–3758. https://doi.org/10.1128/AEM.68.8.3750-3758.2002

Milliken, R. E., and Li, S. (2017). Remote detection of widespread indigenous water in lunar pyroclastic deposits. *Nature Geoscience.* https://doi.org/10.1038/NGEO2993

Mock, T., and Kroon, B. M. A. (2002). Photosynthetic energy conversion under extreme conditions—I: important role of lipids as structural modulators and energy sink under N-limited growth in Antarctic sea ice diatoms. *Phytochemistry*, *61*(1), 41–51. https://doi.org/https://doi.org/10.1016/S0031-9422(02)00216-9

Mogul, R., Barding, G. A., Lalla, S., Lee, S., Madrid, S., Baki, R., ... Walker, J. (2018). Metabolism and Biodegradation of Spacecraft Cleaning Reagents by Strains of Spacecraft-Associated *Acinetobacter*. *Astrobiology*. https://doi.org/10.1089/ast.2017.1814

Moissl-Eichinger, C. (2017). Extremophiles in spacecraft assembly cleanrooms. In *Adaption of Microbial Life to Environmental Extremes: Novel Research Results and Application, Second Edition*. https://doi.org/10.1007/978-3-319-48327-6_10

Moissl, C., Osman, S., La Duc, M. T., Dekas, A., Brodie, E., DeSantis, T., and Venkateswaran, K. (2007). Molecular bacterial community analysis of clean rooms where spacecraft are assembled. *FEMS Microbiology Ecology*, *61*(3), 509–521. https://doi.org/10.1111/j.1574-6941.2007.00360.x

Morgan-Kiss, R. M., Priscu, J. C., Pocock, T., Gudynaite-Savitch, L., and Huner, N. P. A. (2006). Adaptation and Acclimation of Photosynthetic Microorganisms to Permanently Cold Environments. *Microbiology and Molecular Biology Reviews*. https://doi.org/10.1128/MMBR.70.1.222-252.2006

Mori, S., Castoreno, A., Mulligan, M. E., and Lammers, P. J. (2003). Nitrogen status modulates the expression of RNA-binding proteins in cyanobacteria. *FEMS Microbiology Letters*, *227*(2), 203–210. https://doi.org/10.1016/S0378-1097(03)00682-7

Murray, A. E., Kenig, F., Fritsen, C. H., McKay, C. P., Cawley, K. M., Edwards, R., ... Doran, P. T. (2012). Microbial life at -13 C in the brine of an ice-sealed Antarctic lake. *Proceedings of the National Academy of Sciences*. https://doi.org/10.1073/pnas.1208607109

Nickerson, C. A., Ott, C. M., Wilson, J. W., Ramamurthy, R., and Pierson, D. L. (2004). Microbial Responses to Microgravity and Other Low-Shear Environments. *Microbiology and Molecular Biology Reviews*, *68*(2), 345–361. https://doi.org/10.1128/MMBR.68.2.345-361.2004

Nickerson, C. a, Ott, C. M., Mister, S. J., Brian, J., Burns-keliher, L., Pierson, D. L., and Morrow, B. J. (2000). Microgravity as a Novel Environmental Signal Affecting *Salmonella enterica* Serovar *Typhimurium* Virulence Microgravity as a Novel Environmental Signal Affecting *Salmonella enterica* Serovar *Typhimurium* Virulence, *68*(6), 3147–3152. https://doi.org/10.1128/IAI.68.6.3147-3152.2000.Updated

Nuding, D. L., Rivera-Valentin, E. G., Davis, R. D., Gough, R. V., Chevrier, V. F., and Tolbert, M. A. (2014). Deliquescence and efflorescence of calcium perchlorate: An investigation of stable aqueous solutions relevant to mars. *Icarus*. https://doi.org/10.1016/j.icarus.2014.08.036

Nutman, A. P., Bennett, V. C., Friend, C. R. L., Van Kranendonk, M. J., and Chivas, A. R. (2016). Rapid emergence of life shown by discovery of 3,700-million-year-old microbial structures. *Nature*. https://doi.org/10.1038/nature19355

Ojha, L., Wilhelm, M. B., Murchie, S. L., Mcewen, A. S., Wray, J. J., Hanley, J., ... Chojnacki, M. (2015). Spectral evidence for hydrated salts in recurring slope lineae on Mars. *Nature Geoscience*. https://doi.org/10.1038/ngeo2546

Orosei, R., Lauro, S. E., Pettinelli, E., Cicchetti, A., Coradini, M., Cosciotti, B., ... Seu, R. (2018). Radar evidence of subglacial liquid water on Mars. *Science*. https://doi.org/10.1126/science.aar7268

Peters, B., Casciotti, K. L., Samarkin, V. A., Madigan, M. T., Schutte, C. A., and Joye, S. B. (2014). Stable isotope analyses of NO2-, NO3-, and N2O in the hypersaline ponds and soils of the McMurdo Dry Valleys, Antarctica. *Geochimica et Cosmochimica Acta*. https://doi.org/10.1016/j.gca.2014.03.024

Ponder, M. A., Gilmour, S. J., Bergholz, P. W., Mindock, C. A., Hollingsworth, R., Thomashow, M. F., and Tiedje, J. M. (2005). Characterization of potential stress responses in ancient Siberian permafrost psychroactive bacteria. *FEMS Microbiology Ecology*, 53(1), 103–115. https://doi.org/10.1016/j.femsec.2004.12.003

Raymond-Bouchard, I., and Whyte, L. G. (2017). From transcriptomes to metatranscriptomes: Cold adaptation and active metabolisms of psychrophiles from cold environments. In *Psychrophiles: From Biodiversity to Biotechnology: Second Edition*. https://doi.org/10.1007/978-3-319-57057-0_18

Raymond, S. N., Quinn, T., and Lunine, J. I. (2004). Making other earths: Dynamical simulations of terrestrial planet formation and water delivery. *Icarus*. https://doi.org/10.1016/j.icarus.2003.11.019

Robinson, C. K., Wierzchos, J., Black, C., Crits-Christoph, A., Ma, B., Ravel, J., ... Diruggiero, J. (2015). Microbial diversity and the presence of algae in halite endolithic communities are correlated to atmospheric moisture in the hyper-arid zone of the Atacama Desert. *Environmental Microbiology*. https://doi.org/10.1111/1462-2920.12364

Rosado, H., Doyle, M., Hinds, J., and Taylor, P. W. (2010). Low-shear modelled microgravity alters expression of virulence determinants of *Staphylococcus aureus*. *Acta Astronautica*, 66(3–4), 408–413. https://doi.org/10.1016/j.actaastro.2009.06.007

Rosado, H., Stapleton, P., and Taylor, P. (2006). Effect of simulated microgravity on the virulence properties of the opportunistic bacterial pathogen *Staphylococcus aureus*, 1–8. Retrieved from http://discovery.ucl.ac.uk/1350754/

Samarkin, V. A., Madigan, M. T., Bowles, M. W., Casciotti, K. L., Priscu, J. C., McKay, C. P., and Joye, S. B. (2010). Abiotic nitrous oxide emission from the hypersaline Don Juan Pond in Antarctica. *Nature Geoscience*. https://doi.org/10.1038/ngeo847

Seckbach, Joseph and J. Chapman, D. (2010). *Red Algae in the Genomic Age*. https://doi.org/10.1007/978-90-481-3795-4

Siegel, B. Z., McMurty, G., Siegel, S. M., Chen, J., and Larock, P. (1979). Life in the calcium chloride environment of Don Juan Pond, Antarctica [6]. *Nature*. https://doi.org/10.1038/280828a0

Sing, D. K., Pont, F., Aigrain, S., Charbonneau, D., Désert, J. M., Gibson, N., ... Shporer, A. (2011). Hubble Space Telescope transmission spectroscopy of the exoplanet HD189733b: High-altitude atmospheric haze in the optical and near-ultraviolet with STIS. *Monthly Notices of the Royal Astronomical Society.* https://doi.org/10.1111/j.1365-2966.2011.19142.x

Sleep, N. H., Zahnle, K. J., Kasting, J. F., and Morowitz, H. J. (1989). Annihilation of ecosystems by large asteroid impacts on the early Earth. *Nature.* https://doi.org/10.1038/342139a0

Stanish, L. F., Nemergut, D. R., and McKnight, D. M. (2011). Hydrologic processes influence diatom community composition in Dry Valley streams. *Journal of the North American Benthological Society.* https://doi.org/10.1899/11-008.1

Stanish, L. F., O'Neill, S. P., Gonzalez, A., Legg, T. M., Knelman, J., Mcknight, D. M., ... Nemergut, D. R. (2013). Bacteria and diatom co-occurrence patterns in microbial mats from polar desert streams. *Environmental Microbiology.* https://doi.org/10.1111/j.1462-2920.2012.02872.x

Stevenson, A., Burkhardt, J., Cockell, C. S., Cray, J. A., Dijksterhuis, J., Fox-Powell, M., ... Hallsworth, J. E. (2015). Multiplication of microbes below 0.690 water activity: Implications for terrestrial and extraterrestrial life. *Environmental Microbiology.* https://doi.org/10.1111/1462-2920.12598

Ten Kate, I. L. (2018). Organic molecules on Mars. *Science.* https://doi.org/10.1126/science.aat2662

Tixador, R., Richoilley, G., Gasset, G., Templier, J., Bes, J. C., Moatti, N., and Lapchine, L. (1985). Study of minimal inhibitory concentration of antibiotics on bacteria cultivated in vitro in space (Cytos 2 experiment). *Aviation Space and Environmental Medicine*, *56*(8), 748–751.

Toner, J. D., Catling, D. C., and Sletten, R. S. (2017). The geochemistry of Don Juan Pond: Evidence for a deep groundwater flow system in Wright Valley, Antarctica. *Earth and Planetary Science Letters.* https://doi.org/10.1016/j.epsl.2017.06.039

Tosca, N. J., Knoll, A. H., and McLennan, S. M. (2008). Water activity and the challenge for life on early Mars. *Science (New York, N.Y.).* https://doi.org/10.1126/science.1155432

Vago, J., Witasse, O., Svedhem, H., Baglioni, P., Haldemann, A., Gianfiglio, G., ... de Groot, R. (2015). ESA ExoMars program: The next step in exploring Mars. *Solar System Research.* https://doi.org/10.1134/S0038094615070199

Valledor, L., Furuhashi, T., and Weckwerth, W. (2013). Systemic cold stress adaptation of Chlamydomonas reinhardtii, 1–61.

Valvano, M. A., Keith, K. E., and Cardona, S. T. (2005). Survival and persistence of opportunistic *Burkholderia* species in host cells, 99–105. https://doi.org/10.1016/j.mib.2004.12.002

Van Horn, D. J., Wolf, C. R., Colman, D. R., Jiang, X., Kohler, T. J., McKnight, D. M., ... Takacs-Vesbach, C. D. (2016). Patterns of bacterial biodiversity in the glacial meltwater streams of the McMurdo Dry Valleys,

Antarctica. *FEMS Microbiology Ecology.* https://doi.org/10.1093/femsec/fiw148

Volkmann, D. (1988). Microgravity and the organisms. Results of the spacelab mission D1. *Acta Astronautica.* https://doi.org/10.1016/0094-5765(88)90036-7

Voorhies, A. A., and Lorenzi, H. A. (2016). The Challenge of Maintaining a Healthy Microbiome during Long-Duration Space Missions. *Frontiers in Astronomy and Space Sciences*, 3(July), 1–7. https://doi.org/10.3389/fspas.2016.00023

Warren-Rhodes, K. A., Dungan, J. L., Piatek, J., Stubbs, K., Gómez-Silva, B., Chen, Y., and McKay, C. P. (2007). Ecology and spatial pattern of cyanobacterial community island patches in the Atacama Desert, Chile. *Journal of Geophysical Research: Biogeosciences.* https://doi.org/10.1029/2006JG000305

Warren-Rhodes, K. A., Rhodes, K. L., Pointing, S. B., Ewing, S. A., Lacap, D. C., Gómez-Silva, B., McKay, C. P. (2006). Hypolithic cyanobacteria, dry limit of photosynthesis, and microbial ecology in the hyperarid Atacama Desert. *Microbial Ecology.* https://doi.org/10.1007/s00248-006-9055-7

Wierzchos, J., Ascaso, C., and McKay, C. P. (2006). Endolithic Cyanobacteria in Halite Rocks from the Hyperarid Core of the Atacama Desert. *Astrobiology.* https://doi.org/10.1089/ast.2006.6.415

Wierzchos, J., Davila, A. F., Artieda, O., Cámara-Gallego, B., de los Ríos, A., Nealson, K. H., Ascaso, C. (2013). Ignimbrite as a substrate for endolithic life in the hyper-arid Atacama Desert: Implications for the search for life on Mars. *Icarus.* https://doi.org/10.1016/j.icarus.2012.06.009

Wierzchos, J., de los Ríos, A., and Ascaso, C. (2012). Microorganisms in desert rocks: The edge of life on Earth. *International Microbiology.* https://doi.org/10.2436/20.1501.01.170

Wierzchos, J., DiRuggiero, J., Vítek, P., Artieda, O., Souza-Egipsy, V., Škaloud, P., Ascaso, C. (2015). Adaptation strategies of endolithic chlorophototrophs to survive the hyperarid and extreme solar radiation environment of the Atacama Desert. *Frontiers in Microbiology.* https://doi.org/10.3389/fmicb.2015.00934

Williams, R. M. E., Grotzinger, J. P., Dietrich, W. E., Gupta, S., Sumner, D. Y., Wiens, R. C., Moores, J. E. (2013). Martian Fluvial Conglomerates at Gale Crater. *Science.* https://doi.org/10.1126/science.1237317

Williams, R. M. E., and Phillips, R. J. (2001). Morphometric measurements of martian valley networks from Mars Orbiter Laser Altimeter (MOLA) data. *Journal of Geophysical Research E: Planets.* https://doi.org/10.1029/2000JE001409

Wlostowski, A. N., Gooseff, M. N., McKnight, D. M., Jaros, C., and Lyons, W. B. (2016). Patterns of hydrologic connectivity in the McMurdo Dry Valleys, Antarctica: a synthesis of 20 years of hydrologic data. *Hydrological Processes.* https://doi.org/10.1002/hyp.10818

Wouters, J. A., Rombouts, F. M., Kuipers, O. P., de Vos, W. M., and Abee, T. (2001). Chapter 4 The role of cold-shock proteins in low-temperature adaptation. *Cell and Molecular Response to Stress*, *2*, 43–56. https://doi.org/10.1016/S1568-1254(01)80006-1

Zhang, B., Yue, L., Zhou, L., Qi, L., Li, J., and Dong, X. (2017). Conserved TRAM domain functions as an Archaeal cold shock protein via RNA chaperone activity. *Frontiers in Microbiology*, *8*(AUG), 1–11. https://doi.org/10.3389/fmicb.2017.01597

Chapter 2

Are we There Yet? Understanding Interplanetary Microbial Hitchhikers using Molecular Methods

Alexander J. Probst[1]* and Parag Vaishampayan[2]

[1]Group for Aquatic Microbial Ecology (GAME), Environmental Microbiology and Biotechnology, Department of Chemistry, University of Duisburg-Essen, Germany
[2]Biotechnology and Planetary Protection Group, Jet Propulsion Laboratory, NASA, California Institute of Technology, Pasadena, CA, 91109, USA

*alexander.probst@uni-due.de

DOI: https://doi.org/10.21775/9781912530304.02

Abstract
Since the early time of space travel, planetary bodies undergoing chemical or biological evolution have been of particular interest for life detection missions. NASA's and ESA's Planetary Protection offices ensure responsible exploration of the solar system and aim at avoiding inadvertent contamination of celestial bodies with biomolecules or even living organisms. Life forms that have the potential to colonize foreign planetary bodies could be a threat to the integrity of science objectives of life detection missions. While standard requirements for assessing the cleanliness of spacecraft are still based on cultivation approaches, several molecular methods have been applied in the past to elucidate the full breadth of (micro)organisms that can be found on spacecraft and in cleanrooms, where the hardware is assembled. Here, we review molecular assays that have been applied in Planetary Protection research and list their significant advantages and disadvantages. By providing a comprehensive summary of the latest molecular methods yet to be applied in this research area, this article will not only aid in designing technological roadmaps for future Planetary Protection endeavors but also help other disciplines in environmental microbiology that deal with low biomass samples.

From required standards to cutting-edge microbiome profiling

Since the early days of space travel, space-faring nations had been interested in the detection of extraterrestrial life, may it exist. Missions to foreign celestial bodies that are of interest in the context of biological evolution can be equipped with highly sensitive instruments. The integrity of these instruments must be ensured by keeping them clean from terrestrial contaminants during assembly, testing and launching operations. This is one reason why spacecraft are assembled in cleanrooms and undergo rigorous cleaning procedures as a measure of Planetary Protection guidelines (COSPAR, 2002).

In the 1970s, The Viking spacecraft destined to Mars underwent rigorous heat sterilization at the system level for days to ensure sterility of the lander (Puleo et al., 1977). Alternative sterilization methods are necessary to ensure the cleanliness of modern heat-sensitive technical equipment. These alternative methods can only perform surface sterilization and miss the 'embedded bioburden', increasing the risk of contamination of alien celestial bodies with terrestrial life. Consequently, the cleanliness of the spacecraft is continuously monitored throughout its time being on Earth with standardized methodologies. These have been established in the early stages of Planetary Protection and are based on the enumeration of cultivable bacterial endospores per square meter surface of spacecraft or cleanrooms (also termed "standard spore assay"). The main advantage of this cost-effective method is that they are standardized across time and thus comparable between missions. The main disadvantage, however, is that they only target cultivable, heat resistant microorganisms, most of which are in dormant states like endospores. Currently, it is estimated that 0.1%-1% of all microorganisms detected via molecular methods can be cultivated under defined laboratory conditions (Tyson and Banfield, 2005). Moreover, the cultivable spore load of spacecraft surfaces does not correlate with the absolute quantity of microorganisms detectable via molecular methods (Cooper et al., 2011). Consequently, the spore load is only an estimation of spacecraft and cleanroom cleanliness but molecular methods are necessary to determine the actual microbial load and particularly the microbial diversity. Since these modern methods have undergone drastic development over the past years and will continue to be improved in the future, they enable researchers better insight into the microbiome structure and its functional profile associated with spacecraft and cleanrooms, alongside the standard spore assay. A summary of the molecular methods that have been applied in Planetary Protection research and two key methods that will hopefully find application in the near future are depicted in Figure 1.

Detection starts with collection

In order to study microorganisms from spacecraft or cleanrooms, the cells need to be recovered from the respective matrix. Air sampling is fairly straight forward, since there are commercial air samplers available that

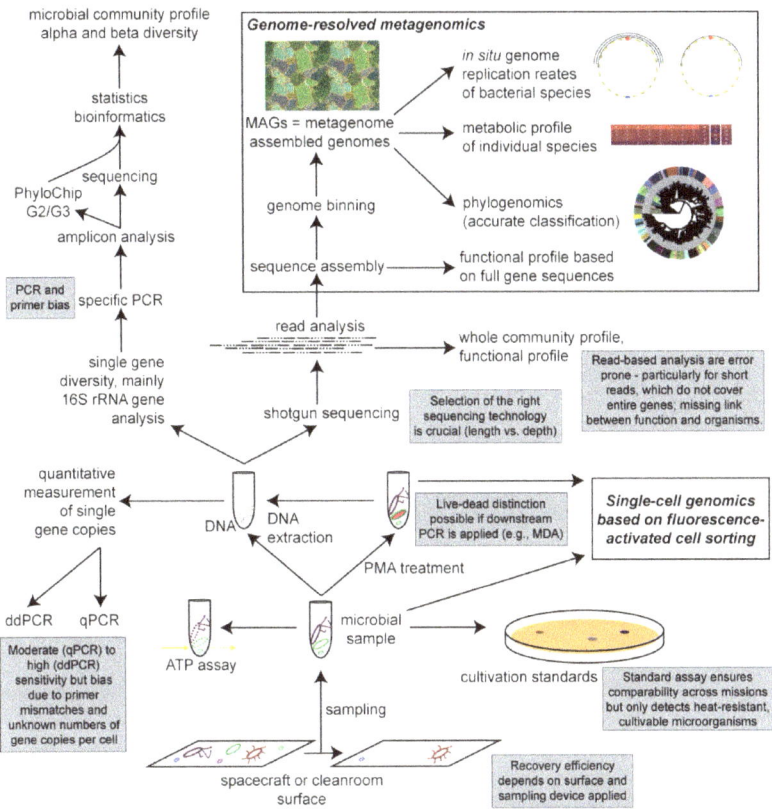

Figure 1. Overview of the molecular methods that have been applied for cleanroom environments and in some cases also for samples from spacecraft hardware. Methods in boxes have not yet been applied but their potential outcomes and limits are discussed in this review. Grey boxes contain comments on the different techniques. Please note that single cell genomics and metagenomics are the only techniques that can result in the detection of viruses because these biological entities do not possess a commonly shared gene that could be used for gene-directed molecular assays.

filter air onto a membrane to catch particles including microorganisms. Retrieving microorganisms from surfaces can be achieved via multiple different methods, whereas wipes and swabs have been the standard for NASA's Planetary Protection over decades (Cooper et al., 2011; La Duc et al., 2012; Ghosh et al., 2010; Probst et al., 2010a, 2010b; Vaishampayan et al., 2010, 2013; Venkateswaran et al., 2001). Other sampling devices (Bargoma et al., 2013; La Duc et al., 2014; Kwan et al., 2011) and their efficiency regarding the retrieval of microorganisms from various surfaces

(Probst et al., 2011) have been investigated and are regularly applied in scientific Planetary Protection studies (La Duc et al., 2004; Mahnert et al., 2015; Moissl-Eichinger et al., 2013; Moissl et al., 2007; Weinmaier et al., 2015). The detection of encapsulated bioburden has proven to be fairly difficult and depending on the matrix, the microbes sometimes cannot be sampled at all (Bauermeister et al., 2014). The evaluation of these techniques has mainly been performed based on endospores or viable bacteria, however, little is known about the recovery efficiency of archaea, eukaryotic cells and viruses from these matrices. In the following paragraphs, we will discuss the analysis of recovered cellular material from any matrix and its downstream analysis and interpretation. An overview of these analyses and the respective methods is provided in Figure 1.

Quantifying contamination
Adenosine triphosphate (ATP) is the main energy currency of all known forms of life. It is fairly stable, for instance, boiling water and alkaline pH over several minutes are necessary to hydrolyze the molecule (Stanley, 1989). Consequently, it is also used as a marker for quantifying microbial activity (Janaszek et al., 1987; Thore et al., 1975; Venkateswaran et al., 2003). Using standardized assays, it is possible to selectively measure the extracellular and intracellular ATP concentration of a sample. This selective ATP quantification technique is used to measure bioburden of spacecraft hardware and associated surfaces in the cleanroom (Figure 1) (Venkateswaran et al., 2003). Total ATP quantification technique was used to rapidly assess cleanliness of spacecraft surfaces in order to assess PP risk (Benardini and Venkateswaran, 2016). However, this quantification is heavily error-prone as i) the sample preparation is dependent on lysozyme-based cell lysis (which does not work for certain microorganisms, e.g. Archaea) and ii) the intracellular ATP concentration depends on the type of microorganisms analyzed and its metabolic state. For example, eukaryotic microorganisms are much larger in cell size and thus usually contain on average many more ATP molecules than prokaryotic cells.

A more accurate way of determining the microbial load of organisms in a cleanroom sample is via quantitative PCR (qPCR; Figure 1). This method has been applied in cleanroom studies to estimate the number of microbes in samples. Usually, this method is based on the amplification of bacterial or archaeal 16S ribosomal RNA genes (rRNA genes or 18S rRNA genes in the case of Eukaryotes) in comparison with standard concentrations. 16S/18S rRNA genes are ubiquitous in all known forms of life on our planet and are a structural component of ribosomes. Targeting individual rRNA genes with qPCR can in theory accurately estimate the number of these genes in a DNA sample. Nevertheless, the method has some disadvantages, which are related to the biology of microorganisms. First of all, the major idea behind bioburden estimation in Planetary Protection is determining the number of microbial cells in a collected sample. However, individual cells can be polyploidic, i.e. have multiple copies of their genomes and thus of their 16S rRNA gene, leading to an overestimation of microbial abundance. Indeed, some organisms have been reported to have hundreds of chromosomes per cell (Ionescu et al., 2017). Although polyploidy usually occurs during high growth rates, which is unlikely to occur in oligotrophic cleanrooms, halophilic archaea have been reported to become polyploidic to use DNA as a phosphate storage (Zerulla et al., 2014). Scientists can currently only speculate about the activity of microbes and their polyploidy in cleanroom environments. Second, some microbial species are known to have multiple copies of 16S rRNA genes encoded on a single genome, adding another uncertainty to the estimation of the microbial load via 16S rRNA gene quantification for bioburden estimation. For instance, one species of *Paenibacillus* isolated from a cleanroom of the European Space Agency is known to have multiple and different copies of 16S rRNA genes (Behrendt et al., 2010). Scientists tried to correct for the bias of targeting multiple copies of 16S rRNA genes of the same chromosome in a metagenomic sample (Moissl-Eichinger, 2011) by considering the average number of bacterial or archaeal 16S rRNA genes per genome (Klappenbach, 2001). However, the average number used was based on the database assembled from sequenced prokaryotes, which was consequently not designed for the diversity of microbes found in spacecraft assembly cleanrooms and much less for the specific diversity of individual samples. As recently revealed, such a process can introduce severe biases and cause a misinterpretation of data (Louca et al., 2018). Third, qPCR, a quantitative PCR approach, is biased toward the primer pair used. In fact, recent investigations regarding the overall diversity that can be reconstructed using metagenomics indicated that conventional 16S rRNA gene primers miss entire clades of organisms (Brown et al., 2015; Eloe-Fadrosh et al., 2016). To what extent these clades, e.g., the Candidate Phyla Radiation of bacteria (Brown et al., 2015), which is estimated to make up nearly one-third of the entire diversity on our planet (Hug et al., 2016), exist in cleanrooms remains unknown due to the lack of genomic information from these ecosystems. Moreover, as any DNA-based method, qPCR is also biased toward the type of DNA extraction method applied. Nevertheless, researchers of the Jet Propulsion Laboratory, Pasadena, have deeply investigated the DNA extraction biases for molecular analyses and have suggested combinations of methods to capture the greatest diversity of organisms in a sample (Cooper et al., 2011). However, to which extent DNA extractions are quantitative for organisms of uncultivated phyla remains to be shown.

Small subunit ribosomal RNA for classifying microbes
Assessing the diversity of microorganisms on spacecraft and in cleanrooms was one of the major foci of research in Planetary Protection over the last two decades. Initiated by Venkateswaran *et al.*, (Venkateswaran et al., 2001) researchers have been probing the diversity of bacteria and archaea using 16S rRNA gene analyses (Figure 1) and found that this diversity is orders of magnitude greater than the cultivable diversity described until that

day (Ghosh et al., 2010; Vaishampayan et al., 2010). With the emergence of next-generation sequencing platforms, this diversity was further explored with some researchers also focusing on the rare biosphere of these ecosystems using deep sequencing of individual samples (Mahnert et al., 2015). Generally, 16S rRNA gene amplicon analyses suffer from the same biases as qPCR techniques, which include DNA extraction, mismatches of primer pairs, and a biased estimation of diversity due to the possibility of multiple 16S rRNA gene copies on a single genome. Moreover, a general PCR bias – the preferred amplification of genes of high abundance and of preferred primer binding – adds another layer of complexity to these analyses. Generally, 16S rRNA gene diversity analyses can be used to infer alpha diversities (e.g., Shannon-Wiener index (Spellerberg and Fedor, 2003) and beta diversities (e.g., changes of microbial communities over time) but existing PCR biases do not allow the estimation of relative abundance of organisms within one sample (changes of relative abundances across samples is indeed possible). Nevertheless, researchers have extensively used 16S rRNA gene amplicon analyses to determine the most abundant taxonomic groups in individual samples from cleanrooms or spacecraft (La Duc et al., 2012; Mahnert et al., 2015; Vaishampayan et al., 2010).

Beyond 16S rRNA gene sequencing, microarray analyses have been used to study the biodiversity changes between samples and over time (Moissl-Eichinger et al., 2015; Vaishampayan et al., 2010, 2013). While PhyloChip G2 (Brodie et al., 2006) and G3 (Hazen et al., 2010) suffer from the ability to accurately call an organism present or absent in a sample, their sensitivity to changes is several orders of magnitude greater than amplicon sequencing due to the amount of PCR product that is hybridized onto the sample (Probst et al., 2014). Although some researchers have looked into using PhyloChip G3 for the detection of microbes that have not been considered in the initial microarray design by compiling new sets of existing probes (Probst et al., 2014), the vastly expanding diversity would require frequent updates of microarray probes. However, due to the drop in sequencing costs and the necessity to accurately call an organism present or absent in a sample have substantially replaced the use of 16S rRNA gene microarray with amplicon sequencing for Planetary Protection research.

Metaomics
Biomass quantities that can be retrieved from cleanrooms and spacecraft surfaces are generally fairly low rendering RNA, protein, or metabolite analyses difficult. However, sequencing of metagenomic DNA extracted from these environments has been successfully applied. Nevertheless, scientists still have to rely on random amplification of the metagenomic DNA (MDA, multi displacement amplification) to create enough biomass for sequencing (Yilmaz et al., 2010) (Figure 1). Randomized amplification of DNA is problematic as it introduced biases to the community, which are

mainly based on the GC content of the template DNA (Probst et al., 2015). Using this approach, scientists have identified a great diversity of microorganisms, including an entire category of biological entities that were previously not reported in cleanrooms: viruses. Weinmaier and co-workers detected signatures of two phage, a Phi29-like virus and an unclassified Siphoviridae, and several viruses associated with humans and other eukaryotes, such as human herpesvirus 4, Cyclovirus TN12, Dragonfly cyclovirus 2, Hypericum japonicum-associated DNA virus, various Fecal-associated gemycircularviruses, and a *Meles meles* fecal virus (Weinmaier et al., 2015)

To overcome the ultralow biomass limitation for generating regular metagenome libraries for sequencing, a larger surface area was sampled. However, if the microbiome is too heterogeneous, this would likely not result in good sequence assemblies and mask the heterogeneous composition of the ecosystem. Hence, a recent study reported the usage of KatharoSeq for generating low-biomass metagenome libraries from cleanroom samples (Minich et al., 2018). However, the metagenomic sample preparation and sequencing technique was not mentioned in detail. Also it remains unclear if amplification steps were involved in this study. Apart from these issues that make KatharoSeq currently not yet attractive for a detailed metagenome study of the functional cleanroom diversity, it seems a very promising technology. Another interesting approach is the generation of regular metagenome libraries from few nanograms using emulsion PCR (Blow et al., 2008) or from biomass as small as a few femtograms (Rinke et al., 2016). Rinke *et al.* applied this protocol to several samples from different environments and produced reliable metagenome data after removing duplicate reads from the samples. This technology should find application in future Planetary Protection endeavors and might enable researchers to go beyond metagenomics of cleanroom samples and produce the first metagenome libraries from spacecraft hardware.

Live-dead distinction
In nature, microorganisms can have several different states of existence: They can either be alive and metabolically active, in a dormant state (e.g., as an endospore), viable but non-culturable (VBNC) or they can simply be dead (Barer and Harwood, 1999; Oliver, 2005, 2010; Xu et al., 1982). Even though distinguishing viable microbial cells from dead cells will have a paramount effect on ecological inferences in cleanrooms and on spacecraft, very few microbial diversity studies take this into account. A lack of understanding about the viability of a microbial population could have serious and sometimes grave consequences, since this is the portion of the community that contributes or affects the ecosystem. This is of particular importance for microbial diversity analyses in food and medical device manufacturing. Estimates of the viable microbial population in the spacecraft assembly cleanroom would help Planetary Protection engineers to calculate the viable microbial bioburden and in turn the Planetary

Protection risk for forward contamination. In-depth understanding of the viable microbes present on the cleanroom environment will also guide the development of more effective bioburden reduction techniques.

Propidium iodide (PI) represents one of the most commonly used fluorescent dyes to determine cell viability by probing the membrane integrity of microorganisms. PI can penetrate only compromised cell membranes and intercalate with DNA resulting in a red fluorescence (excitation 493 nm and emission 636 nm). It is a component of the LIVE/ DEAD BacLight Bacterial Viability Kits (ThermoFisher, USA) along with SYTO 9 (Boulos et al., 1999). The latter stains all cells with a different color resulting in a green (viable) versus red (dead, overshadowing the green) differentiation of dead from viable cells for epifluorescence microscopy, flow cytometry, and fluorometry techniques (Williams et al., 1998). Andreas Nocker and colleagues further developed this technique and replaced PI with propidium monoazide (PMA), which also intercalates into DNA of membrane-compromised cells (Fittipaldi et al., 2012; Nocker et al., 2006, 2007). However, the azide group can be photoactivated resulting in a covalent bond between DNA and the fluorescence dye. Followed by DNA-extraction and PCR reaction (e.g., 16S rRNA gene PCR), this method is selective for viable microorganisms, since the DNA of the PMA-tagged cells can no longer be amplified due to a steric hindrance of the DNA polymerase in binding.

Vaishampayan *et al.* reported pre-PCR propidium monoazide (PMA) treatment of samples followed by downstream 16S rRNA gene analyses (via qPCR, pyrosequencing and PhyloChip DNA microarray) to understand the diversity and distribution of the viable bacterial population in spacecraft assembly cleanrooms (Vaishampayan et al., 2010) (Figure 1). Their results demonstrate a substantially lower bioburden of viable cells compared to total cells and a very limited diversity of living microorganisms. One step further, Weinmaier *et al.* published the first viability-linked metagenomic analysis of cleanroom environments resulting in many novel findings including viruses (see above). Subsequently, Mahnert *et al.* published another PMA-based viability study on spacecraft assembly cleanrooms reporting the effect of cleanroom maintenance on microbial diversity and abundance.

Another advantage of PMA is its application in reduction of contaminants during sample processing to ensure the cleanliness of the reagents applied. Here, PMA is used to remove contaminating extracellular DNA present in almost all commercial PCR reagents (Salter et al., 2014), particularly while using low biomass samples such as cleanrooms. PCR reagents can be treated with PMA to exclude contaminating DNA from amplification during the PCR reaction. Thus treatment of both environmental samples and PCR reagents could improve the detection of viable cells from low biomass samples (Schnetzinger et al., 2013). Several other cultivation-independent techniques such as stable isotope labeling (Dumont and Murrell, 2005; Fischer and Pusch, 1999), respiration detection (Winding et al., 1994), BONCAT (Hatzenpichler et al., 2014) , isothermal microcalorimetary (IMC) (Rong et al., 2007) were recently reviewed (Emerson et al., 2017). These techniques would, however, necessitate the establishment of mesocosms of cleanroom populations and thus result in skewing the community structure. Nevertheless, these techniques could be useful for hypothesis testing regarding the metabolic activity of organisms in response to a certain substrate.

Open questions

Planetary Protection research, in general, has been very descriptive by cataloging the microbial diversity in cleanroom environments and associated spacecraft hardware. Little effort has been performed in understanding the ecology of these built environments. First and foremost, a general understanding of the entire breadth of organisms and viruses in these ecosystem needs to be established. Most of the assays are geared towards the detection of Bacteria, although Archaea (Moissl-Eichinger, 2011; Moissl-Eichinger et al., 2015) and even Eukarya (La Duc et al., 2012) and viruses (Weinmaier et al., 2015) have also been detected in these ecosystems. It is obvious, that current sampling techniques of surfaces can recover these organisms but the actual efficiency has only been established for a few bacterial strains (Bargoma et al., 2013; Probst et al., 2010b).

More importantly, the general nature of the assembly of the ecosystem of the cleanrooms needs to be deciphered. Based on the current understanding, the ecosystem of cleanroom facilities has substantial selective pressures on microbes. These pressures arise from the harsh cleaning procedures and environmental conditions that are maintained within these facilities limiting the survival of organisms. They also result in little nutrient availability (oligotrophy) posing a challenge for most microbes to thrive in these ecosystems, although a recent study reported the growth of bacterial species from a cleanroom on cleaning reagents (Mogul et al., 2018). In theory, only a small portion of the detectable microbiome thrives in these environments and other microbes are random contaminants in cleanrooms. This enables the assumption that cleanroom microbiomes do neither follow a deterministic model nor a pure stochastic ecosystem assembly (Dumbrell et al., 2010; Hubbell, 2001; Langenheder and Székely, 2011; Ofiteru et al., 2010). Indeed, multiple factors like biogeography of skin microbiome, soil composition of the surrounding ecosystem, weather influences on soil and hardware entering the cleanroom serve as sources for microbial dispersal and suggest a stochastic model for the inactive community. In contrast, the active community, microbes that might grow and increase in cell numbers in this oligotrophic environment, should in theory follow a deterministic model for ecosystem assembly.

A few studies have looked into beta diversity changes of the microbiomes in spacecraft assembly cleanrooms (Moissl et al., 2007; Vaishampayan et al., 2010) to understand the temporal or spatial differences of the microbial communities. Moissl and co-workers concluded that the surrounding ecosystem of the cleanroom buildings substantially impacts the detected biodiversity (Moissl et al., 2007). Particularly 16S rRNA genes of microbes putatively originating from soil were detected. This conclusion is based on studying geographically distinct cleanroom facilities. However, the conclusion is questionable considering that a) the cleanrooms had different maintenance procedures (e.g. particulate filtering) and b) different people were working in these cleanrooms. The human microbiome shows highly significant variations between human beings (Kolde et al., 2018; Morgan et al., 2013); ergo, different workers would ultimately mean the transport of different microbiomes into the cleanrooms (e.g. by shedding skin particles) but linking signatures of microbes from cleanrooms to those of the workers has not been performed yet. A first attempt has been done for Archaea, for which it has been shown that certain 16S rRNA genes of Thaumarchaeota are also present on human skin (Probst et al., 2013). Other studies have tried to identify contamination routes and followed microbial signatures from outside the cleanroom (e.g., changing room) into the actual cleanroom (Mahnert et al., 2015) as a first attempt to identify how these ecosystems assemble over time.

In the future, rigorous source tracking of microbes involving sampling the workers' microbiome would need to be performed to understand how human beings impact the cleanroom microbiome. For instance, it is unclear if human skin particles can serve as nutrients for microorganisms that survive under the harsh cleanroom conditions. At the same time, contamination routes of microbes on hardware entering the cleanroom and from the surrounding ecosystem and would need to be explored in detail, which would involve sampling the outside of the assembly facilities. At the same time, the active microbiome in these facilities needs to be identified by going beyond simple live/dead distinction. Activity measurements linked to phylogeny are necessary to understand the active portion of microbes that might assemble via deterministic processes. Measuring activity of microbes in an ecosystem is generally hard to achieve (see below for calculating *in situ* replication rates) without bringing the microorganisms into an enrichment culture. However, enrichment can be used to test important hypotheses like the growth of microbes on cleaning agents as performed recently. Here, Mogul et al 2018 showed that spacecraft cleaning reagents may serve as nutrient sources under oligotrophic conditions (Mogul et al., 2018). The researchers demonstrated that spacecraft associated *Acinetobacter* strains, one of the dominating and recurring microbial species, can grow on ethanol (ethyl alcohol), 2-propanol (isopropyl alcohol), or Kleenol 30 (floor detergent) under minimal conditions in the laboratory. Results of this study enable speculation about the survival and dynamics of the active microorganisms in spacecraft-associated environments suggesting a partially deterministic ecosystem assembly.

Next steps in microbiome profiling

Planetary Protection research has mostly been lacking behind several years compared to the state-of-the-art in environmental microbiology. For instance, cloning and sequencing of 16S rRNA genes from environmental samples was published in 1990 (Giovannoni et al., 1990), yet it took more than ten years until the technique was applied to cleanroom environments by Venkateswaran et al. (2001). However, NASA's Jet Propulsion Laboratory Planetary Protection research group has generally been employing cutting edge technology ever since to study cleanroom diversity, e.g., 16S rRNA gene microarrays called PhyloChip, next generation sequencing and shotgun metagenome sequencing (La Duc et al., 2009, 2012; Vaishampayan et al., 2010; Weinmaier et al., 2015).

There have been several technical advances in microbiome research in the recent years that will prove useful in the near future for Planetary Protection research and that have not yet been applied. Some of these advances regard 16S rRNA gene analyses, including the usage of long-read sequencing for a fairly accurate micro-diversity measure. For instance, PacBio sequencing of circularized 16S rRNA gene amplicons can detect microdiversity of bacteria in environmental samples (Singer et al., 2017). Another 16S rRNA gene-based technology is its accurate and very sensitive quantifications using digital droplet PCR (ddPCR) (Hu et al., 2014; Lin et al., 2017; White et al., 2009); Multiple 16S rRNA genes can occur per genome and the restricted diversity cannot be represented by an overall correction for this phenomenon as performed earlier (see above). We suggest that a comprehensive database of genomes from cleanrooms and their relative distribution based on amplicon sequencing could be used to qualitatively correct for the occurrence of multiple 16S rRNA genes in genomes. Such a genome database could be generated either from public reference genomes matching cleanroom 16S rRNA gene data or directly by resolving population genomes via environmental genomics. Particularly the latter is a major advance in the field of environmental microbiology but was already introduced in 2004 (Tyson et al., 2004) and has made substantial advantages since (Sieber et al., 2018).

The low biomass of cleanroom environments has so far been very challenging in producing high-quality metagenomes and has so far not enabled researchers to perform genome-resolved metagenomics (Figure 1). However, there are multiple promising techniques available that will help to achieve this goal as outline above. Having population genomes from cleanrooms and/or spacecraft surfaces at hand would not only enable the above-mentioned correction of 16S rRNA gene surveys but also bolster the understanding of the metabolic diversity of microorganisms in cleanrooms. Moreover, the known diversity of mobile genetic elements in these environments would be greatly enhanced leading to important insights for pharmaceutical sciences. A recent bioinformatics technology that enables researchers to calculate genome replication rates of microorganisms from sequencing reads (Korem et al., 2015) would provide further insight into the ecology of cleanrooms by deciphering which organism actively replicates under these harsh conditions. During this procedure non-MDA biased metagenomic sequence reads are mapped to microbial genomes and the relative abundance of sequence information of the origin of replication is compared to the terminus of replication. The difference in the relative abundance can be interpreted as the presence of replication forks running from the origin to the terminus of replication. The result is an average of the entire population of a representative genome and can also be applied to genomes from metagenomes (Brown et al., 2016).

Last but not least, a really elegant and probably the most feasible approach for generating genomes from cleanroom samples would be the use of single-cell genomics, which has in the past lead to the discovery of several hundred novel lineages from environmental samples (Rinke et al., 2013). This approach is based on the sorting of particles that have been stained with a DNA-intercalating dye, and might even be combined with PMA-treatment of samples to selectively sort living microorganisms (Figure 1). After sorting, the cell is lysed and the DNA is amplified using an MDA approach, followed by sequencing (Rinke et al., 2014). Interestingly, this approach can, in theory, also result in the detection of viruses if these are integrated as prophages, attached to their host or have infected the host. Compared to other techniques, this method is particularly useful for low-biomass samples as retrieved from cleanroom environments or even spacecraft hardware and could provide substantial information on the metabolism of microorganisms that reside there and even resolve their strain distribution (Blainey, 2013).

Molecular methods have provided substantial insight into the cultivable and not-yet-cultivable microbiome of cleanrooms and spacecraft surfaces over the past two decades. These methods enabled researchers multiple break throughs including the detection of viruses, identifying contamination routes and potential sources of contaminants, deciphering the viable microbiome and investigating the dynamics of the microbiome in these ecosystems. Although the ecosystem assembly of cleanroom facilities has not yet been deciphered, there are already novel technologies on the horizon that are just waiting to be applied by Planetary Protection researchers to fully understand these ecosystems and minimize the risk of microbial hitchhikers on spacecraft destined to foreign planetary bodies.

Acknowledgements
Funding by the Ministerium für Kultur und Wissenschaft des Landes Nordrhein-Westfalen ("Nachwuchsgruppe Dr. Alexander Probst") is acknowledged. Part of the work was carried out at the Jet Propulsion

Laboratory (California Institute of Technology, Pasadena) under a contract with the National Aeronautics and Space Administration.

References

Alexander J. Probst, Pek Yee Lum, Bettina John, Eric A. Dubinsky, Yvette M. Piceno, Lauren M. Tom, Gary L. Andersen, Z.H. and T.Z.D. (2014). Microarray of 16S rRNA Gene Probes for Quantifying Population Differences Across Microbiome Samples. In Microarrays: Current Technology, Innovations and Applications, Zhili He, ed. (Caister Academic Press, U.K.), p.

Barer, M.R., and Harwood, C.R. (1999). Bacterial viability and culturability. Adv. Microb. Physiol. *41*, 93–137.

Bargoma, E., La Duc, M.T., Kwan, K., Vaishampayan, P., and Venkateswaran, K. (2013). Differential recovery of phylogenetically disparate microbes from spacecraft-qualified metal surfaces. Astrobiology *13*. https://doi.org/10.1089/ast.2012.0917

Bauermeister, A., Mahnert, A., Auerbach, A., Böker, A., Flier, N., Weber, C., Probst, A.J., Moissl-Eichinger, C., and Haberer, K. (2014). Quantification of Encapsulated Bioburden in Spacecraft Polymer Materials by Cultivation-Dependent and Molecular Methods. PLoS One *9*, e94265. https://doi.org/10.1371/journal.pone.0094265

Behrendt, U., Schumann, P., Stieglmeier, M., Pukall, R., Augustin, J., Spröer, C., Schwendner, P., Moissl-Eichinger, C., and Ulrich, A. (2010). Characterization of heterotrophic nitrifying bacteria with respiratory ammonification and denitrification activity - Description of Paenibacillus uliginis sp. nov., an inhabitant of fen peat soil and Paenibacillus purispatii sp. nov., isolated from a spac. Syst. Appl. Microbiol. *33*, 328–336. https://doi.org/10.1016/j.syapm.2010.07.004

Benardini, J.N., and Venkateswaran, K. (2016). Application of the ATP assay to rapidly assess cleanliness of spacecraft surfaces: a path to set a standard for future missions. AMB Express *6*, 113. https://doi.org/10.1186/s13568-016-0286-9

Blainey, P.C. (2013). The future is now: single-cell genomics of bacteria and archaea. FEMS Microbiol. Rev. *37*, 407–427. https://doi.org/10.1111/1574-6976.12015

Blow, M.J., Zhang, T., Woyke, T., Speller, C.F., Krivoshapkin, A., Yang, D.Y., Derevianko, A., and Rubin, E.M. (2008). Identification of ancient remains through genomic sequencing. Genome Res. *18*, 1347–1353. https://doi.org/10.1101/gr.076091.108

Boulos, L., Prévost, M., Barbeau, B., Coallier, J., and Desjardins, R. (1999). LIVE/DEAD®BacLight™: application of a new rapid staining method for direct enumeration of viable and total bacteria in drinking water. J. Microbiol. Methods *37*, 77–86. https://doi.org/10.1016/S0167-7012(99)00048-2

Brodie, E.L., DeSantis, T.Z., Joyner, D.C., Baek, S.M., Larsen, J.T., Andersen, G.L., Hazen, T.C., Richardson, P.M., Herman, D.J., Tokunaga, T.K., et al. (2006). Application of a high-density oligonucleotide microarray approach to study bacterial population dynamics during uranium reduction and reoxidation. Appl. Environ. Microbiol. 72, 6288–6298. https://doi.org/10.1128/AEM.00246-06

Brown, C.T., Hug, L.A., Thomas, B.C., Sharon, I., Castelle, C.J., Singh, A., Wilkins, M.J., Wrighton, K.C., Williams, K.H., and Banfield, J.F. (2015). Unusual biology across a group comprising more than 15% of domain Bacteria. Nature 523, 208–211. https://doi.org/10.1038/nature14486

Brown, C.T., Olm, M.R., Thomas, B.C., and Banfield, J.F. (2016). Measurement of bacterial replication rates in microbial communities. Nat. Biotechnol. 34, 1256–1263. https://doi.org/10.1038/nbt.3704

Cooper, M., La Duc, M.T., Probst, A., Vaishampayan, P., Stam, C., Benardini, J.N., Piceno, Y.M., Andersen, G.L., and Venkateswaran, K. (2011). Comparison of innovative molecular approaches and standard spore assays for assessment of surface cleanliness. Appl. Environ. Microbiol. 77, 5438–5444. https://doi.org/10.1128/AEM.00192-11

COSPAR (2002). Planetary Protection Policy, October 2002, as amended, March 2008. In Planetary Protection Policy, October 2002, as Amended, March 2008, Committee of Space Research 2002, (Houston, TX: COSPAR), p. http://www.cosparhq.org/scistr/PPPolicy.htm.

Crespo, B.G., Wallhead, P.J., Logares, R., and Pedrós-Alió, C. (2016). Probing the Rare Biosphere of the North-West Mediterranean Sea: An Experiment with High Sequencing Effort. PLoS One 11, e0159195. https://doi.org/10.1371/journal.pone.0159195

La Duc, M.T., Kern, R., and Venkateswaran, K. (2004). Microbial monitoring of spacecraft and associated environments. In Microbial Ecology, 47(2): 150-158. https://doi.org/10.1007/s00248-003-1012-0

La Duc, M.T., Osman, S., Vaishampayan, P., Piceno, Y., Andersen, G., Spry, J. a, and Venkateswaran, K. (2009). Comprehensive census of bacteria in clean rooms by using DNA microarray and cloning methods. Appl. Environ. Microbiol. 75, 6559–6567. https://doi.org/10.1128/AEM.01073-09

La Duc, M.T., Vaishampayan, P., Nilsson, H.R., Torok, T., and Venkateswaran, K. (2012). Pyrosequencing-derived bacterial, archaeal, and fungal diversity of spacecraft hardware destined for Mars. Appl. Environ. Microbiol. 78, 5912–5922. https://doi.org/10.1128/AEM.01435-12

La Duc, M.T., Venkateswaran, K., and Conley, C.A. (2014). A Genetic Inventory of Spacecraft and Associated Surfaces. Astrobiology 14, 15–23. https://doi.org/10.1007/s00248-003-1012-0

Dumbrell, A.J., Nelson, M., Helgason, T., Dytham, C., and Fitter, A.H. (2010). Relative roles of niche and neutral processes in structuring a soil microbial community. ISME J. 4, 337–345. https://doi.org/10.1038/ismej.2009.122

Dumont, M.G., and Murrell, J.C. (2005). Stable isotope probing — linking microbial identity to function. Nat. Rev. Microbiol. 3, 499–504. https://doi.org/10.1038/nrmicro1162

Eloe-Fadrosh, E.A., Ivanova, N.N., Woyke, T., and Kyrpides, N.C. (2016). Metagenomics uncovers gaps in amplicon-based detection of microbial diversity. Nat. Microbiol. *1*, 15032. https://doi.org/10.1038/nmicrobiol.2015.32

Emerson, J.B., Adams, R.I., Román, C.M.B., Brooks, B., Coil, D.A., Dahlhausen, K., Ganz, H.H., Hartmann, E.M., Hsu, T., Justice, N.B., et al. (2017). Schrödinger's microbes: Tools for distinguishing the living from the dead in microbial ecosystems. Microbiome *5*. https://doi.org/10.1186/s40168-017-0285-3

Fischer, H., and Pusch, M. (1999). Use of the [(14)C]leucine incorporation technique to measure bacterial production in river sediments and the epiphyton. Appl. Environ. Microbiol. *65*, 4411–4418.

Fittipaldi, M., Nocker, A., and Codony, F. (2012). Progress in understanding preferential detection of live cells using viability dyes in combination with DNA amplification. J. Microbiol. Methods *91*, 276–289. https://doi.org/10.1016/j.mimet.2012.08.007

Ghosh, S., Osman, S., Vaishampayan, P., and Venkateswaran, K. (2010). Recurrent isolation of extremotolerant bacteria from the clean room where phoenix spacecraft components were assembled. Astrobiology *10*. https://doi.org/10.1089/ast.2009.0396

Giovannoni, S.J., Britschgi, T.B., Moyer, C.L., and Field, K.G. (1990). Genetic diversity in Sargasso Sea bacterioplankton. Nature *345*, 60–63. https://doi.org/10.1038/345060a0

Hatzenpichler, R., Scheller, S., Tavormina, P.L., Babin, B.M., Tirrell, D.A., and Orphan, V.J. (2014). In situ visualization of newly synthesized proteins in environmental microbes using amino acid tagging and click chemistry. Environ. Microbiol. *16*, 2568–2590. https://doi.org/10.1111/1462-2920.12436

Hazen, T.C., Dubinsky, E.A., DeSantis, T.Z., Andersen, G.L., Piceno, Y.M., Singh, N., Jansson, J.K., Probst, A., Borglin, S.E., Fortney, J.L., et al. (2010). Deep-sea oil plume enriches indigenous oil-degrading bacteria. Science *330*, 204–208. https://doi.org/10.1126/science.1195979

Hu, W., Chen, R., Zhang, C., An, Z., Wang, B., and Ping, Y. (2014). [Species identification and absolute quantification of biological samples by droplet digital PCR]. Fa Yi Xue Za Zhi *30*, 342–345.

Hubbell, S.P. (2001). The unified neutral theory of biodiversity and biogeography (Princeton University Press).

Hug, L.A., Baker, B.J., Anantharaman, K., Brown, C.T., Probst, A.J., Castelle, C.J., Butterfield, C.N., Hernsdorf, A.W., Amano, Y., Ise, K., et al. (2016). A new view of the tree of life. Nat. Microbiol. *1*, 16048. https://doi.org/10.1038/nmicrobiol.2016.48

Ionescu, D., Bizic-Ionescu, M., De Maio, N., Cypionka, H., and Grossart, H.P. (2017). Community-like genome in single cells of the sulfur bacterium Achromatium oxaliferum. Nat. Commun. https://doi.org/10.1016/0306-2619(86)90021-8

Janaszek, W., Aleksandrowicz, J., and Sitkiewicz, D. (1987). The use of the firefly bioluminescent reaction for the rapid detection and counting of mycobacterium BCG. J. Biol. Stand. *15*, 11–16.

Klappenbach, J.A. (2001). rrndb: the Ribosomal RNA Operon Copy Number Database. Nucleic Acids Res. https://doi.org/10.1093/nar/29.1.181

Kolde, R., Franzosa, E.A., Rahnavard, G., Hall, A.B., Vlamakis, H., Stevens, C., Daly, M.J., Xavier, R.J., and Huttenhower, C. (2018). Host genetic variation and its microbiome interactions within the Human Microbiome Project. Genome Med. *10*, 6. https://doi.org/10.1186/s13073-018-0515-8

Korem, T., Zeevi, D., Suez, J., Weinberger, A., Avnit-Sagi, T., Pompan-Lotan, M., Matot, E., Jona, G., Harmelin, A., Cohen, N., et al. (2015). Growth dynamics of gut microbiota in health and disease inferred from single metagenomic samples. Science *349*, 1101–1106. https://doi.org/10.1126/science.aac4812

Kwan, K., Cooper, M., La Duc, M.T., Vaishampayan, P., Stam, C., Benardini, J.N., Scalzi, G., Moissl-Eichinger, C., and Venkateswaran, K. (2011). Evaluation of procedures for the collection, processing, and analysis of biomolecules from low-biomass surfaces. Appl. Environ. Microbiol. *77*, 2943–2953. https://doi.org/10.1128/AEM.02978-10

Langenheder, S., and Székely, A.J. (2011). Species sorting and neutral processes are both important during the initial assembly of bacterial communities. ISME J. *5*, 1086–1094. https://doi.org/10.1038/ismej.2010.207

Lin, J., Su, G., Su, W., and Zhou, C. (2017). [Progress in digital PCR technology and application]. Sheng Wu Gong Cheng Xue Bao *33*, 170–177. https://doi.org/10.13345/j.cjb.160269

Louca, S., Doebeli, M., and Parfrey, L.W. (2018). Correcting for 16S rRNA gene copy numbers in microbiome surveys remains an unsolved problem. Microbiome *6*, 41. https://doi.org/10.1186/s40168-018-0420-9

Mahnert, A., Vaishampayan, P., Probst, A.J., Auerbach, A., Moissl-Eichinger, C., Venkateswaran, K., and Berg, G. (2015). Cleanroom maintenance significantly reduces abundance but not diversity of indoor microbiomes. PLoS One *10*. https://doi.org/10.1371/journal.pone.0134848

Minich, J.J., Zhu, Q., Janssen, S., Hendrickson, R., Amir, A., Vetter, R., Hyde, J., Doty, M.M., Stillwell, K., Benardini, J., et al. (2018). KatharoSeq Enables High-Throughput Microbiome Analysis from Low-Biomass Samples. MSystems *3*. https://doi.org/10.1128/mSystems.00218-17

Mogul, R., Barding, G.A., Lalla, S., Lee, S., Madrid, S., Baki, R., Ahmed, M., Brasali, H., Cepeda, I., Gornick, T., et al. (2018). Metabolism and Biodegradation of Spacecraft Cleaning Reagents by Strains of Spacecraft-Associated *Acinetobacter*. Astrobiology. https://doi.org/10.1089/ast.2017.1814

Moissl-Eichinger, C. (2011). Archaea in artificial environments: Their presence in global spacecraft clean rooms and impact on planetary protection. ISME J. *5*, 209–219. https://doi.org/10.1038/ismej.2010.124

Moissl-Eichinger, C., Pukall, R., Probst, A.J., Stieglmeier, M., Schwendner, P., Mora, M., Barczyk, S., Bohmeier, M., and Rettberg, P. (2013). Lessons Learned from the Microbial Analysis of the Herschel Spacecraft during Assembly, Integration, and Test Operations. Astrobiology. https://doi.org/10.1089/ast.2013.1024

Moissl-Eichinger, C., Auerbach, A.K., Probst, A.J., Mahnert, A., Tom, L., Piceno, Y., Andersen, G.L., Venkateswaran, K., Rettberg, P., Barczyk, S., et al. (2015). Quo vadis? Microbial profiling revealed strong effects of cleanroom maintenance and routes of contamination in indoor environments. Sci. Rep. *5*, 9156. https://doi.org/10.1038/srep09156

Moissl, C., Osman, S., La Duc, M.T., Dekas, A., Brodie, E., DeSantis, T., and Venkateswaran, K. (2007). Molecular bacterial community analysis of clean rooms where spacecraft are assembled. FEMS Microbiol. Ecol. https://doi.org/10.1111/j.1574-6941.2007.00360.x

Morgan, X.C., Segata, N., and Huttenhower, C. (2013). Biodiversity and functional genomics in the human microbiome. Trends Genet. *29*, 51–58. https://doi.org/10.1016/j.tig.2012.09.005

Nocker, A., Cheung, C.Y., and Camper, A.K. (2006). Comparison of propidium monoazide with ethidium monoazide for differentiation of live vs. dead bacteria by selective removal of DNA from dead cells. J Microbiol Methods *67*, 310–320. https://doi.org/S0167-7012(06)00125-4 [pii]10.1016/j.mimet.2006.04.015

Nocker, A., Sossa-Fernandez, P., Burr, M.D., and Camper, A.K. (2007). Use of Propidium Monoazide for Live/Dead Distinction in Microbial Ecology. Appl. Environ. Microbiol. *73*, 5111–5117. https://doi.org/10.1128/AEM.02987-06

Ofiteru, I.D., Lunn, M., Curtis, T.P., Wells, G.F., Criddle, C.S., Francis, C.A., and Sloan, W.T. (2010). Combined niche and neutral effects in a microbial wastewater treatment community. Proc. Natl. Acad. Sci. U. S. A. *107*, 15345–15350. https://doi.org/10.1073/pnas.1000604107

Oliver, J.D. (2005). The viable but nonculturable state in bacteria. J. Microbiol. *43 Spec No*, 93–100.

Oliver, J.D. (2010). Recent findings on the viable but nonculturable state in pathogenic bacteria. FEMS Microbiol. Rev. *34*, 415–425. https://doi.org/10.1111/j.1574-6976.2009.00200.x

Probst, A., Vaishampayan, P., Osman, S., Moissl-Eichinger, C., Andersen, G.L., and Venkateswaran, K. (2010a). Diversity of anaerobic microbes in spacecraft assembly clean rooms. Appl Env. Microbiol *76*, 2837–2845. https://doi.org/AEM.02167-09 [pii]10.1128/AEM.02167-09

Probst, A., Facius, R., Wirth, R., and Moissl-Eichinger, C. (2010b). Validation of a Nylon-Flocked-Swab Protocol for Efficient Recovery of Bacterial Spores from Smooth and Rough Surfaces. Appl. Environ. Microbiol. *76*, 5148–5158. https://doi.org/10.1128/AEM.00399-10

Probst, A., Facius, R., Wirth, R., Wolf, M., and Moissl-Eichinger, C. (2011). Recovery of *Bacillus* Spore Contaminants from Rough Surfaces: a Challenge to Space Mission Cleanliness Control. Appl. Environ. Microbiol. 77, 1628–1637. https://doi.org/10.1128/AEM.02037-10

Probst, A.J., Auerbach, A.K., and Moissl-Eichinger, C. (2013). Archaea on human skin. PLoS One 8, e65388. https://doi.org/10.1371/journal.pone.0065388

Probst, A.J., Birarda, G., Holman, H.-Y.N., DeSantis, T.Z., Wanner, G., Andersen, G.L., Perras, A.K., Meck, S., Völkel, J., Bechtel, H.A., et al. (2014). Coupling Genetic and Chemical Microbiome Profiling Reveals Heterogeneity of Archaeome and Bacteriome in Subsurface Biofilms That Are Dominated by the Same Archaeal Species. PLoS One 9, e99801. https://doi.org/10.1371/journal.pone.0099801

Probst, A.J., Weinmaier, T., DeSantis, T.Z., Santo Domingo, J.W., and Ashbolt, N. (2015). New Perspectives on Microbial Community Distortion after Whole-Genome Amplification. PLoS One 10, e0124158. https://doi.org/10.1371/journal.pone.0124158

Puleo, J.R., Fields, N.D., Bergstrom, S.L., Oxborrow, G.S., Stabekis, P.D., and Koukol, R. (1977). Microbiological profiles of the Viking spacecraft. Appl. Environ. Microbiol. 33, 379–384.

Rinke, C., Schwientek, P., Sczyrba, A., Ivanova, N.N., Anderson, I.J., Cheng, J.-F., Darling, A., Malfatti, S., Swan, B.K., Gies, E.A., et al. (2013). Insights into the phylogeny and coding potential of microbial dark matter. Nature 499, 431–437. https://doi.org/10.1038/nature12352

Rinke, C., Lee, J., Nath, N., Goudeau, D., Thompson, B., Poulton, N., Dmitrieff, E., Malmstrom, R., Stepanauskas, R., and Woyke, T. (2014). Obtaining genomes from uncultivated environmental microorganisms using FACS–based single-cell genomics. Nat. Protoc. 9, 1038–1048. https://doi.org/10.1038/nprot.2014.067

Rinke, C., Low, S., Woodcroft, B.J., Raina, J.-B., Skarshewski, A., Le, X.H., Butler, M.K., Stocker, R., Seymour, J., Tyson, G.W., et al. (2016). Validation of picogram- and femtogram-input DNA libraries for microscale metagenomics. PeerJ 4, e2486. https://doi.org/10.7717/peerj.2486

Rong, X.-M., Huang, Q.-Y., Jiang, D.-H., Cai, P., and Liang, W. (2007). Isothermal Microcalorimetry: A Review of Applications in Soil and Environmental Sciences. Pedosphere 17, 137–145. https://doi.org/10.1016/S1002-0160(07)60019-8

Salter, S.J., Cox, M.J., Turek, E.M., Calus, S.T., Cookson, W.O., Moffatt, M.F., Turner, P., Parkhill, J., Loman, N.J., and Walker, A.W. (2014). Reagent and laboratory contamination can critically impact sequence-based microbiome analyses. BMC Biol. 12. https://doi.org/10.1186/s12915-014-0087-z

Schnetzinger, F., Pan, Y., and Nocker, A. (2013). Use of propidium monoazide and increased amplicon length reduce false-positive signals in quantitative PCR for bioburden analysis. Appl. Microbiol. Biotechnol. 97, 2143–2162. https://doi.org/10.1007/s00253-013-4711-6

Sieber, C.M.K., Probst, A.J., Sharrar, A., Thomas, B.C., Hess, M., Tringe, S.G., and Banfield, J.F. (2018). Recovery of genomes from metagenomes via a dereplication, aggregation and scoring strategy. Nat. Microbiol. *3*, 836–843. https://doi.org/10.1038/s41564-018-0171-1

Singer, E., Wagner, M., and Woyke, T. (2017). Capturing the genetic makeup of the active microbiome in situ. ISME J. *11*, 1949–1963. https://doi.org/10.1038/ismej.2017.59

Spellerberg, I.F., and Fedor, P.J. (2003). A tribute to Claude Shannon (1916-2001) and a plea for more rigorous use of species richness, species diversity and the 'Shannon-Wiener' Index. Glob. Ecol. Biogeogr. *12*, 177–179. https://doi.org/10.1046/j.1466-822X.2003.00015.x

Stanley, P.E. (1989). A review of bioluminescent ATP techniques in rapid microbiology. J. Biolumin. Chemilumin. *4*, 375–380. https://doi.org/10.1002/bio.1170040151

Thore, A., Anséhn, S., Lundin, A., and Bergman, S. (1975). Detection of bacteriuria by luciferase assay of adenosine triphosphate. J. Clin. Microbiol. *1*, 1–8.

Tyson, G.W., and Banfield, J.F. (2005). Cultivating the uncultivated: a community genomics perspective. Trends Microbiol. *13*, 411–415. https://doi.org/10.1016/j.tim.2005.07.003

Tyson, G.W., Chapman, J., Hugenholtz, P., Allen, E.E., Ram, R.J., Richardson, P.M., Solovyev, V. V., Rubin, E.M., Rokhsar, D.S., and Banfield, J.F. (2004). Community structure and metabolism through reconstruction of microbial genomes from the environment. Nature *428*, 37–43. https://doi.org/10.1038/nature02340

Vaishampayan, P., Osman, S., Andersen, G., and Venkateswaran, K. (2010). High-density 16S microarray and clone library-based microbial community composition of the Phoenix spacecraft assembly clean room. Astrobiology *10*, 499–508. https://doi.org/10.1089/ast.2009.0443

Vaishampayan, P., Probst, A.J., La Duc, M.T., Bargoma, E., Benardini, J.N., Andersen, G.L., and Venkateswaran, K. (2013). New perspectives on viable microbial communities in low-biomass cleanroom environments. ISME J. *7*. https://doi.org/10.1038/ismej.2012.114

Venkateswaran, K., Satomi, M., Chung, S., Kern, R., Koukol, R., Basic, C., and White, D. (2001). Molecular microbial diversity of a spacecraft assembly facility. Syst. Appl. Microbiol. https://doi.org/10.1078/0723-2020-00018

Venkateswaran, K., Hattori, N., La Duc, M.T., and Kern, R. (2003). ATP as a biomarker of viable microorganisms in clean-room facilities. J. Microbiol. Methods *52*, 367–377.

Weinmaier, T., Probst, A.J., La Duc, M.T., Ciobanu, D., Cheng, J.-F., Ivanova, N., Rattei, T., and Vaishampayan, P. (2015). A viability-linked metagenomic analysis of cleanroom environments: eukarya, prokaryotes, and viruses. Microbiome *3*. https://doi.org/10.1186/s40168-015-0129-y

White, R. a, Blainey, P.C., Fan, H.C., and Quake, S.R. (2009). Digital PCR provides sensitive and absolute calibration for high throughput sequencing. BMC Genomics *10*, 116. https://doi.org/10.1186/1471-2164-10-116

Williams, S.C., Hong, Y., Danavall, D.C.A., Howard-Jones, M.H., Gibson, D., Frischer, M.E., and Verity, P.G. (1998). Distinguishing between living and nonliving bacteria: Evaluation of the vital stain propidium iodide and its combined use with molecular probes in aquatic samples. J. Microbiol. Methods *32*, 225–236. https://doi.org/10.1016/S0167-7012(98)00014-1

Winding, A., Binnerup, S.J., and Sørensen, J. (1994). Viability of indigenous soil bacteria assayed by respiratory activity and growth. Appl. Environ. Microbiol. *60*, 2869–2875.

Xu, H.S., Roberts, N., Singleton, F.L., Attwell, R.W., Grimes, D.J., and Colwell, R.R. (1982). Survival and viability of nonculturableEscherichia coli andVibrio cholerae in the estuarine and marine environment. Microb. Ecol. *8*, 313–323. https://doi.org/10.1007/BF02010671

Yilmaz, S., Allgaier, M., and Hugenholtz, P. (2010). Multiple displacement amplification compromises quantitative analysis of metagenomes. Nat. Methods *7*, 943–944. https://doi.org/10.1038/nmeth1210-943

Zerulla, K., Chimileski, S., Nather, D., Gophna, U., Papke, R.T., and Soppa, J. (2014). DNA as a phosphate storage polymer and the alternative advantages of polyploidy for growth or survival. PLoS One. https://doi.org/10.1371/journal.pone.0094819

Chapter 3

Detection of Organic Matter and Biosignatures in Space Missions

Zita Martins[1]*

[1]Centro de Química Estrutural (CQE), Departamento de Engenharia Química, Instituto Superior Técnico (IST), Universidade de Lisboa, Portugal

*zita.martins@tecnico.ulisboa.pt

DOI: https://doi.org/10.21775/9781912530304.03

Abstract
Carbon-based compounds are widespread throughout the Universe, including abiotic molecules that are the components of the life as we know it. This article reviews the space missions that have aimed to detect organic matter and biosignatures in planetary bodies of our solar system. While to date there was only one life-detection space mission, i.e., the Viking mission to Mars, several past and present space missions have searched for organic matter, paving the way for the future detection of signatures of extra-terrestrial life. This review also reports on the *in-situ* analysis of organic matter and sample-return missions from primitive bodies, i.e. comets and asteroids, providing crucial information on the conditions of the early solar system as well as on the building blocks of life delivered to the primitive Earth.

Introduction
Organic matter, i.e. matter composed of carbon-based compounds is widespread throughout the Universe, including the interstellar (ISM) and circumstellar medium (around 75% of the nearly 200 molecules detected to date contain at least one carbon atom; Agúndez et al., 2018), and solar system bodies, such as comets (Crovisier and Bockelée-Morvan, 1999; Bockelée-Morvan et al., 2004; Sandford et al., 2006; Crovisier et al., 2009; Mumma and Charnley, 2011; Cochran et al., 2015), asteroids (Pinilla-Alonso et al., 2013; Trigo-Rodriguez et al., 2014; Martins, 2019), micrometeorites, interplanetary dust particles (IDPs) and Ultra-Carbonaceous Antarctic Micrometeorites (UCAMMs) (Clemett et al., 1993; Clemett et al., 1998; Brinton et al., 1998; Flynn et al., 2003; Flynn et al., 2004; Glavin et al., 2004; Keller et al., 2004; Matrajt et al., 2004; Matrajt et

al., 2005; Duprat et al., 2010; Dartois et al., 2013), and several planets and satellites (Lorenz et al., 2008; Postberg et al., 2018). Organic matter was exogenously delivered to the primitive Earth, by comets, asteroids and their fragments during the late heavy bombardment (Schidlowski, 1988; Chyba and Sagan, 1992; Schopf, 1993). Furthermore, the potential existence of extra-terrestrial life in planetary bodies of our solar system, e.g. Mars, and some of the icy moons of Jupiter and Saturn, makes it even more crucial to search for organic molecules representative of signatures of life in these places. This review describes the efforts of past, present and future space missions to detect organic matter and biosignatures in solar system bodies.

Detection of organic matter in solar system bodies by space missions

Mars

The 1976 Viking spacecrafts were the first, and to date the only life-detection mission to a solar system body that landed intact and carried out measurements. Mars regolith samples were pyrolyzed at 500°C for 30 seconds, followed by the analysis of the evolved volatiles using a gas chromatography-mass spectrometer (GC-MS). Organic compounds were not detected above a threshold level of a few parts-per-billion (ppb) in two surface samples from the *Chryse Planitia* region of Mars (Biemann et al., 1976). This was later explained by two possible reasons: technological problems and searching in the "wrong location". Regarding the first point, a study years later showed that amino acids present in several million bacterial cells per gram of soil would not have been detected by the Viking GC-MS (Glavin et al., 2001). Regarding the second point, it is now established that the search for biosignatures should not target the surface of Mars, as there are several processes at work on the surface and subsurface that destroy organic molecules, i.e., UV radiation (Stoker and Bullock, 1997; ten Kate et al., 2005, 2006; Garry et al., 2006; Poch et al., 2013; Poch et al., 2014; Poch et al., 2015; dos Santos et al., 2016; Fornaro et al., 2018; Laurent et al., 2019), cosmic rays (Pavlov et al., 2012; Crandall et al. 2017), and oxidation reactions that will likely produce non-volatile products (Benner et al., 2000). Indeed, oxidized species (perchlorates) were detected by the Wet Chemistry Laboratory (WCL) of the Phoenix spacecraft, and the Thermal Evolved Gas Analyser detected the release of oxygen, which is consistent with the thermal decomposition of perchlorate (Hecht et al., 2009; Kounaves et al., 2010). Due to these results, the GC-MS data from the Viking mission was re-analysed, and ≤0.1% perchlorate at the Viking landing sites was suggested (Navarro-González et al., 2010). The Mars Science Laboratory's (MSL) Sample Analysis at Mars (SAM) instrument indicated the presence of oxychlorine phases (i.e. likely perchlorate and chlorate) in the Gale Crater (Leshin et al., 2013; Archer et al., 2014; Ming et al, 2014). Chlorinated hydrocarbons, such as chlorobenzene (150 - 300 parts per billion by weight (ppbw)) and C_2 to C_4 dichloroalkanes (up to 70 ppbw) were detected by the SAM instrument in the Sheepbed mudstone at Yellowknife Bay (Glavin et al.,

2013; Ming et al, 2014; Freissinet et al., 2015). Chlorobenzene was also identified in the re-analysis of the Viking GC-MS data sets (Guzman et al., 2018). It was concluded that the chlorinated hydrocarbons were the reaction products of Martian chlorine and organic carbon from Martian sources, or exogenous sources (e.g. meteorites, comets, or IDPs) (Freissinet et al., 2015). More recently, organic matter (e.g. thiophenic, aromatic, and aliphatic compounds) was found preserved in lacustrine mudstones at the base of the Murray formation at Pahrump Hills, Gale crater, by the SAM instrument suite onboard the Curiosity rover (Eigenbrode et al., 2018). In 2020, the ExoMars mission, a collaborative project between the European Space Agency (ESA) and Roscosmos will launch a rover to search for biosignatures on the subsurface of Mars. It will obtain samples at various depths, from 0 down to 2 meters (Vago et al., 2017). In the meantime, the ExoMars Trace Gas Orbiter (TGO) is already in orbit around Mars, performing its science activities since April 2018. It will determine the presence of methane and other trace gases in the Martian atmosphere, providing evidence for possible biological activity on the Red Planet (Liuzzi et al., 2019). The potential presence of methane and its origin (suggested to be originated by different processes, e.g. biological or by early Mars serpentinization) in the atmosphere of Mars has been the subject of a long discussion in the scientific community, which hopefully will be solved by the ExoMars TGO data (Krasnopolsky et al., 2004; Formisano et al., 2004; Mumma et al., 2009; Fonti et al., 2015; Webster et al., 2015; Webster et al., 2018).

Dwarf planets
The dwarf planet Pluto and its satellites were visited in July 2015 by the New Horizons spacecraft, which flew past them. The New Horizons team analysed the colours and chemical compositions of the surfaces of Pluto and the satellite Charon by analysing the reflectance and absorption spectra, which distinguishes the various cryogenic surface ices via their characteristic spectroscopic features. Charon has a dark red spot on its polar cap, similar to the colours on the surface of Pluto. An absorption around 2.3 μm on the surface of Pluto was indicative of hydrocarbons heavier than CH_4. In fact, it has been attributed to tholin-like organic macromolecules produced by energetic radiation processing of hydrocarbons. Similar process is hypothesised to happen on Charon (Stern et al., 2015; Grundy et al., 2016a; Grundy et al., 2016b). More recently, Telfer and co-authors (Telfer et al., 2018) analysed images taken by the New Horizons spacecraft and identified dunes in the Sputnik Planitia region on Pluto. In addition, they showed that the wavelength of the dunes could be formed by the deposition of sand-sized grains of methane ice in moderate Pluto winds.

Another dwarf planet, Ceres has been visited by NASA's Dawn spacecraft. Aliphatic organic material was identified on its surface based on the near-infrared (NIR) reflectance spectra obtained by the Visible and InfraRed

(VIR) spectrometer by the Dawn mission (De Sanctis et al., 2017). Further analysis showed that Ceres exhibits a stronger 3.4 μm aliphatic organic absorption (De Sanctis et al., 2017) than those observed in lab spectra of carbonaceous chondrites (Kaplan et al., 2018). In addition, spectral models require 45% to 65% meteorite-derived insoluble organic matter (IOM) to fit the spectra from Ceres (Kaplan et al., 2018).

Comets
The first *in-situ* chemical analyses of comet dust particles were performed on comet 1P/Halley by the PIA and PUMA mass spectrometers on-board the Giotto and Vega space missions (Kissel et al., 1986a; Kissel et al., 1986b). Most of the comet Halley dust particles were rich in light elements such as carbon, hydrogen, oxygen and nitrogen, suggesting that the cometary dust included organic material (Clark et al., 1987; Langevin et al., 1987; Lawler & Brownlee, 1992; Fomenkova et al., 1994) and consisted of a chondritic core with an organic mantle composed of highly unsaturated compounds (Kissel and Krueger, 1987). The very low surface albedo, obtained by the Halley Multicolour Camera (HMC) from Giotto would lead to a strong solar energy absorption, and therefore to a total gas and dust production about an order of magnitude higher than actually observed. This implied the presence of dark non-volatile organic compounds covering the surface ice of comet Halley (Keller et al., 1986).

The Stardust mission performed a fly-by of comet 81P/Wild 2, and later returned samples to Earth. The Cometary and Interstellar Dust Analyzer (CIDA) instrument, a time-of-flight mass spectrometer on the Stardust spacecraft obtained spectra during the flyby of Comet 81P/Wild 2 and confirmed the predominance of organic matter in cometary particles (Kissel et al., 2004). Samples of cometary particles collected in the coma of comet 81P/Wild 2 were returned to Earth, and a preliminary examination showed that the non-volatile portion of this comet is a heterogeneous and unequilibrated assortment of materials, with both presolar and solar system origin (Brownlee et al., 2006). In addition, the presence of deuterium and ^{15}N excesses in some of the comet particles also suggests an interstellar/ protostellar chemical heritage for the cometary organics (Sandford et al., 2006). Further analysis showed that the organic matter of these particles was very rich in both oxygen and nitrogen and contained a diverse set of both aromatic and non-aromatic compounds (Sandford et al., 2006; Cody et al., 2008). Amorphous and organic carbon were dominant in the particles of this comet, with a heterogeneous distribution of the carbonaceous phases (which contained carbonyl functional groups) within these particles (Matrajt et al., 2008). Glycine, the smallest amino acid, as well as methylamine and ethylamine were detected for the first time in particles of comet 81P/Wild 2 (Sandford et al., 2006). The carbon isotopic composition of glycine (29 ± 6‰) strongly indicated a non-terrestrial origin for this compound. Nevertheless, some terrestrial contamination may have occurred, as the δ^{13}C value of -25 ± 2‰ for ε-amino-*n*-caproic acid (EACA) present in the cometary samples indicated terrestrial contamination during the curation process (Elsila et al., 2009). More recent analysis indicated that samples of comet 81P/Wild 2 had N/C atomic ratios spanning over a wide range from 0.002 to 0.18 (De Gregorio et al., 2011). They contained refractory material, dominated by IOM and highly aromatic organic matter (De Gregorio et al., 2011). This refractory organic component is more closely related to that of IDPs than to that of carbonaceous chondrites (Clemett et al., 2010).

The 67P/Churyumov-Gerasimenko comet was analysed *in-situ* by the Rosetta space mission. The COmetary Secondary Ion Mass Analyzer (COSIMA) instrument onboard the Rosetta mission analysed the composition of the dust particles of this comet, showing that they contained solid organic matter, which shared some similarities with the IOM extracted from carbonaceous chondrites (Fray et al., 2016). The N/C atomic ratios of 27 cometary particles collected in the environment of the 67P/Churyumov–Gerasimenko comet ranged from 0.018 to 0.06, with an averaged value of 0.035 ± 0.011, which is in agreement with the values obtained for comet 1P/Halley (Fray et al., 2017). The Visible, Infrared and Thermal Imaging Spectrometer (VIRTIS) instrument on-board the Rosetta spacecraft indicated the presence of carbon-bearing compounds on the nucleus of the comet 67P/Churyumov-Gerasimenko, in particular non-volatile organic macromolecular materials (Capaccioni et al., 2015). Furthermore, several carbon-rich species were detected on comet 67P/Churyumov-Gerasimenko, i.e. alcohols, carbonyls, amines, nitriles, amides, isocyanates, the polymer polyoxymethylene (Goesmann et al., 2015; Wright et al., 2015), phosphorus and glycine (Altwegg et al., 2016).

Asteroids
Space missions to primitive asteroids are crucial to determine the conditions of the history of the solar system and the evolution of organic molecules relevant to the origin of life on Earth. The Hayabusa spacecraft from the Japan Aerospace Exploration Agency (JAXA) successfully collected surface particles from the near-Earth asteroid 25143 Itokawa, returning these samples to Earth to be analysed (Nakamura et al., 2011). The particles were classified in 4 categories (Yada et al., 2014). Particles of categories 1 and 2 were regolith of Itokawa (Ebihara et al., 2011; Nagao et al., 2011; Nakamura et al., 2011; Noguchi et al., 2011; Tsuchiyama et al., 2011; Yurimoto et al., 2011), composed mainly of silicate minerals, and no indigenous carbonaceous matter (such as amino acids or IOM) was identified in these particles (Naraoka et al., 2012). Category 3 contained carbonaceous particles, as carbon was shown to be dominant in these particles, based on their chemical composition using field emission scanning electron microscope (FE-SEM) with energy dispersion spectrometer (EDS). Small peaks corresponding to nitrogen and oxygen, and trace amounts of fluorine and sulphur were detected by SEM/EDS (Uesugi et al., 2014). Three carbonaceous category 3 particles were further

analysed by time of flight-secondary ion mass spectrometry (ToF-SIMS). The homogenous organic carbon distribution was associated with nitrogen, silicon, and/or fluorine. As this distribution is different to the distribution of carbon-rich extra-terrestrial samples, it was suggested that these three particles could be debris of silicon rubber and fluorinated compounds, i.e. they were man-made artefacts and not indigenous organic matter (Naraoka et al., 2015). In particular, it was suggested that this was the result of the degradation of a polyimide/polyamide resin (Kitajima et al., 2015). Indeed, the nature and origin of the same three particles is unsure as H, C, and N isotopic compositions indicate that they were terrestrial within the experimental error (Ito et al., 2014). Finally, category 4 particles were manmade artefacts from the sample catcher (Yada et al., 2014).

Hayabusa2 is a sample return mission from the near-Earth C-type asteroid 162173 Ryugu. The space mission has reached its target in 2018, has successfully landed in February 2019, and will arrive on Earth with samples in 2020 (Wada et al., 2018). A near-infrared spectrum of this asteroid, from 0.85 to 2.2 µm, is consistent with a carbonaceous-type classification, which is related to carbonaceous meteorites (Pinilla-Alonso et al., 2013). Also, in the year 2018, another asteroid sample return mission, OSIRIS-Rex has reached the B-type asteroid 101955 Bennu. It will return with samples to Earth for analysis in 2023 (Lauretta et al., 2017).

Jovian and Saturnian moons
The Saturnian moon Titan has a dense atmosphere composed mainly of nitrogen, methane and hydrogen, and traces of other hydrocarbons, which were confirmed by space missions as Pioneer 11, Voyager 1 and Cassini-Huygens (Coustenis et al., 2003; Coustenis, 2005; Flasar et al., 2005; Coustenis et al., 2006; Coustenis et al., 2007; Teanby et al., 2009; Waite et al., 2007; Cui et al., 2009; Niemann et al., 2010). The latter flew by Titan in 2004, and in 2005 the Huygens probe landed on the surface of Titan. Lakes, rivers, and seas found on the surface of this satellite were mainly composed of methane, ethane and propane (Lorenz et al., 2008; Clark et al., 2010; Stephan et al., 2010; Wall et al., 2010). Furthermore, Titan has a methane cycle that includes evaporation of methane from lakes, followed by partly conversion into ethane. Subsequent condensation of these two compounds leads to the formation of clouds that generate precipitation (i.e. rain or aerosols) (Poch et al., 2012; Dhingra et al., 2019).

Iapetus is a satellite of Saturn and contains hydrocarbons and other organic compounds (e.g. cycloalkanes, olefinic compounds, CH_3OH, and N-substituted polycyclic aromatic hydrocarbon (PAHs)) on its leading and trailing hemispheres as detected by the Cassini Visual and Infrared Mapping Spectrometer (VIMS) instrument (Cruikshank et al., 2014). The aromatic hydrocarbons had a stronger signal than the aliphatic hydrocarbons in Iapetus. Hyperion and Phoebe, another two Saturnian moons also exhibited spectral absorptions indicative of C-H in aromatic and aliphatic hydrocarbons (Dalton et al., 2012; Cruikshank et al., 2014).

The Saturnian moon Enceladus contains a global water ocean below its icy crust and above a rocky core (less et al., 2014; Thomas et al., 2016). Ice grains and vapour, originating from the subsurface ocean were injected into space, and the plumes contained simple organic compounds with molecular masses mostly below 50 atomic mass units (Waite et al., 2006), as well as complex macromolecular organic material with molecular masses above 200 atomic mass units, as measured by the Cassini spacecraft (Postberg et al., 2018). Several icy solar system bodies, such as the moons Ganymede, Callisto, and Europa are also thought to have subsurface oceans, which makes them potentially habitable environments that may host extra-terrestrial life (Carr et al., 1998; Khurana et al., 1998; Zimmer et al., 2000; Spohn and Schubert, 2003; Kuskov et al., 2005; Hansen et al., 2006; less et al., 2014; Vance et al., 2014;). Such locations are ideal to detect signatures of extra-terrestrial life potentially present in the oceans of icy worlds. While the subsurface oceans of these icy worlds have not been directly analysed by space missions, the Jupiter Icy Moon Explorer (JUICE) by ESA will happen in the next decade and will visit three moons of the Jovian system: Ganymede, Callisto, and Europa. One of its objectives is to determine the content of organic molecules connected to life in the moon Europa (Grasset et al., 2013).

Conclusions

This review outlines the past, present and future space missions that aim to detect organic matter and biosignatures in planetary bodies of our solar system. Destination of past space missions included Mars, dwarf planets (Pluto and Ceres), comets, asteroids, and Jovian and Saturnian moons (Titan, Iapetus, Hyperion, Phoebe, and Enceladus). While organic matter is widespread throughout the solar system, to date no space mission has successfully detected any signatures of extra-terrestrial life. Several future life-detection missions are currently planned, including the ExoMars and JUICE space missions. The asteroid sample return missions Hayabusa2 and OSIRIS-Rex will provide crucial information about the conditions of the early solar system and the prebiotic molecules relevant to the first living organisms on Earth. Overall data resulting from *in-situ* and sample return missions from several planetary bodies play a central role in our understanding of the formation, evolution and distribution of organic matter in the solar system.

Acknowledgements

This work was financed by FEDER - Fundo Europeu de Desenvolvimento Regional funds through the COMPETE 2020 - Operational Programme for Competitiveness and Internationalisation (POCI), and by Portuguese funds through FCT - Fundação para a Ciência e a Tecnologia in the framework of the project POCI-01-0145-FEDER-029932 (PTDC/FIS-AST/29932/2017).

The author would like to kindly acknowledge the constructive comments of the Reviewer that improved the quality of this manuscript.

References

Altwegg, K., Balsiger, H., Bar-Nun, A., Berthelier, J.-J., Bieler, A., Bochsler, P., Briois, C., Calmonte, U., Combi, M. R., Cottin, H., De Keyser, J., Dhooghe, F., Fiethe, B., Fuselier, S. A., Gasc, S., Gombosi, T. I., Hansen, K. C., Haessig, M., Jackel, A., Kopp, E., Korth, A., Le Roy, L., Mall, U., Marty, B., Mousis, O., Owen, T., Reme, H., Rubin, M., Semon, T., Tzou, C.-Y., Waite, J. Hunter, Wurz, P. (2016). Prebiotic chemicals - amino acid and phosphorus - in the coma of comet 67P/Churyumov-Gerasimenko. Science Advances 2, e1600285-e1600285. https://dx.doi.org/10.1126/sciadv.1600285

Archer, P. D., Franz, H. B., Sutter, B., Arevalo, R. D., Coll, P., Eigenbrode, J. L., Glavin, D. P., Jones, J. J., Leshin, L. A., Mahaffy, P. R., McAdam, A. C., McKay, C. P., Ming, D. W., Morris, R. V., Navarro-González, R., Niles, P. B., Pavlov, A., Squyres, S. W., Stern, J. C., Steele, A., Wray, J. J. (2014). Abundances and implications of volatile-bearing species from evolved gas analysis of the Rocknest aeolian deposit, Gale Crater, Mars. Journal of Geophysical Research: Planets 119, 237-254. https://dx.doi.org/10.1002/2013JE004493

Benner, S. A., Devine, K. G., Matveeva, L. N., Powell, D. H. (2000). The missing organic molecules on Mars. Proceedings of the National Academy of Science 97, 2425-2430. https://dx.doi.org/10.1073/pnas.040539497

Biemann, K., Oro, J., Toulmin, P. III, Orgel, L. E., Nier, A. O., Anderson, D. M., Simmonds, P. G., Flory, D., Diaz, A. V., Rushneck, D. R., Biller, J. A. (1976). Search for organic and volatile inorganic compounds in two surface samples from the Chryse Planitia region of Mars. Science 194, 72-76. https://dx.doi.org/10.1126/science.194.4260.72

Bockelée-Morvan, D., Crovisier, J., Mumma, M. J., Weaver, H. A. (2004). The composition of cometary volatiles. In Comets II, M.C. Festou, H.U. Keller, and H.A. Weaver, eds. (Tucson: University of Arizona Press), pp. 391-423.

Brinton, K. L. F., Engrand, C., Glavin, D. P., Bada, J. L., Maurette, M. (1998). A Search for Extraterrestrial Amino Acids in Carbonaceous Antarctic Micrometeorites. Origins of Life and Evolution of the Biosphere 28, 413-424.

Brownlee, D., et al. (2006). Comet 81P/Wild 2 Under a Microscope. Science 314, 1711-1716. https://dx.doi.org/10.1126/science.1135840

Capaccioni, F. (2015). The organic-rich surface of comet 67P/Churyumov-Gerasimenko as seen by VIRTIS/Rosetta. Science 347, aaa0628-1-aaa0628-4. https://dx.doi.org/10.1126/science.aaa0628

Carr, M. H., Belton, M. J. S., Chapman, C. R., Davies, M. E., Geissler, P., Greenberg, R., McEwen, A. S., Tufts, B. R., Greeley, R., Sullivan, R., Head, J. W., Pappalardo, R. T., Klaasen, K. P., Johnson, T. V., Kaufman, J., Senske, D., Moore, J., Neukum, G., Schubert, G., Burns, J. A.,

Thomas, P., Veverka, J. (1998). Evidence for a subsurface ocean on Europa. Nature 391, 363-365. https://dx.doi.org/10.1038/34857

Chyba, C., Sagan, C. (1992). Endogenous production, exogenous delivery and impact-shock synthesis of organic molecules: an inventory for the origins of life. Nature 355, 125-132. https://dx.doi.org/10.1038/355125a0

Clark, B. C., Mason, L. W., Kissel, J. (1987). Systematics of the CHON and Other Light Element Particle Populations in Comet p/ Halley. Astronomy and Astrophysics 187, 779-784.

Clark, R. N., Curchin, J. M., Barnes, J. W., Jaumann, R., Soderblom, L., Cruikshank, D. P., Brown, R. H., Rodriguez, S., Lunine, J., Stephan, K., Hoefen, T. M., Le Mouélic, S., Sotin, C., Baines, K. H., Buratti, B. J., Nicholson, P. D. (2010). Detection and mapping of hydrocarbon deposits on Titan. Journal of Geophysical Research 115, E1005-1- E1005-28. https://dx.doi.org/10.1029/2009JE003369

Clemett, S. J., Maechling, C. R., Zare, R. N., Swan, P. D., Walker, R. M. (1993). Identification of Complex Aromatic Molecules in Individual Interplanetary Dust Particles. Science 262, 721-725. https://dx.doi.org/10.1126/science.262.5134.721

Clemett, S. J., Chillier, X. D. F., Gillette, S., Zare, R. N., Maurette, M., Engrand, C., Kurat, G. (1998). Observation of Indigenous Polycyclic Aromatic Hydrocarbons in `Giant' carbonaceous Antarctic Micrometeorites. Origins of Life and Evolution of the Biosphere 28, 425-448.

Clemett, S. J., Sandford, S. A., Nakamura-Messenger, K., Hörz, F., McKay, D. S. (2010). Complex aromatic hydrocarbons in Stardust samples collected from comet 81P/Wild 2. Meteoritics and Planetary Science 45, 701-722. https://dx.doi.org/10.1111/j.1945-5100.2010.01062.x

Cochran, A. L., Levasseur-Regourd, A.-C., Cordiner, M., Hadamcik, E., Lasue, J., Gicquel, A., Schleicher, D. G., Charnley, S. B., Mumma, M. J., Paganini, L., Bockelée-Morvan, D., Biver, N., Kuan, Y.-J. (2015). The Composition of Comets. Space Science Reviews 197, 9-46. https://dx.doi.org/10.1007/s11214-015-0183-6

Cody, G. D., Ade, H., O'D. Alexander, C. M., Araki, T., Butterworth, A., Fleckenstein, H., Flynn, G., Gilles, M. K., Jacobsen, C., Kilcoyne, A. L. D., Messenger, K., Sandford, S. A., Tyliszczak, T., Westphal, A. J., Wirick, S., Yabuta, H. (2008). Quantitative organic and light-element analysis of comet 81P/Wild 2 particles using C-, N-, and O-μ-XANES. Meteoritics & Planetary Science 43, 353-365. https://dx.doi.org/10.1111/j.1945-5100.2008.tb00627.x

Coustenis, A. (2005). Formation and Evolution of Titan's Atmosphere. Space Science Reviews 116, 171-184. https://dx.doi.org/10.1007/s11214-005-1954-2

Coustenis, A., Salama, A. Schulz, B., Ott, S., Lellouch, E., Encrenaz, T. h., Gautier, D., Feuchtgruber, H. (2003). Titan's atmosphere from ISO mid-infrared spectroscopy. Icarus 161, 383-403. https://dx.doi.org/10.1016/S0019-1035(02)00028-3

Coustenis, A., Negrão, A., Salama, A., Schulz, B., Lellouch, E., Rannou, P., Drossart, P., Encrenaz, T., Schmitt, B., Boudon, V., Nikitin, A. (2006). Titan's 3-micron spectral region from ISO high-resolution spectroscopy. Icarus 180, 176-185. https://dx.doi.org/10.1016/j.icarus.2005.08.007

Coustenis, A., Achterberg, R. K., Conrath, B. J., Jennings, D. E., Marten, A., Gautier, D., Nixon, C. A., Flasar, F. M., Teanby, N. A., Bézard, B., Samuelson, R. E., Carlson, R. C., Lellouch, E., Bjoraker, G. L., Romani, P. N., Taylor, F. W., Irwin, P. G. J., Fouchet, T., Hubert, A., Orton, G. S., Kunde, V. G., Vinatier, S., Mondellini, J., Abbas, M. M., Courtin, R. (2007). The composition of Titan's stratosphere from Cassini/CIRS mid-infrared spectra. Icarus 189, 35-62. https://dx.doi.org/10.1016/j.icarus.2006.12.022

Crandall, P. B., Góbi, S., Gillis-Davis, J., Kaiser, R. I. (2017). Can perchlorates be transformed to hydrogen peroxide (H_2O_2) products by cosmic rays on the Martian surface?. Journal of Geophysical Research: Planets 122, 1880-1892. https://doi.org/10.1002/2017JE005329

Crovisier, J., Bockelée-Morvan, D. (1999). Remote Observations of the Composition of Cometary Volatiles. Space Science Reviews 90, 19-32. https://dx.doi.org/10.1023/A:1005217224240

Crovisier, J., Biver, N., Bockelée-Morvan, D., Boissier, J., Colom, P., Lis, D. C. (2009). The chemical diversity of comets. Earth Moon Planets 105, 267–272. https://dx.doi.org/10.1007/s11038-009-9293-z

Cruikshank, D. P., Dalle Ore, C. M., Clark, R. N., Pendleton, Y. J. (2014). Aromatic and aliphatic organic materials on Iapetus: Analysis of Cassini VIMS data. Icarus 233, 306-315. https://dx.doi.org/10.1016/j.icarus.2014.02.011

Cui, J., Yelle, R. V., Vuitton, V., Waite, J. H., Kasprzak, W. T., Gell, D. A., Niemann, H. B., Müller-Wodarg, I. C. F., Borggren, N., Fletcher, G. G., Patrick, E. L., Raaen, E., Magee, B. A. (2009). Analysis of Titan's neutral upper atmosphere from Cassini Ion Neutral Mass Spectrometer measurements. Icarus 200, 581-615. https://dx.doi.org/10.1016/j.icarus.2008.12.005

Dalton, J. B. III, Shirley, J. H., Kamp, L. W. (2012). Europa's icy bright plains and dark linea: Exogenic and endogenic contributions to composition and surface properties. Journal of Geophysical Research 117, E03003-1-E03003-16. https://dx.doi.org/10.1029/2011JE003909

Dartois, E., Engrand, C., Brunetto, R., Duprat, J., Pino, T., Quirico, E., Remusat, L., Bardin, N., Briani, G., Mostefaoui, S., Morinaud, G., Crane, B., Szwec, N., Delauche, L., Jamme, F., Sandt, Ch., Dumas, P. (2013). UltraCarbonaceous Antarctic micrometeorites, probing the Solar System beyond the nitrogen snow-line. Icarus 224, 243-252. https://dx.doi.org/10.1016/j.icarus.2013.03.002

de Gregorio, B. T., Stroud, R. M., Cody, G. D., Nittler, L. R., David Kilcoyne, A. L., Wirick, S. (2011) Correlated microanalysis of cometary organic grains returned by Stardust. Meteoritics & Planetary Science 46, 1376-1396. https://dx.doi.org/10.1111/j.1945-5100.2011.01237.x

De Sanctis, M. C., Ammannito, E., McSween, H. Y., Raponi, A., Marchi, S., Capaccioni, F., Capria, M. T., Carrozzo, F. G., Ciarniello, M., Fonte, S., Formisano, M., Frigeri, A., Giardino, M., Longobardo, A., Magni, G., McFadden, L. A., Palomba, E., Pieters, C. M., Tosi, F., Zambon, F., Raymond, C. A., Russell, C. T. (2017). Localized aliphatic organic material on the surface of Ceres. Science 355, 719-722. https://dx.doi.org/10.1126/science.aaj2305

Dhingra, R. D. et al. (2019). Observational evidence for summer rainfall at Titan's north pole. Geophysical Research Letters https://dx.doi.org/10.1029/2018GL080943

dos Santos, R., Patel, M., Cuadros, J., Martins, Z. (2016) Influence of mineralogy on the preservation of amino acids under simulated Mars conditions, Icarus 277, 342–353. https://doi.org/10.1016/j.icarus.2016.05.029

Duprat, J., Dobrică, E., Engrand, C., Aléon, J., Marrocchi, Y., Mostefaoui, S., Meibom, A., Leroux, H., Rouzaud, J.-N., Gounelle, M., Robert, F. (2010) Extreme Deuterium Excesses in Ultracarbonaceous Micrometeorites from Central Antarctic Snow. Science 328, 742-745. https://dx.doi.org/10.1126/science.1184832

Ebihara, M., Sekimoto, S., Shirai, N., Hamajima, Y., Yamamoto, M., Kumagai, K., Oura, Y., Ireland, T. R., Kitajima, F., Nagao, K., Nakamura, T., Naraoka, H., Noguchi, T., Okazaki, R., Tsuchiyama, A., Uesugi, M., Yurimoto, H., Zolensky, M. E., Abe, M., Fujimura, A., Mukai, T., Yada, Y. (2011). Neutron Activation Analysis of a Particle Returned from Asteroid Itokawa. Science 333, 1119-1121. https://dx.doi.org/10.1126/science.1207865

Eigenbrode, J. L., Summons, R. E., Steele, A., Freissinet, C., Millan, M., Navarro-González, R., Sutter, B., McAdam, A. C., Franz, H. B., Glavin, D. P., Archer, P. D., Mahaffy, P. R., Conrad, P. G., Hurowitz, J. A., Grotzinger, J. P., Gupta, S., Ming, D. W., Sumner, D. Y., Szopa, C., Malespin, C., Buch, A., Coll, P. (2018). Organic matter preserved in 3-billion-year-old mudstones at Gale crater, Mars. Science 360, 1096-1101. https://dx.doi.org/10.1126/science.aas9185

Elsila, J. E., Glavin, D. P., Dworkin, J. P. (2009). Cometary glycine detected in samples returned by Stardust. Meteoritics & Planetary Science 44, 1323-1330. https://dx.doi.org/10.1111/j.1945-5100.2009.tb01224.x

Flasar, F. M., Achterberg, R. K., Conrath, B. J., Gierasch, P. J., Kunde, V. G., Nixon, C. A., Bjoraker, G. L., Jennings, D. E., Romani, P. N., Simon-Miller, A. A., Bézard, B., Coustenis, A., Irwin, P. G. J., Teanby, N. A., Brasunas, J., Pearl, J. C., Segura, M. E., Carlson, R. C., Mamoutkine, A., Schinder, P. J., Barucci, A., Courtin, R., Fouchet, T., Gautier, D., Lellouch, E., Marten, A., Prangé, R., Vinatier, S., Strobel, D. F., Calcutt, S. B., Read, P. L., Taylor, F. W., Bowles, N., Samuelson, R. E., Orton, G. S., Spilker, L. J., Owen, T. C., Spencer, J. R., Showalter, M. R., Ferrari, C., Abbas, M. M., Raulin, F., Edgington, S., Ade, P., Wishnow, E. H. (2005). Titan's Atmospheric Temperatures, Winds, and Composition. Science 308, 975-978. https://dx.doi.org/10.1126/science.1111150

Flynn, G. J., Keller, L. P., Feser, M., Wirick, S., Jacobsen, C. (2003). The origin of organic matter in the solar system: evidence from the interplanetary dust particles. Geochimica et Cosmochimica Acta 67, 4791-4806. https://dx.doi.org/10.1016/j.gca.2003.09.001

Flynn, G. J., Keller, L. P., Jacobsen, C., Wirick, S. (2004). An assessment of the amount and types of organic matter contributed to the Earth by interplanetary dust. Advances in Space Research 33, 57-66. https://dx.doi.org/10.1016/j.asr.2003.09.036

Fomenkova, N. M., Chang, S., Mukhin, L. M. (1994). Carbonaceous components in the comet Halley dust. Geochimica et Cosmochimica Acta 58, 4503-4512. https://dx.doi.org/10.1016/0016-7037(94)90351-4

Fonti, S., Mancarella, F., Liuzzi, G., Roush, T. L., Chizek Frouard, M., Murphy, J., Blanco, A. (2015). Revisiting the identification of methane on Mars using TES data. Astronomy & Astrophysics 581, A136-1-A136-11. https://dx.doi.org/10.1051/0004-6361/201526235

Formisano, V., Atreya, S., Encrenaz, T., Ignatiev, N., Giuranna, M. (2004). Detection of Methane in the Atmosphere of Mars. Science 306, 1758-1761. https://dx.doi.org/10.1126/science.1101732

Fornaro, T., Boosman, A., Brucato, J. R., ten Kate, I. L., Siljeström, S., Poggiali, G., Steele, A., Hazen, R. M. (2018) UV irradiation of biomarkers adsorbed on minerals under Martian-like conditions: Hints for life detection on Mars. Icarus 313, 38-60. https://doi.org/10.1016/j.icarus.2018.05.001

Fray, N., Bardyn, A., Cottin, H., Altwegg, K., Baklouti, D., Briois, C., Colangeli, L., Engrand, C., Fischer, H., Glasmachers, A., Grün, E., Haerendel, G., Henkel, H., Höfner, H., Hornung, K., Jessberger, E. K., Koch, A., Krüger, H., Langevin, Y., Lehto, H., Lehto, K., Le Roy, L., Merouane, S., Modica, P., Orthous-Daunay, F.-R., Paquette, J., Raulin, F., Rynö, J., Schulz, R., Silén, J., Siljeström, S., Steiger, W., Stenzel, O., Stephan, T., Thirkell, L., Thomas, R., Torkar, K., Varmuza, K., Wanczek, K.-P., Zaprudin, B., Kissel, J., Hilchenbach, M. (2016). High-molecular-weight organic matter in the particles of comet 67P/Churyumov-Gerasimenko. Nature 538, 72-74. https://dx.doi.org/10.1038/nature19320

Fray, N., Bardyn, A., Cottin, H., Baklouti, D., Briois, C., Engrand, C., Fischer, H., Hornung, K., Isnard, R., Langevin, Y., Lehto, H., Le Roy, L., Mellado, E. M., Merouane, S., Modica, P., Orthous-Daunay, F.-R., Paquette, J., Rynö, J., Schulz, R., Silén, J., Siljeström, S., Stenzel, O., Thirkell, L., Varmuza, K., Zaprudin, B., Kissel, J., Hilchenbach, M. (2017). Nitrogen-to-carbon atomic ratio measured by COSIMA in the particles of comet 67P/Churyumov-Gerasimenko. Monthly Notices of the Royal Astronomical Society 469, S506-S516. https://dx.doi.org/10.1093/mnras/stx2002

Freissinet, C., Glavin, D. P., Mahaffy, P. R., Miller, K. E., Eigenbrode, J. L., Summons, R. E., Brunner, A. E., Buch, A., Szopa, C., Archer, P. D., Jr., Franz, H. B., Atreya, S. K., Brinckerhoff, W. B., Cabane, M., Coll, P., Conrad, P. G., Des Marais, D. J., Dworkin, J. P., Fairén, A. G., François, P., Grotzinger, J. P., Kashyap, S., ten Kate, I. L., Leshin, L. A., Malespin,

C. A., Martin, M. G., Martin-Torres, J. F., McAdam, A. C., Ming, D. W., Navarro-González, R., Pavlov, A. A., Prats, B. D., Squyres, S. W., Steele, A., Stern, J. C., Sumner, D. Y., Sutter, B., Zorzano, M.-P., MSL Science Team (2015). Organic molecules in the Sheepbed Mudstone, Gale Crater, Mars. Journal of Geophysical Research: Planets 120, 495-514. https://dx.doi.org/10.1002/2014JE004737

Garry, J. R. C., ten Kate, I. L., Martins, Z., Nørnberg, P., Ehrenfreund, P. (2006) Analysis and survival of amino acids in Martian regolith analogs. Meteoritics & Planetary Science 41, 391-405. https://doi.org/10.1111/j.1945-5100.2006.tb00470.x

Glavin, D. P., Schubert, M., Botta, O., Kminek, G., Bada, J. L. (2001). Detecting pyrolysis products from bacteria on Mars. Earth and Planetary Science Letters 185, 1-5. https://dx.doi.org/10.1016/S0012-821X(00)00370-8

Glavin, D. P., Matrajt, G., Bada, J. L. (2004). Re-examination of amino acids in Antarctic micrometeorites. Advances in Space Research 33, 106-113. https://dx.doi.org/10.1016/j.asr.2003.02.011

Glavin, D. P., Freissinet, C., Miller, K. E., Eigenbrode, J. L., Brunner, A. E., Buch, A., Sutter, B., Archer, P. D., Atreya, S. K., Brinckerhoff, W. B., Cabane, M., Coll, P., Conrad, P. G., Coscia, D., Dworkin, J. P., Franz, H. B., Grotzinger, J. P., Leshin, L. A., Martin, M. G., McKay, C., Ming, D. W., Navarro-González, R., Pavlov, A., Steele, A., Summons, R. E., Szopa, C., Teinturier, S., Mahaffy, P. R. (2013). Evidence for perchlorates and the origin of chlorinated hydrocarbons detected by SAM at the Rocknest aeolian deposit in Gale Crater. Journal of Geophysical Research: Planets 118, 1955-1973. https://dx.doi.org/10.1002/jgre.20144

Goesmann, F., Rosenbauer, H., Bredehöft, J. H., Cabane, M., Ehrenfreund, P., Gautier, T.; Giri, C., Krüger, H., Le Roy, L., MacDermott, A. J., McKenna-Lawlor, S., Meierhenrich, U. J., Caro, G. M. M., Raulin, F., Roll, R., Steele, A., Steininger, H., Sternberg, R., Szopa, C., Thiemann, W., Ulamec, S. (2015). Organic compounds on comet 67P/Churyumov-Gerasimenko revealed by COSAC mass spectrometry. Science 349, aab0689-1- aab0689-4. https://dx.doi.org/10.1126/science.aab0689

Grasset, O., Dougherty, M. K., Coustenis, A., Bunce, E. J., Erd, C., Titov, D., Blanc, M., Coates, A., Drossart, P., Fletcher, L. N., Hussmann, H., Jaumann, R., Krupp, N., Lebreton, J.-P., Prieto-Ballesteros, O., Tortora, P., Tosi, F., Van Hoolst, T. (2013). JUpiter ICy moons Explorer (JUICE): An ESA mission to orbit Ganymede and to characterise the Jupiter system. Planetary and Space Science 78, 1-21. https://dx.doi.org/10.1016/j.pss.2012.12.002

Grundy, W. M. et al. (2016a). The formation of Charon's red poles from seasonally cold-trapped volatiles. Nature 539, 65-68. https://dx.doi.org/10.1038/nature19340

Grundy, W. M., Binzel, R. P., Buratti, B. J., Cook, J. C., Cruikshank, D. P., Dalle Ore, C. M., Earle, A. M., Ennico, K., Howett, C. J. A., Lunsford, A. W., Olkin, C. B., Parker, A. H., Philippe, S., Protopapa, S., Quirico, E., Reuter, D. C., Schmitt, B., Singer, K. N., Verbiscer, A. J., Beyer, R. A.,

Buie, M. W., Cheng, A. F., Jennings, D. E., Linscott, I. R., Parker, J. Wm., Schenk, P. M., Spencer, J. R., Stansberry, J. A., Stern, S. A., Throop, H. B., Tsang, C. C. C., Weaver, H. A., Weigle, G. E., Young, L. (2016b). A.Surface compositions across Pluto and Charon. Science 351, aad9189-1-aad9189-8. https://dx.doi.org/10.1126/science.aad9189

Guzman, M., McKay, C. P., Quinn, R. C., Szopa, C., Davila, A. F., Navarro-González, R., Freissinet, C. (2018). Identification of Chlorobenzene in the Viking Gas Chromatograph-Mass Spectrometer Data Sets: Reanalysis of Viking Mission Data Consistent With Aromatic Organic Compounds on Mars. Journal of Geophysical Research: Planets 123, 1674-1683. https://dx.doi.org/10.1029/2018JE005544

Hansen, C. J., Esposito, L., Stewart, A. I. F., Colwell, J., Hendrix, A., Pryor, W., Shemansky, D., West, R. (2006). Enceladus' Water Vapor Plume. Science 311, 1422-1425. https://dx.doi.org/10.1126/science.1121254

Hecht, M. H., Kounaves, S. P., Quinn, R. C., West, S. J., Young, S. M. M., Ming, D. W., Catling, D. C., Clark, B. C., Boynton, W. V., Hoffman, J., DeFlores, L. P., Gospodinova, K., Kapit, J., Smith, P. H. (2009). Detection of Perchlorate and the Soluble Chemistry of Martian Soil at the Phoenix Lander Site. Science 325, 64-67. https://dx.doi.org/10.1126/science.1172466

Iess, L., Stevenson, D. J., Parisi, M., Hemingway, D., Jacobson, R. A., Lunine, J. I., Nimmo, F., Armstrong, J. W., Asmar, S. W., Ducci, M., Tortora, P. (2014). The Gravity Field and Interior Structure of Enceladus. Science 344, 78-80. https://dx.doi.org/10.1126/science.1250551

Ito, M., Uesugi, M., Naraoka, H., Yabuta, H., Kitajima, F., Mita, H., Takano, Y., Karouji, Y., Yada, T., Ishibashi, Y., Okada, T., Abe, M. (2014). H, C, and N isotopic compositions of Hayabusa category 3 organic samples. https://dx.doi.org/10.1186/1880-5981-66-91

Kaplan, H. H., Milliken, R. E., Alexander, C. M. O'D. (2018). New Constraints on the Abundance and Composition of Organic Matter on Ceres. Geophysical Research Letters 45, 5274-5282. https://dx.doi.org/10.1029/2018GL077913

Keller, H. U., Arpigny, C., Barbieri, C., Bonnet, R. M., Cazes, S., Coradini, M., Cosmovici, C. B., Delamere, W. A., Huebner, W. F., Hughes, D. W., Jamar, C., Malaise, D., Reitsema, H. J., Schmidt, H. U., Schmidt, W. K. H., Seige, P., Whipple, F. L., Wilhelm, K. (1986). First Halley multicolour camera imaging results from Giotto. Nature 321, 320-326. https://dx.doi.org/10.1038/321320a0

Keller, L. P., Messenger, S., Flynn, G. J., Clemett, S., Wirick, S., Jacobsen, C. (2004). The nature of molecular cloud material in interplanetary dust. Geochimica et Cosmochimica Acta 68, 2577-2589. https://dx.doi.org/10.1016/j.gca.2003.10.044

Khurana, K. K., Kivelson, M. G., Stevenson, D. J., Schubert, G., Russell, C. T., Walker, R. J., Polanskey, C. (1998). Induced magnetic fields as evidence for subsurface oceans in Europa and Callisto. Nature 395, 777-780. https://dx.doi.org/10.1038/27394

Kissel, J., Krueger, F. R. (1987). The organic component in dust from comet Halley as measured by the PUMA mass spectrometer on board Vega 1. Nature 326, 755-760. https://dx.doi.org/10.1038/326755a0

Kissel, J., Brownlee, D. E., Buchler, K., Clark, B. C., Fechtig, H., Grun, E., Hornung, K., Igenbergs, E. B., Jessberger, E. K., Krueger, F. R., Kuczera, H., McDonnell, J. A. M., Morfill, G. M., Rahe, J., Schwehm, G. H., Sekanina, Z., Utterback, N. G., Volk, H. J., Zook, H. A. (1986a). Composition of comet Halley dust particles from Giotto observations. Nature 321, 336-337. https://dx.doi.org/10.1038/321336a0

Kissel, J., Sagdeev, R. Z., Bertaux, J. L., Angarov, V. N., Audouze, J., Blamont, J. E., Buchler, K., Evlanov, E. N., Fechtig, H., Fomenkova, M. N., von Hoerner, H., Inogamov, N. A., Khromov, V. N., Knabe, W., Krueger, F. R., Langevin, Y., Leonasv, B., Levasseur-Regourd, A. C., Managadze, G. G., Podkolzin, S. N., Shapiro, V. D., Tabaldyev, S. R., Zubkov, B. V. (1986b). Composition of comet Halley dust particles from VEGA observations. Nature 321, 280-282. https://dx.doi.org/10.1038/321280a0

Kissel, J., Krueger, F. R., Silén, J., Clark, B. C. (2004). The Cometary and Interstellar Dust Analyzer at Comet 81P/Wild 2. Science 304, 1774-1776. https://dx.doi.org/10.1126/science.1098836

Kitajima, F., Uesugi, M., Karouji, Y., Ishibashi, Y., Yada, T., Naraoka, H., Abe, M., Fujimura, A., Ito, M., Yabuta, H., Mita, H., Takano, Y., Okada, T. (2015). A micro-Raman and infrared study of several Hayabusa category 3 (organic) particles. Earth, Planets and Space 67, 20-31. https://dx.doi.org/10.1186/s40623-015-0182-6

Kounaves, S. P., Hecht, M. H., Kapit, J., Gospodinova, K., DeFlores, L., Quinn, R. C., Boynton, W. V., Clark, B. C., Catling, D. C., Hredzak, P., Ming, D. W., Moore, Q., Shusterman, J., Stroble, S., West, S. J., Young, S. M. M. (2010). Wet Chemistry experiments on the 2007 Phoenix Mars Scout Lander mission: Data analysis and results. Journal of Geophysical Research 115, E00E10-1- E00E10-16. https://dx.doi.org/10.1029/2009JE003424

Krasnopolsky, V. A., Maillard, J. P., Owen, T. C. (2004). Detection of methane in the martian atmosphere: evidence for life?. Icarus 172, 537-547. https://dx.doi.org/10.1016/j.icarus.2004.07.004

Kuskov, O. L., Kronrod, V. A. (2005). Internal structure of Europa and Callisto. Icarus 177, 550-569. https://dx.doi.org/10.1016/j.icarus.2005.04.014

Langevin, Y., Kissel, J., Bertaux, J.-L., Chassefiere, E. (1987). First statistical analysis of 5000 mass spectra of cometary grains obtained by PUMA 1 (Vega 1) and PIA (Giotto) impact ionization mass spectrometers in the compressed modes. Astronomy and Astrophysics 187, 761-766.

Laurent, B., Cousins, C. R., Pereira, M. F. C., Martins, Z. (2019) Effects of UV-organic interaction and Martian conditions on the survivability of organics. Icarus 323, 33-39. https://doi.org/10.1016/j.icarus.2019.01.020

Lauretta, D. S., Balram-Knutson, S. S., Beshore, E., Boynton, W. V., Drouet d'Aubigny, C., DellaGiustina, D. N., Enos, H. L., Golish, D. R., Hergenrother, C. W., Howell, E. S., Bennett, C. A., Morton, E. T., Nolan,

M. C., Rizk, B., Roper, H. L., Bartels, A. E., Bos, B. J., Dworkin, J. P., Highsmith, D. E., Lorenz, D. A., Lim, L. F., Mink, R., Moreau, M. C., Nuth, J. A., Reuter, D. C., Simon, A. A., Bierhaus, E. B., Bryan, B. H., Ballouz, R., Barnouin, O. S., Binzel, R. P., Bottke, W. F., Hamilton, V. E., Walsh, K. J., Chesley, S. R., Christensen, P. R., Clark, B. E., Connolly, H. C., Crombie, M. K., Daly, M. G., Emery, J. P., McCoy, T. J., McMahon, J. W., Scheeres, D. J., Messenger, S., Nakamura-Messenger, K., Righter, K., Sandford, S. A. (2017). OSIRIS-REx: Sample Return from Asteroid (101955) Bennu. Space Science Reviews 212, 925-984. https://dx.doi.org/10.1007/s11214-017-0405-1.

Lawler, M. E., Brownlee, D. E. (1992). CHON as a component of dust from Comet Halley. Nature 359, 810-812. https://dx.doi.org/10.1038/359810a0

Leshin, L. A. et al. (2013). Volatile, Isotope, and Organic Analysis of Martian Fines with the Mars Curiosity Rover. Science 341, 1238937-1-1238937-9. https://dx.doi.org/10.1126/science.1238937

Lorenz, R. D., Mitchell, K. L., Kirk, R. L., Hayes, A. G., Aharonson, O., Zebker, H. A., Paillou, P., Radebaugh, J., Lunine, J. I., Janssen, M. A., Wall, S. D., Lopes, R. M., Stiles, B., Ostro, S., Mitri, G., Stofan, E. R. (2008). Titan's inventory of organic surface materials. Geophysical Research Letters 35, L02206-1-L02206-6. https://dx.doi.org/10.1029/2007GL032118

Liuzzi, G., Villanueva, G. L., Mumma, M. J., Smith, M. D., Daerden, F., Ristic, B., Thomas, I., Vandaele, A. C., Patel, M. R., Lopez-Moreno, J.-J., Bellucci, G., NOMAD Team (2019). Methane on Mars: New insights into the sensitivity of CH4 with the NOMAD/ExoMars spectrometer through its first in-flight calibration. Icarus 321, 671-690. https://dx.doi.org/10.1016/j.icarus.2018.09.021

Matrajt, G., Pizzarello, S., Taylor, S., Brownlee, D. (2004). Concentration and variability of the AIB amino acid in polar micrometeorites: Implications for the exogenous delivery of amino acids to the primitive Earth. Meteoritics & Planetary Science 39, 1849-1858. https://dx.doi.org/10.1111/j.1945-5100.2004.tb00080.x

Matrajt, G., Muñoz Caro, G. M., Dartois, E., D'Hendecourt, L., Deboffle, D., Borg, J. (2005). FTIR analysis of the organics in IDPs: Comparison with the IR spectra of the diffuse interstellar medium. Astronomy and Astrophysics 433, 979-995. https://dx.doi.org/10.1051/0004-6361:20041605

Matrajt, G., Ito, M., Wirick, S., Messenger, S., Brownlee, D. E., Joswiak, D., Flynn, G., Sandford, S., Snead, C., Westphal, A. (2008). Carbon investigation of two Stardust particles: A TEM, NanoSIMS, and XANES study. Meteoritics & Planetary Science 43, 315-334. https://dx.doi.org/10.1111/j.1945-5100.2008.tb00625.x

Martins, Z. (2019) Organic Molecules in Meteorites and Their Astrobiological Significance. In Handbook of Astrobiology, Kolb, V (ed.): CRC Press. pp. 177-194.

Ming, D. W., et al. (2014). Volatile and Organic Compositions of Sedimentary Rocks in Yellowknife Bay, Gale Crater, Mars. Science 343, 1245267-1-1245267-15. https://dx.doi.org/10.1126/science.1245267

Mumma, M. J., Charnley, S. B. (2011). The Chemical Composition of Comets—Emerging Taxonomies and Natal Heritage. Annual Review of Astronomy and Astrophysics 49, 471-524. https://dx.doi.org/10.1146/annurev-astro-081309-130811

Mumma, M. J., Villanueva, G. L., Novak, R. E., Hewagama, T., Bonev, B. P., DiSanti, M. A., Mandell, A. M., Smith, M. D. (2009). Strong Release of Methane on Mars in Northern Summer 2003. Science 323, 1041-1045. https://dx.doi.org/10.1126/science.1165243

Nagao, K., Okazaki, R., Nakamura, T., Miura, Y. N., Osawa, T., Bajo, K.-i., Matsuda, S., Ebihara, M., Ireland, T. R., Kitajima, F., Naraoka, H., Noguchi, T., Tsuchiyama, A., Yurimoto, H., Zolensky, M. E., Uesugi, M., Shirai, K., Abe, M., Yada, T., Ishibashi, Y., Fujimura, A., Mukai, T., Ueno, M., Okada, T., Yoshikawa, M., Kawaguchi, J. (2011). Irradiation History of Itokawa Regolith Material Deduced from Noble Gases in the Hayabusa Samples. Science 333, 1128-1131. https://dx.doi.org/10.1126/science.1207785

Nakamura, T., Noguchi, T., Tanaka, M., Zolensky, M. E., Kimura, M., Tsuchiyama, A., Nakato, A., Ogami, T., Ishida, H., Uesugi, M., Yada, T., Shirai, K., Fujimura, A., Okazaki, R., Sandford, S. A., Ishibashi, Y., Abe, M., Okada, T., Ueno, M., Mukai, T., Yoshikawa, M., Kawaguchi, J. (2011). Itokawa Dust Particles: A Direct Link Between S-Type Asteroids and Ordinary Chondrites. Science 333, 1113-1116. https://dx.doi.org/10.1126/science.1207758

Naraoka, H. et al. (2012). Preliminary organic coumpound analysis of microparticles returned from Asteroid 25143 Itokawa by the Hayabusa mission. Geochimcal Journal 46, 61-72.

Naraoka, H., Aoki, D., Fukushima, K., Uesugi, M., Ito, M., Kitajima, F., Mita, H., Yabuta, H., Takano, Y., Yada, T., Ishibashi, Y., Karouji, Y., Okada, T., Abe, M. (2015). ToF-SIMS analysis of carbonaceous particles in the sample catcher of the Hayabusa spacecraft. Earth, Planets and Space 67, 67-1-67-9. https://dx.doi.org/10.1186/s40623-015-0224-0

Navarro-González, R., Vargas, E., de la Rosa, J., Raga, A. C., McKay, C. P. (2010). Reanalysis of the Viking results suggests perchlorate and organics at midlatitudes on Mars. Journal of Geophysical Research 115, E12010-1-E12010-11. https://dx.doi.org/10.1029/2010JE003599

Niemann, H. B., Atreya, S. K., Demick, J. E., Gautier, D., Haberman, J. A., Harpold, D. N., Kasprzak, W. T., Lunine, J. I., Owen, T. C., Raulin, F. (2010). Composition of Titan's lower atmosphere and simple surface volatiles as measured by the Cassini-Huygens probe gas chromatograph mass spectrometer experiment. Journal of Geophysical Research 115, E12006-1- E12006-22. https://dx.doi.org/10.1029/2010JE003659

Noguchi, T., Nakamura, T., Kimura, M., Zolensky, M. E., Tanaka, M., Hashimoto, T., Konno, M., Nakato, A., Ogami, T., Fujimura, A., Abe, M., Yada, T., Mukai, T., Ueno, M., Okada, T., Shirai, K., Ishibashi, Y.,

Okazaki, R. (2011). Incipient Space Weathering Observed on the Surface of Itokawa Dust Particles. https://dx.doi.org/10.1126/science.1207794

Pavlov, A. A., Vasilyev, G., Ostryakov, V. M., Pavlov, A. K., Mahaffy, P. (2012) Degradation of the organic molecules in the shallow subsurface of Mars due to irradiation by cosmic rays. Geophysical Research Letters 39, L13202. https://doi.org/10.1029/2012GL052166

Pinilla-Alonso, N., Lorenzi, V., Campins, H., de Leon, J., Licandro, J. (2013). Near-infrared spectroscopy of 1999 JU3, the target of the Hayabusa 2 mission. Astronomy & Astrophysics 552, 79-81. https://dx.doi.org/10.1051/0004-6361/201221015

Poch, O., Coll, P., Buch, A., Ramírez, S. I., Raulin, F. (2012). Production yields of organics of astrobiological interest from H2O-NH3 hydrolysis of Titan's tholins. Planetary and Space Science 61, 114-123. https://dx.doi.org/10.1016/j.pss.2011.04.009

Poch, O., Noblet, A., Stalport, F., Correia, J. J., Grand, N., Szopa, C., Coll, P. (2013). Chemical evolution of organic molecules under Mars-like UV radiation conditions simulated in the laboratory with the "Mars organic molecule irradiation and evolution" (MOMIE) setup. Planetary and Space Science 85, 188-197. https://doi.org/10.1016/j.pss.2013.06.013

Poch, O., Kaci, S., Stalport, F., Szopa, C., Coll, P. (2014). Laboratory insights into the chemical and kinetic evolution of several organic molecules under simulated Mars surface UV radiation conditions. Icarus 242, 50-63. https://doi.org/10.1016/j.icarus.2014.07.014

Poch, O., Jaber, M., Stalport, F., Nowak, S., Georgelin, T., Lambert, J.-F., Szopa, C., Coll, P. (2015). Effect of Nontronite Smectite Clay on the Chemical Evolution of Several Organic Molecules under Simulated Martian Surface Ultraviolet Radiation Conditions. Astrobiology 15, 221-237. https://doi.org/10.1089/ast.2014.1230

Postberg, F., Khawaja, N.; Abel, B.; Choblet, G.; Glein, C. R.; Gudipati, M. S.; Henderson, B. L.; Hsu, H.-W.; Kempf, S.; Klenner, F.; Moragas-Klostermeyer, G.; Magee, B.; Nölle, L.; Perry, M.; Reviol, R.; Schmidt, J.; Srama, R.; Stolz, F.; Tobie, G.; Trieloff, M.; Waite, J. H. (2018). Macromolecular organic compound s from the depths of Enceladus. Nature 558, 564-568. https://dx.doi.org/10.1038/s41586-018-0246-4

Sandford, S. A., Aléon, J., Alexander, C. M. O.'D., Araki, T., Bajt, S., Baratta, G. A., Borg, J., Bradley, J. P., Brownlee, D. E., Brucato, J. R., Burchell, M. J., Busemann, H., Butterworth, A., Clemett, S. J., Cody, G., Colangeli, L., Cooper, G., D'Hendecourt, L., Djouadi, Z., Dworkin, J. P., Ferrini, G., Fleckenstein, H., Flynn, G. J., Franchi, I. A., Fries, M., Gilles, M. K., Glavin, D. P., Gounelle, M., Grossemy, F., Jacobsen, C., Keller, L. P., Kilcoyne, A. L. D., Leitner, J., Matrajt, G., Meibom, A., Mennella, V., Mostefaoui, S., Nittler, L. R., Palumbo, M. E., Papanastassiou, D. A., Robert, F., Rotundi, A., Snead, C. J., Spencer, M. K., Stadermann, F. J., Steele, A., Stephan, T., Tsou, P., Tyliszczak, T., Westphal, A. J., Wirick, S., Wopenka, B., Yabuta, H., Zare, R. N., Zolensky, M. E. (2006). Organics Captured from Comet 81P/Wild 2 by the Stardust Spacecraft. Science 314, 1720-1724. https://dx.doi.org/10.1126/science.1135841

Schidlowski, M. (1988). A 3,800-million-year isotopic record of life from carbon in sedimentary rocks. Nature 333, 313-318. https://dx.doi.org/10.1038/333313a0

Schopf, J. W. (1993). Microfossils of the Early Archean Apex Chert: New Evidence of the Antiquity of Life. Science 260, 640-646. https://dx.doi.org/10.1126/science.260.5108.640

Spohn, T., Schubert, G. (2003). Oceans in the icy Galilean satellites of Jupiter?. Icarus 161, 456-467.   https://dx.doi.org/10.1016/S0019-1035(02)00048-9

Stephan, K., Jaumann, R., Brown, R. H., Soderblom, J. M., Soderblom, L. A., Barnes, J. W., Sotin, C., Griffith, C. A., Kirk, R. L., Baines, K. H., Buratti, B. J., Clark, R. N., Lytle, D. M., Nelson, R. M., Nicholson, P. D. (2010). Specular reflection on Titan: Liquids in Kraken Mare. Geophysical Research Letters 37, L07104-1-L07104-5. https://dx.doi.org/10.1029/2009GL042312

Stern, S. A., et al. (2015). The Pluto system: Initial results from its exploration by New Horizons. Science 350, aad1815-1-aad1815-10. https://dx.doi.org/10.1126/science.aad1815

Stoker, C. R., Bullock, M. A. (1997) Organic degradation under simulated martian conditions. J. Geophys. Res. 102, 10881–10888. https://doi.org/10.1029/97JE00667

Teanby, N. A., Irwin, P. G. J., de Kok, R., Jolly, A., Bézard, B., Nixon, C. A., Calcutt, S. B. (2009). Titan's stratospheric C2N2, C3H4, and C4H2 abundances from Cassini/CIRS far-infrared spectra. Icarus 202, 620-631. https://dx.doi.org/10.1016/j.icarus.2009.03.022

Telfer, M. W., Parteli, E. J. R., Radebaugh, J., Beyer, R. A., Bertrand, T., Forget, F., Nimmo, F., Grundy, W. M., Moore, J. M., Stern, S. A., Spencer, J., Lauer, T. R., Earle, A. M., Binzel, R. P., Weaver, H. A., Olkin, C. B., Young, L. A., Ennico, K., Runyon, K. (2018). Dunes on Pluto. Science 360, 992-997. https://dx.doi.org/10.1126/science.aao2975

ten Kate, I. L., Garry, J. R. C., Peeters, Z., Quinn, R., Foing, B., Ehrenfreund, P. (2005). Amino acid photostability on the martian surface. Meteorit. Planet. Sci. 40, 1185–1193. https://doi.org/10.1111/j.1945-5100.2005.tb00183.x

ten Kate, I.L., Garry, J. R. C., Peeters, Z., Foing, B., Ehrenfreund, P., (2006) The effects of martian near surface conditions on the photochemistry of amino acids. Planet. Space Sci. 54, 296–302. https://doi.org/10.1016/j.pss.2005.12.002

Thomas, P. C., Tajeddine, R., Tiscareno, M. S., Burns, J. A., Joseph, J., Loredo, T. J., Helfenstein, P., Porco, C. (2016). Enceladus's measured physical libration requires a global subsurface ocean. Icarus 264, 37-47. https://dx.doi.org/10.1016/j.icarus.2015.08.037

Trigo-Rodríguez, J. M., Moyano-Cambero, C. E., Llorca, J., Fornasier, S., Barucci, M. A., Belskaya, I., Martins, Z., Rivkin, A. S., Dotto, E., Madiedo, J. M., Jacinto, A.-A. (2014). UV to far-IR reflectance spectra of carbonaceous chondrites - I. Implications for remote characterization of dark primitive asteroids targeted by sample-return missions. Monthly

Notices of the Royal Astronomical Society 437, 227-240. https://dx.doi.org/10.1093/mnras/stt1873

Tsuchiyama, A., Uesugi, M., Matsushima, T., Michikami, T., Kadono, T., Nakamura, T., Uesugi, K., Nakano, T., Sandford, S. A., Noguchi, R., Matsumoto, T., Matsuno, J., Nagano, T., Imai, Y., Takeuchi, A., Suzuki, Y., Ogami, T., Katagiri, J., Ebihara, M., Ireland, T. R., Kitajima, F., Nagao, K., Naraoka, H., Noguchi, T., Okazaki, R., Yurimoto, H., Zolensky, M. E., Mukai, T., Abe, M., Yada, T., Fujimura, A., Yoshikawa, M., Kawaguchi, J. (2011). Three-Dimensional Structure of Hayabusa Samples: Origin and Evolution of Itokawa Regolith. Science 333, 1125-1128. https://dx.doi.org/10.1126/science.1207807

Uesugi, M., Naraoka, H., Ito, M., Yabuta, H., Kitajima, F., Takano, Y., Mita, H., Ohnishi, I., Kebukawa, Y., Yada, T., Karouji, Y., Ishibashi, Y., Okada, T., Abe, M. (2014). Sequential analysis of carbonaceous materials in Hayabusa-returned samples for the determination of their origin. Earth, Planets and Space 66, 102-1-102-11. https://dx.doi.org/10.1186/1880-5981-66-102

Vago, J. L. et al. (2017). Habitability on Early Mars and the Search for Biosignatures with the ExoMars Rover. Astrobiology 17, 471-510. https://dx.doi.org/10.1089/ast.2016.1533

Vance, S., Bouffard, M., Choukroun, M., Sotin, C. (2014). Ganymede's internal structure including thermodynamics of magnesium sulfate oceans in contact with ice. Planetary and Space Science 96, 62-70. https://dx.doi.org/10.1016/j.pss.2014.03.011

Wada, K., Grott, M., Michel, P., Walsh, K.J., Barucci, A.M., Biele, J., Blum, J., Ernst, C.M., Grundmann, J.T., Gundlach, B., Hagermann, A., Hamm, M., Jutzi, M., Kim, M.-J., Kührt, E., Le Corre, L., Libourel, G., Lichtenheldt, R., Maturilli, A., Messenger, S.R., Michikami, T., Miyamoto, H., Mottola, S., Müller, T., Nakamura, A.M., Nittler, L.R., Ogawa, K., Okada, T., Palomba, E., Sakatani, N., Schröder, S.E., Senshu, H., Takir, D., Zolensky, M.E. (2018). Asteroid Ryugu before the Hayabusa2 encounter. Progress in Earth and Planetary Science 5, 82 1-30. https://dx.doi.org/10.1186/s40645-018-0237-y

Waite, J. H.; Combi, M. R.; Ip, W.-H.; Cravens, T. E.; McNutt, R. L.; Kasprzak, W.; Yelle, R.; Luhmann, J.; Niemann, H.; Gell, D.; Magee, B.; Fletcher, G.; Lunine, J.; Tseng, W.-L. (2006). Cassini Ion and Neutral Mass Spectrometer: Enceladus Plume Composition and Structure. Science 311, 1419-1422. https://dx.doi.org/10.1126/science.1121290

Waite, J. H., Young, D. T., Cravens, T. E., Coates, A. J., Crary, F. J., Magee, B., Westlake, J. (2007). The Process of Tholin Formation in Titan's Upper Atmosphere. Science 316, 870-875. https://dx.doi.org/10.1126/science.1139727

Wall, S., Hayes, A., Bristow, C., Lorenz, R., Stofan, E., Lunine, J., Le Gall, A., Janssen, M., Lopes, R., Wye, L., Soderblom, L., Paillou, P., Aharonson, O., Zebker, H., Farr, T., Mitri, G., Kirk, R., Mitchell, K., Notarnicola, C., Casarano, D., Ventura, B. (2010). Active shoreline of Ontario Lacus, Titan: A morphological study of the lake and its surroundings. Geophysical Research Letters 37, L05202-1-L05202-5. https://dx.doi.org/10.1029/2009GL041821

Webster, C. R., et al. (2015). Mars methane detection and variability at Gale crater. Science 347, 415-417. https://dx.doi.org/10.1126/science.1261713

Webster, C. R., et al. (2018). Background levels of methane in Mars' atmosphere show strong seasonal variations. Science 360, 1093-1096. https://dx.doi.org/10.1126/science.aaq0131

Wright, I. P., Sheridan, S., Barber, S. J., Morgan, G. H., Andrews, D. J., Morse, A. D. (2015). CHO-bearing organic compounds at the surface of 67P/Churyumov-Gerasimenko revealed by Ptolemy. Science 349, aab0673-1-aab0673-3. https://dx.doi.org/10.1126/science.aab0673

Yada, T., Fujimura, A., Abe, M., Nakamura, T., Noguchi, T., Okazaki, R., Nagao, K., Ishibashi, Y., Shirai, K., Zolensky, M. E., Sandford, S., Okada, T., Uesugi, M., Karouji, Y., Ogawa, M., Yakame, S., Ueno, M., Mukai, T., Yoshikawa, M., Kawaguchi, J. (2014). Hayabusa-returned sample curation in the Planetary Material Sample Curation Facility of JAXA. Meteoritics & Planetary Science 49, 135-153. https://dx.doi.org/10.1111/maps.12027

Yurimoto, H., et al. (2011). Oxygen Isotopic Compositions of Asteroidal Materials Returned from Itokawa by the Hayabusa Mission. Science 333, 1116-1119. https://dx.doi.org/10.1126/science.1207776

Zimmer, C., Khurana, K. K., Kivelson, M. G. (2000). Subsurface Oceans on Europa and Callisto: Constraints from Galileo Magnetometer Observations. Icarus 147, 329-347. https://dx.doi.org/10.1006/icar.2000.6456

Chapter 4

Microbial Life in Impact Craters

Charles S. Cockell[1*], Gordon Osinski[2], Haley Sapers[3], Alexandra Pontefract[4] and John Parnell[5]

[1] School of Physics and Astronomy, James Clerk Maxwell Building, King's Buildings, University of Edinburgh, Edinburgh, UK, EH9 3FD, UK
[2] Department of Earth Sciences/Physics and Astronomy, 1151 Richmond St., University of Western Ontario, London, Ontario, N6A 5B7, Canada
[3] California Institute of Technology, Mail code 100-23, Pasadena, CA 91125, USA
[4] Earth, Atmospheric and Planetary Sciences, Massachusetts Institute of Technology
Greene Building, MIT, Cambridge, MA 02139, USA
[5] School of Geosciences, University of Aberdeen, King's College, Aberdeen, AB24 3FX, UK

*e-mail: c.s.cockell@ed.ac.uk

DOI: https://doi.org/10.21775/9781912530304.04

Abstract
Asteroid and comet impacts are known to have caused profound disruption to multicellular life, yet their influence on habitats for microorganisms, which comprise the majority of Earth's biomass, is less well understood. Of particular interest are geological changes in the target lithology at and near the point of impact that can persist for billions of years. Deep subsurface and surface-dwelling microorganisms are shown to gain advantages from impact-induced fracturing of rocks. Deleterious changes are associated with impact-induced closure of pore spaces in rocks. Superimposed on these long-term geological changes are post-impact alterations such as changes in the hydrological system in and around a crater. The close coupling between geological changes and the conditions for microorganisms yields a synthesis of the fields of microbiology and impact cratering. We use these data to discuss how craters can be used in the search for life beyond Earth.

Introduction
The collision of asteroids or comets with the surface of Earth yields large quantities of energy that cause profound changes to the geology of the target region (Melosh, 1989). The production of a fireball, blast waves,

dust-loading of the atmosphere and changes in atmospheric chemistry (Toon et al., 1997; Kring, 2003) also perturb the environment. These discoveries have focused attention on the deleterious consequences of impact events to the surface biota, focus reinforced by research on the impact thought to have been responsible for the end Mesozoic (Cretaceous–Palaeogene) extinctions (Alvarez et al., 1980; Schulte et al. 2010). Although impacts are known to have caused changes in some microbial communities; for example, the turnover in calcareous nanoplankton flora across the Cretaceous–Palaeogene boundary (Pospichel, 1996; Bown, 2005), less is known about the effects of impacts on habitats for microorganisms, particularly changes associated with geological processes. Impact events are unique among Earth-system perturbations because they are the only extraterrestrial mechanism capable of delivering a localised pulse of destructive energy into an ecosystem.

Although impacts can potentially cause global-scale environmental perturbations, asteroid and comet impacts have had a more common influence at the local scale (Kring, 1997; Adushkin and Nemchinov, 1994). An event that could cause the global changes observed at the Cretaceous–Palaeogene boundary is thought to occur at an average rate of about once per 100 million years, whereas an event causing the formation of a 1 km diameter crater with its associated environmental effects, such as those associated with the Barringer (Meteor) crater (Kring, 1997), will occur approximately once every few thousand years. Both global- and local-scale impact events will result in permanent changes to the target geology. Thus, the more common local impacts are as important in understanding how impacts alter rocky habitats for microorganisms as large events associated with global environmental perturbations.

Understanding the link between impact processes and microbiology is scientifically important from a number of perspectives. First, it gives a more complete picture of how the biosphere is shaped by geological perturbations and processes. Second, it yields insights into how the astronomical environment specifically (i.e. asteroid and comets) can shape our planet's biosphere. Third, we gain a more complete view of the role of impacts in shaping the biosphere and its subsequence evolution beyond the simplistic view that impacts 'destroy life'. Fourth, as no solar system-forming process is likely to occur without leaving behind debris, we might reasonably suggest that impact events on the surfaces of rocky planets are a universal process. Thus, understanding their influence on planetary habitability and a planetary biota provides us with a universe-scale picture of how the cosmic environment shapes the conditions for life alongside our understanding of endogenous processes within planets (such as volcanism, plate tectonics etc.) that influence planetary habitability and life.

In this paper, we review the state of knowledge on the association of microorganisms with impact craters. We highlight some of the major features of impact cratering that influence the subsequent microbial colonisation of craters. We also suggest how the study of impact-generated habitats might influence efforts to test the hypothesis of life on other planetary bodies.

Geological changes caused by impact of relevance to microbiology

The effects of meteorite impacts on microbiology may be defined by the sequence of geological changes that follow from such events. Meteorite impacts are characterised by an initial contact and compression stage (when the impactor makes contact with the target lithology or water), followed by an excavation phase during which a bowl-shaped "transient" crater cavity is formed; for diameters exceeding 2–4 km on Earth subsequent modification can occur, resulting in the formation of a central peak and/or peak ring, depending on the magnitude of the event (Melosh, 1989).

The kinetic energy of such an event is enormous due to the high velocity of objects intersecting with Earth's orbit (the mean impact velocity with the Earth is ~21 km/s; Stuart and Binzel, 2004). This energy will be released in large part as heat. The longevity of this thermal excursion will depend, *inter alia*, upon the target lithology, the availability of water, and local climate.

The Haughton impact structure - a 23-km diameter structure in Nunavut, Canadian High Arctic, formed about 23.5 million years ago (Young et al., 2013) - is thought to have hosted a hydrothermal system for thousands of years with initial temperatures estimated at ~650–700°C (Osinski et al., 2001; Osinski et al., 2005a), evidenced by a sequence of high temperature hydrothermal minerals such as carbonates, sulfides, and quartz followed by the precipitation of cooler temperature minerals such as sulfates. The hydrothermal system in the smaller 4-km-diameter Kärdla crater, Estonia, is also estimated to have lasted for several thousand years (Versh et al., 2005). By contrast, a lifetime of one to two million years is estimated for the hydrothermal systems of larger craters such as the 250-km-diameter Sudbury structure, Canada (Abramov and Kring, 2004) and the 170-km-diameter Chicxulub structure, Mexico (Abramov and Kring, 2007). The physicochemical environment in these hydrothermal systems will depend upon the local target lithology; however, geochemical evidence from a number of craters suggests that neutral or alkaline pH values are usual for post-impact hydrothermal systems (Naumov, 2005).

The energy delivered into the target area during contact and compression will, in addition to vaporizing the impactor, deform, heat, and metamorphose rocks (French, 2004). An important effect of shock processing is that the target rocks can become highly fractured and experience large increases in porosity. Shock-induced alteration of the target rocks at the Vredefort (Reimold and Gibson, 1996) and Sudbury (Ames et al., 2002) impact structures is still clearly evident today, 2.0 Ga and 1.8 Ga after they formed, showing that these changes can persist over geological timescales.

The formation of a crater cavity during the excavation and modification stages will influence the local hydrological cycle through the disruption or alteration of flow paths and also potentially through the formation of long-lived water bodies that host aquatic microbial ecosystems. Approximately half of the impact craters known on Earth today with clear surface expressions host some type of intra-crater water body. Lakes formed in the intra-crater cavity of land-based craters can persist for hundreds of millions of years. For example, the Lac Couture impact structure in Quebec hosts a lake approximately 430 Ma after the crater's formation; although more recent glaciations make it unlikely that a lake has persisted for this entire duration. However, these features are eventually subject to a breach of the crater rim and/or infilling of the crater (this is, of course, not relevant for impacts in the marine environment), making them, in most cases, more short-lived than changes to the target geology.

These impact-induced geological processes influence the microbiology of the crater and are the backdrop against which microbial communities are re-established. Although the given sequence and duration of these impact environments (i.e. hydrothermal system, lakes, exposed impact-altered rocks) vary from crater to crater, they are a common theme in understanding the distribution of microbial communities within craters.

The microbiology of impact structures

How are microbial processes linked to these geological changes associated with impact? Data on the microbiota of impact craters is sparse and so we examine this question by reviewing and synthesizing data from two "type" localities - the Haughton impact structure, Canada, and the Chesapeake impact structure, USA (Table 1). Other craters are discussed where data are available.

Proceeding chronologically through the formation of the impact crater environment, we propose that the first major geological effect on microbiology is that of the shock-induced heating resulting from the transfer of kinetic energy from the projectile to the target. The energy associated with this process could sterilize extant microbial communities (e.g., Sleep 1989) or severely alter the ecology (e.g. Abramov and Mojzsis 2009).

Hydrothermal minerals at the Haughton structure (Osinski et al., 2001), such as quartz, record post-impact temperatures that locally exceeded the known upper temperature limit for microorganisms (>122°C) (Takai et al., 2008). Similar observations have been made in the deep subsurface: drilling of the Chesapeake impact structure has shown that at 1397 to

Table 1. Location, size and age of impact structures in the text of this paper*.

Name of crater	Country	Size (km)	Age (Myr)
Barringer	Arizona, USA	1.186	0.049 ± 0.003
Chesapeake	Virginia, USA	90	35.5 ± 0.3
Chicxulub	Yucatan, Mexico	170	64.98 ± 0.05
Clearwater lakes	Quebec, Canada	26 and 36	290 ± 20
El'gygytgyn	Russia	18	3.5 ± 0.5
Haughton	Nunavut, Canada	24	22.5 ± 2.0
Kärdla	Estonia	4	~455
Lonar	India	1.83	0.052 ± 0.006
Mjølnir	Norway	40	142.0 ± 2.6
New Quebec	Quebec, Canada	3.44	1.4 ± 0.1
Ries	Germany	24	14.6 ± 0.2
Rochechouart	France	23	203 ± 3.0
Siljan	Sweden	52	376.8 ± 1.7
Sudbury	Ontario, Canada	250	1850 ± 3
Tswaing	South Africa	1.13	0.22 ± 0.05
Vredefort	South Africa	300	2023 ± 4

* Data from the Earth impact database maintained by the Planetary and Space Science Centre, University of New Brunswick, Canada.

1424 m depth, the rock consists of 20-30% glassy impact melt clasts (Horton et al. 2009) suggesting an average temperature at the time of deposition of greater than 350°C (Malinconico et al., 2009). These data show that impact-induced sterilization can occur in surface and deep subsurface microbial communities.

As the target begins to cool, the next phase of thermally controlled microbiology can begin, where habitats for heat-loving (thermophilic and hyperthermophilic) microorganisms become available. The formation of hydrothermally-induced habitats is less well documented. Direct evidence for microbial colonisation of an impact-induced hydrothermal system is reported for the deep subsurface of the Chesapeake impact structure, where mineralised microbial structures have been observed (Glamoclija et al., 2009) and from the Siljan impact structure, Sweden where mineralized microbial biofilms associated with a low temperature hydrothermal system have been reported (Hode et al., 2009). Lipid distributions in hydrothermally precipitated gypsum in the Haughton structure have been suggested to be associated with the impact hydrothermal system (Bowden and Parnell, 2007) in contrast to lipids from present-day microorganisms (Parnell et al., 2004). In Haughton specifically, the hydrothermally associated mineral assemblages are complex, including diverse mineral states of iron and sulfur. They serve as a record of geochemical conditions that varied spatially and temporally, pointing to the existence of a range of microbial habitats with different nutrient and energy availabilities, and preservation potentials (Izawa et al., 2011).

Direct evidence for post-impact colonization of a hydrothermal system might be found in isotopic signatures. Large sulfur isotope excursions in hydrothermally deposited sufides in the Haughton impact structure are suggested to be evidence for post-impact colonization by sulfate-reducing microorganisms (Parnell et al., 2010). Similarly sulfate isotope excursions associated with impact-induced fractures are associated with the Rochechouart impact structure in France and have been hypothesized to be evidence for post-impact microbial sulfate reduction (Simpson et al., 2017).

The influence of the post-hydrothermal phase of impact cratering on biology is much better constrained, since these hydrologic systems can be observed in many craters today. The impact-induced disruption to the local geology and the formation of a crater cavity leads to the hypothesis that changes in the hydrological cycles, and therefore nutrient and redox couple supply, will influence the abundance, distribution and diversity of microbial communities. This is most clearly represented by the microbial ecosystems of intra-crater lakes. Many craters today host lakes whose microbiota have been examined; for example, the Tswaing (Ashton, 1999; Schoeman and Ashton, 1982; Ashton and Shoeman, 1983; Ashton and Schoeman, 1988), Clearwater (Maltais and Vincent, 1997), New Quebec (Bouchard, 1989; Gronlund et al., 1990), El'gygytgyn (Cremer and Wagner, 2003), and Lonar (Wani et al., 2006; Surakasi et al., 2007; Joshi et al., 2008; Antony et al., 2014) impact lakes. Microorganisms associated with impact crater lakes are also preserved in the fossil record. Fossil remains of ancient lake algal bioherms have been reported at the Ries impact structure (Riding, 1979). Few of these ecosystems are directly influenced by the fact that the cavity is impact-induced. In the case of New Quebec, the steep littoral zone that is generally a feature of crater cavities, and therefore the low biomass of peripheral photosynthesis communities, has been implicated in the low overall productivity of the lake (Gronlund et al., 1990), although a steep littoral zone is not a unique property of cavities created by impacts.

The most profound influences on a microbiota caused by an impact event, due to their potential longevity, would be expected to arise from permanent changes to the target rock, where the physical and chemical alteration of target materials by the impact would be hypothesized to change both the availability and physicochemical characteristics of habitats for microorganisms, persisting over geological timescales.

Investigations at the Haughton impact structure, situated in what is today a polar desert (Cockell et al., 2001) have provided a particularly lucid test of this hypothesis. Evidence for the influence of impact processes on microbial colonization patterns is found in altered distributions of endolithic communities (organisms that live within the rock interstices) within shocked sandstone and gneiss lithologies (Cockell et al., 2002; Cockell et al., 2003; Pontefract et al. 2014; Pontefract et al. 2016).

Of particular interest are the shocked gneiss communities, as gneissic rocks do not generally provide suitable substrates for endolithic, especially cryptoendolithic communities, although weathering rinds can be colonized by phototrophs (chemical and biological rock weathering itself may be accelerated by impact-induced changes in the target material; Leroux, 2005; Cockell et al., 2007). The gneiss clasts within the crater, associated with the carbonate-rich melt rocks (Metzler et al., 1988; Osinski et al., 2005b; Osinski et al., 2005c), reveal an observed increase in microbial biomass and species diversity with increasing shock metamorphism (Pontefract et al., 2014, 2016); Figure 1. This is most pronounced within the photosynthetic community, where phototrophs such as *Gloeocapsa* and *Chroococcidiopsis* morphotypes are abundant within the endolithic environment, and are also found inhabiting the underside of rocks in the arctic (Cockell and Stokes, 2004; 2006). These cryptoendolithic microorganisms are limited to the interiors of rocks where pore size and permeability is sufficient to allow for the growth of the organisms throughout the rock interstices and the movement of nutrients and redox couples (Figure 2).

The cryptoendolithic colonization of gneissic rocks within the crater, specifically the enhanced colonization by phototrophs, can be explained by the effects of impact fracturing and bulking, which increased the porosity of the rocks (pore spaces of 1 micron and greater are increased in surface area by 25-fold) and the translucence of the rocks (Singleton et al. 2011). The penetration of photosynthetically active radiation through the bulk material was increased by an order of magnitude, primarily because of the formation of vesicles and fractures (Cockell et al., 2002). The organisms inhabit the upper ~5 mm of the rock substrate (Figure 1a). Although other geological events such as volcanism and earthquakes can fracture rocks, the gneissic rocks at Haughton show how impacts can cause a systemic increase in permeability, porosity and translucence throughout the rock matrix, leading to a direct cause-effect relationship between impact and the colonization of impact-shocked rocks by a specific group of microorganisms – phototrophs. This accounts for the lack of reports of cryptoendolithic communities within unweathered crystalline rocks in other geological settings. The community composition between low shock and

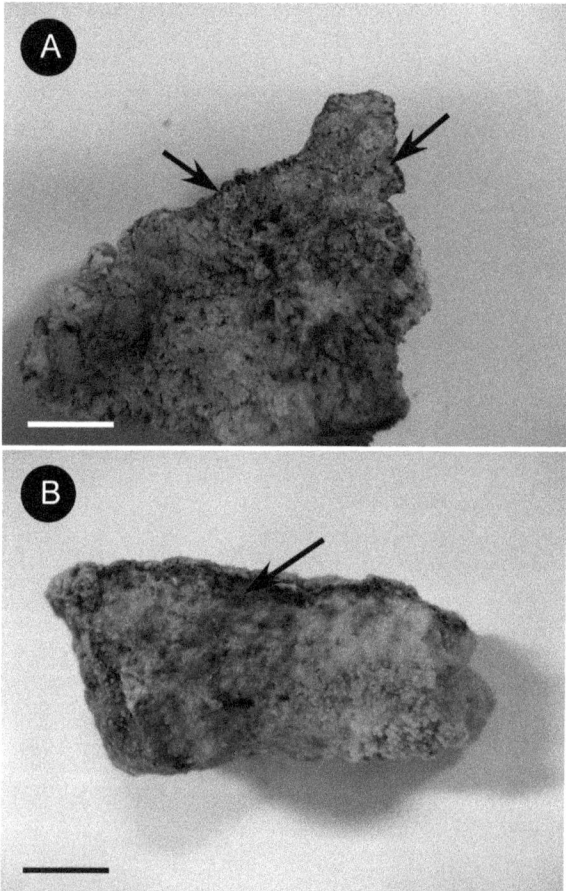

Figure 1. Cryptoendoliths associated with impact-altered rocks. Location of organisms is indicated with arrows. a) Cryptoendolithic phototrophs associated with impact-altered gneiss (shocked to ~30 GPa) from the Haughton impact structure, Nunavut, Canadian High Arctic, b) similar communities associated with vesiculated and shocked sandstones (~10-20 GPa) from the same impact structure. Scale bar 1 cm.

high shock is also different in non-phototrophic taxa (Pontefract et al. 2016).

Detailed examination of shock gneisses, whereby shock levels were estimated from petrographic analysis and correlated to microbial colonisation allowed for the quantification of the optimal shock range in which the habitat is improved for colonisation. This was found to be between 55 and 65 GPa (Pontefract et al., 2014). This study examined the depth distribution of organisms into shocked gneissses and using confocal

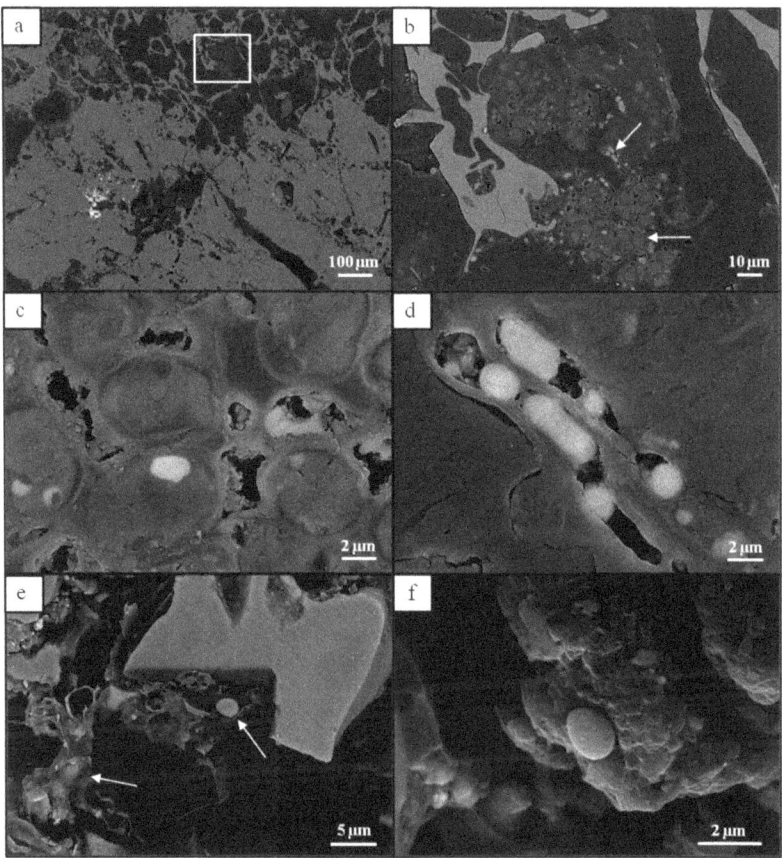

Figure 2. (A-B) SEM-BSE (Scanning Electron Microscope-Backscattered electrons) micrographs showing a cryptoendolithic community in gneiss shocked to ~50GPa. (C) Higher magnification SEM-BSE image of lower arrow in B. (D) Higher magnification SEM-BSE micrograph of upper arrow in B. Light patches visible in B-D indicate OsO4 staining of thylakoids. (E) SEM micrograph showing exopolymeric substance (left arrow) and a cell within the EPS, revealed using FIB milling. (Pontefract et al., unpublished data).

microscopy the direct link between the fractures, pore spaces, and microbial colonisation was demonstrated.

Microbial diversity within impact-shocked rocks is also correlated with shock metamorphism. In the Haughton impact structure, species diversity was found to increase in gneisses exposed to higher shock pressures (Pontefract et al., 2016), with distinct populations based on the level of porosity and light transmission within the rock. In each population of organisms studied, Actinobacteria were the most abundant phylum, and it was not until shock pressures of 55 GPa were attained that phototrophic bacteria began to represent a significant portion (>10%) of the microbial community. The study found that these effects on microbial diversity are likely caused by a variety of influences resulting from the impact processing of the gneisses, such as changes in surface area available for growth and changes in fluid flow and nutrient availability, and potentially the effects of changed primary production from the greater abundance of phototrophs observed at higher shock levels. The study demonstrates that impact events can change the microbial diversity that can be supported by the target location.

Although the shocked rocks at Haughton are chemically varied (Metzler et al., 1988), it was possible to examine the effects of impact shock on major and trace elements. Pontefract et al. (2012) showed that major cation abundances expressed as oxides were lower in highly shocked crystalline rocks compared to lower shocked rocks. Concentrations peaked at shock level 3 (~10-30 GPa), which could be an artefact of the fact that unshocked or very low-shocked rocks sourced outside the crater do not exactly match the pre-impact basement below the crater. These patterns were not observed for trace elements, which could be explained by these elements being associated with more stable phases within the rock. Similar patterns were not observed in sedimentary rocks. In some cases, the concentration of some elements such as phosphorus increased, which they attributed to possible hydrothermal mobilisation of phosphorus or enrichment by organisms on account of enhanced colonisation. The authors note that the formation of glasses by high shock pressures may enhance element availability to a biota since glasses are more easily dissolved by organic acids than crystalline rocks. Thus, the physical changes in the rock may change elemental availability.

The formation of glasses by impact events may offer entirely new habitats for life. Tubular structures in volcanic glasses have long been associated with microbial boring (e.g., ,Staudigel et al., 1995, 1998; Torsvik et al., 1998; Thorseth et al., 1991) although specific occurrences remain controversial and the means of their formation remains a matter of discussion. Tubular features have been observed in impact glass–bearing breccias from the Ries impact structure (Sapers et al., 2014). These features have associated with them organic features observed using Fourier transform infrared (FTIR) spectroscopy. These features raise the possibility that impact glasses may provide habitats for microorganisms capable of active boring. Furthermore, if they are biogenic, they demonstrate potential as biosignatures of life in preserved impact glasses.

Though the impact-induced changes in rocky microbial habitats discussed above have been shown to have a linear correlation (though a more complex model may be supported by the data) with increasing shock metamorphism (Pontefract et al. 2014), not all rock types conform to this type of relationship. Investigations on the effects of shock on the colonization of sandstones within the Haughton impact structure reveal complexities in the correlation between shock level and colonization associated with the rock type (Osinski, 2008; Cockell and Osinski, 2007) (Figure 1b, 3 and 4). As the sandstones are exposed in the same locations and experience the same environmental conditions as the shocked gneisses, they offer a comparison to the crystalline rocks. Whereas unshocked sandstones are generally porous and suitable for endolithic colonization (Friedmann, 1982; Wessels and Büdel, 1995; Weber et al., 1996; Colwell et al., 1997; Büdel et al., 2004; Blackhurst et al., 2005; Omelon et al., 2006), crystalline rocks are not. A comparison between the two lithologies yields insights into effects of impact on rocky microbial habitats with quite different initial conditions for microbial colonization.

At low shock pressures (<5.5 GPa), sandstones were found to suffer from pore collapse (Osinski, 2008), which impeded colonization by a laboratory-cultivated cyanobacterium *Chroococcidiopsis* sp. and the Gram-positive bacterium *Bacillus subtilis*, consistent with expectations from previous geological observations of shocked sandstones (Kieffer, 1971; Kieffer et al., 1976; Osinski, 2008). However, at higher shock pressures, the formation of

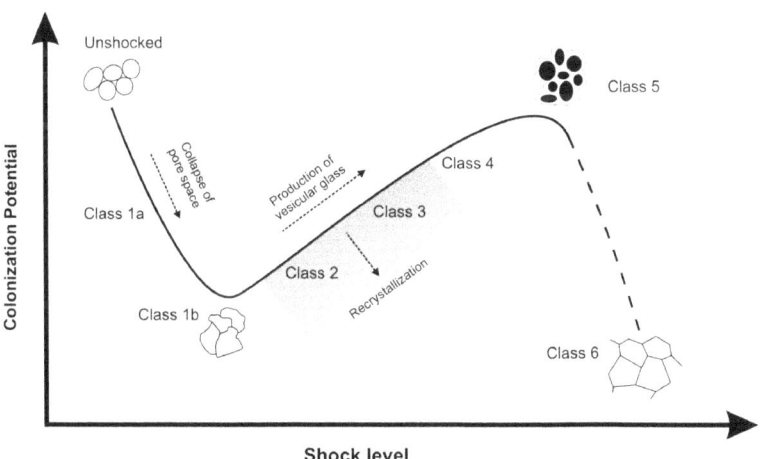

Figure 3. Schematic diagram illustrating the potential for colonization of sandstones with respect to shock level. 'Colonisation potential' means the accessibility of the rock interior for colonisation. This general schema can be modified to any type of material, but illustrates some of the complexities that influence colonisation potential. Low shock (class 1b; ~3-5.5 GPa, porosity ~2-5%) was found to close pore spaces and retarded colonisation. Beyond these shock pressures and up to ~20-30 GPa (up to class 5; 30-36% porosity), production of vesicles in the glass improved permeability and colonisation, although recrystallization in some samples was noted to reduce accessibility to organisms. At shock pressures greater than ~35GPa (class 6), recrystallization was observed to an extent that microorganisms could no longer colonise these materials (taken from Cockell and Osinski, 2007).

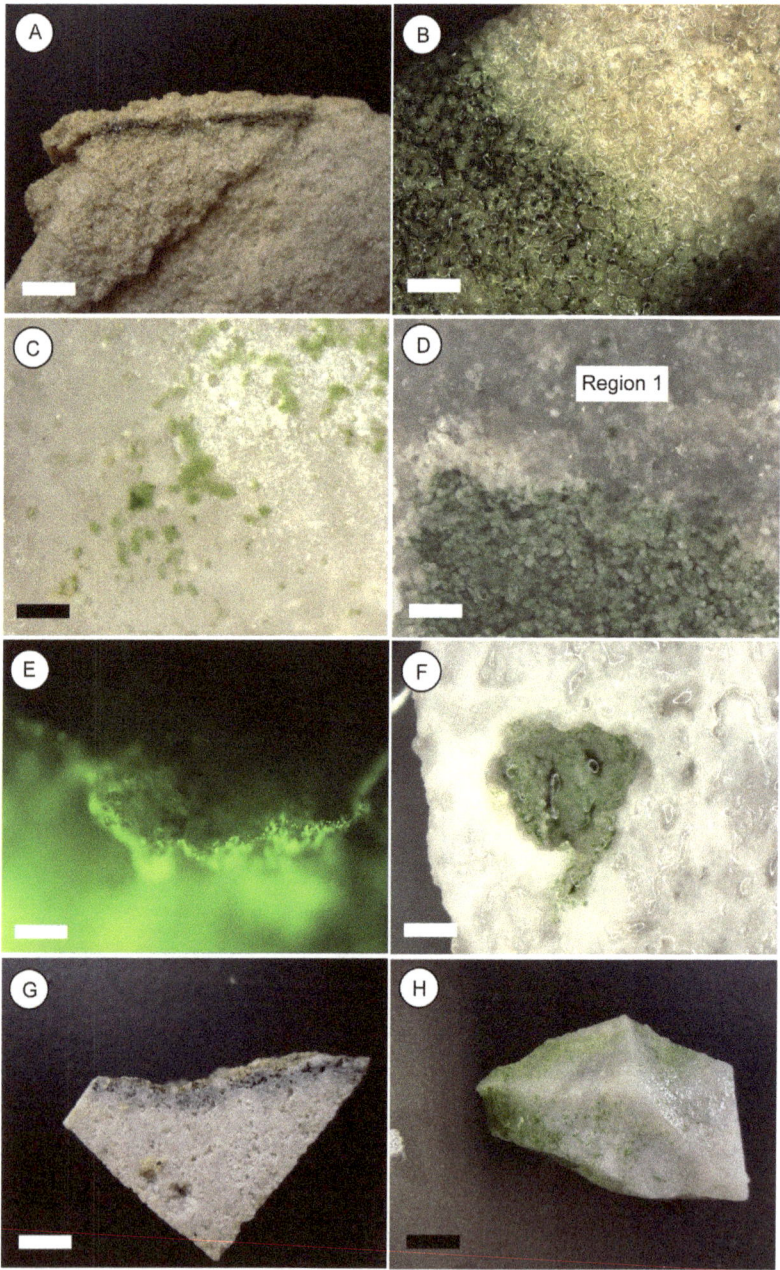

Figure 4. Effects of impact shock on colonization of sandstones. a) Natural cryptoendolithic colonization of Beacon sandstones from the Dry Valleys of Antarctica (see, e.g., Friedmann, 1982 for a description). Black layer is a lichen layer within rock interstices. Scale bar 1 cm; b) Colonization front moving through the interstices of Beacon sandstone (Antarctica) inoculated with *Chroococcidiopsis* sp. 029 as a positive control. Scale bar 0.2 cm; c) Epilithic colonization of class 1b rock. No crypto- or chasmoendolithic colonization of the material is possible. Scale bar 0.5 cm; d) Impeded colonization front of *Chroococcidiopsis* in sandstone (class 2) blocked by 'Region 1'. Scale bar 0.2 cm; e) Impeded colonization front of *Bacillus subtilis* (seen here as green, stained with Syto 9 DNA binding dye) in sandstone [similar location to (d)]. Scale bar 0.2 cm; f) Chasmoendolithic colonization of vesicles by *Chroococcidiopsis* in sandstone (class 4). Scale bar 0.5 cm; g) Natural cryptoendolithic colonization of class 3 shocked sandstone from Haughton by *Gloeocapsa* sp. Scale bar 1 cm; h) Epilithic colonization of class 6 rock. No crypto- or chasmoendolithic colonization of the material is possible. Scale bar 0.5 cm (images from Cockell and Osinski, 2007).

vesicles in the rock, caused by irregular patterns of melting, improved colonization. Natural cyanobacterial colonization of sandstones shocked to 10-20 GPa is observed in field samples (Figure 1b). At yet higher shock pressures (~35 GPa) cryptoendolithic colonization may again be impeded by recrystallization and the formation of solid glassy material. Although a systematic pattern can be discerned as illustrated in Figure 3, shock processes are heterogeneous, even at centimeter scales, leading to irregular colonization patterns depending on where the shock wave has generated compressed pores, glass and/or vesicles (Figure 4d). The data show that impacts cause quite specific increases or decreases in colonization potential that are linked to specific shock levels.

In summary, impacts can, in the same target location, cause geomicro-biological conditions to be reversed. Low porosity crystalline rocks that are typically difficult for microorganisms to colonize can be improved as habitats due to fracturing and vesicularisation. By contrast, high porosity rocks, such as sandstones, can, at certain shock pressures, be impoverished with respect to their accessibility to a biota.

Impact events can generate new habitats by mobilising fluids that precipitate new minerals, for example in hydrothermal systems. In the Haughton impact structure, the colonization of hydrothermally deposited calcium sulfate (selenite) was demonstrated, whereby microorganisms colonise the cleavage planes in selenite (Parnell et al., 2004). As the material is transparent to photosynthetically active radiation, the spaces can act as habitats for phototrophs.

Across a wide range of rock types, impact events fracture materials, generating new habitats for chasmoendoliths that inhabit the fractures connected to the surface of rocks. These effects are less subtle than the changes in internal permeability described above for gneiss and sandstone and relate more to the general shattering of target rocks. The fracturing of carbonates is observed in Haughton that provide habitats for a diversity of microorganisms, for example in ejecta blocks (Cockell et al., 2003). Fractures are also produced in exposed outcrops of gypsum that harbor a diversity of phototrophs and heterotrophs (Cockell et al. 2010).

In addition to the physical effects of shock metamorphism, the thermal effects on any indigenous organic carbon in the crater is a constraint. This carbon contributes to what is available for reprocessing as biomass by subsequent colonisers. However, the carbon becomes more thermally mature due to the energy of impact, as measured in the Haughton Crater (Parnell et al. 2005, Lindgren et al. 2009), and less amenable to reprocessing.

A large quantity of the diversity and biomass of microbial life on the Earth resides in the subsurface (Whitman et al. 1998; Horsfield et al., 2007; Kallmeyer et al., 2012; Magnabosco et al., 2018). The drilling of the Chesapeake Bay impact structure (Gohn et al., 2008) using microbiological contamination control (Gronstal et al., 2009) allowed the changes in the deep subsurface caused by impact events to be examined (Cockell et al., 2009). The crater was formed in the Late Eocene and is approximately 35 million years old (Koeberl et al. 1996; Powars and Bruce, 1999; Poag et al., 2004; Hortson et al., 2005). The buried structure has the form of an inverted sombrero and a 1.6 km core was collected within the elliptical moat approximately 9 km from its centre (Gohn et al., 2008). The microbiology of the crater core (Figure 5) can be broadly split into three zones (Cockell et al., 2009). The upper zone (127-867 m depth) exhibits a logarithmic decline in cell numbers similar to other deep subsurface environments, with cell numbers declining to below detection limits within the middle parts of the tsunami resurge deposits that filled the crater cavity following the impact. The steeper decline in cell numbers with depth associated with this zone compared to the rate of reduction in cell numbers with depth in other subsurface sites (Parkes et al., 1994; D'Hondt et al., 2004) may be a general feature of the terrestrial biosphere. Alternatively, it may be specifically linked to a change in geochemical conditions, such as the increasing salinity of the water within the crater with depth, which is a result of the inundation by seawater at the time of impact along with the redistribution of a subsurface brine layer thought to be pre-impact in origin (Sanford, 2003; Sanford et al., 2009).

A second zone within the crater (867-1397 m depth) exhibits cell counts below detection limits, sections where cells were not cultivatable and DNA could not be recovered. The hydrological data suggest that neither the introduced seawater nor the pre-impact briny water within the section has been flushed since impact. This interval may have remained biologically impoverished since impact.

More direct evidence for an impact-induced effect on the habitat for microorganisms was found in the third microbiological zone in the crater

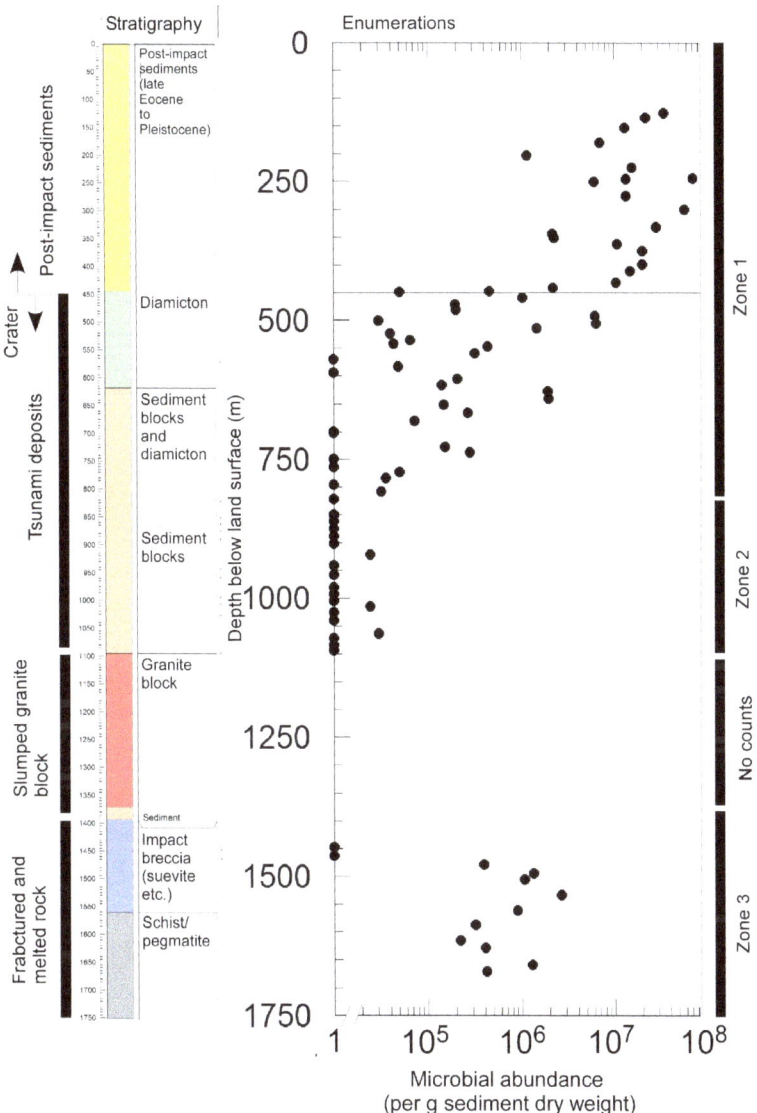

Figure 5. Enumeration of microorganisms with depth through the Chesapeake Bay impact structure. The lithological sequence through the post-impact sediments and the crater is shown on the left. Enumerations (right) can be split into three distinct zones. Zone 1 is a region with a logarithmic decline in the post-impact sediments and the upper part of the impact tsunami deposits. Zone 2 is a region with enumerations below detection limits and Zone 3 is a region of fractured rock below a granite megablock. The detection limit of enumerations was taken as 10^4 cells/g.

(below 1397 m depth) associated with a rise in microbial abundance beneath a granite block that slumped into the crater cavity in the final stages of the impact event. The lower part of this zone is associated with fractured schist-pegmatite rock. Impact breccia veins and dykes that contain shock-deformed rocks are locally present, often associated with fracture networks.

Although we do not have enumeration data for unaltered material immediately outside the crater at a corresponding depth, cell abundances three orders of magnitude lower were observed in samples collected from cores taken northwest of the crater rim in the Coastal Plain (Chapelle et al., 1987). The layered sections in this core represent the geologic units of the original target material and provide the only comparison of unaltered material. These data suggest that the impact and post-impact processes contributed to an increase in the porosity of this region of the crater and played a role in the formation of fracture networks through which microorganisms could have migrated. The presence of breccia dykes within the material suggests that dilatancy, or the opening of fractures, occurred during their emplacement, which would have contributed to biological recolonization. In addition, compaction of crater fill material over time resulted in an upward advection of fluids that may have enhanced microbial recolonization.

The data obtained from the deep subsurface of the Chesapeake structure show that impacts can change the conditions for microbial communities for many millions of years after the event. The deep melt-rich sections of the crater provide evidence that the shock wave and its associated thermal excursion sterilized the deep subsurface, but the fracture networks and advection later allowed recolonization and movement of both redox couples and nutrients into the units of the crater cavity, improving the environment for life.

The data from the Haughton and Chesapeake impact structures, when considered together, show that despite general observable changes in geology caused by impact, particularly rock fracturing (Gurov and Gurova, 1983; Henkel, 1992; Pilkington and Grieve, 1992; Pesonen et al., 1999; Plado et al., 2000; Salminen, 2004; Kumar, 2005; Kumar and Kring, 2008) whose biological effects were noted in earlier work (Cockell and Lee, 2002; Cockell et al., 2005), the particular geological effects of asteroid and comet impacts allow us to discriminate their influence on habitats for microbial communities compared to other geological agents that fracture or compress rocks.

Successional changes in the biota
The changes described above can be presented within the context of classical ideas of ecological succession, which allows for the suggestion of some major successional phases associated with impact craters. A first

phase, a phase of thermal biology, is associated with the hydrothermal systems established in the impact crater and around it (Cockell and Lee, 2002). It will last for as long as the hydrothermal systems persist. In some cases, where the heating is insufficient, for example in small impacts, the influence of the heating on the local biota will be inconsequential. These systems are analogous in some ways to colonisation of deep-sea hydrothermal vents (Sylvan et al., 2012; Christakis et al., 2017; Patwardhan et al., 2018). However, impact systems would be different in two major respects. First, impacts can occur indiscriminately in any lithology. Thus, the geochemistry of the fluids generated, and the subsequent conditions for life, will depend uniquely on the flow regime and geology of the target site. Of course, impact can occur into land, which will greatly differ in terms of fluid circulation and the cooling and diluting influences of the ocean in the case of hydrothermal vents. Second, the temperature regimes in impact craters will depend on the scale of the impact and the distribution of heat, which itself will depend on the target lithology. In general, unlike hydrothermal vents, which are localised point sources of heat, impact events are distributed sources of heat which gradually cool. The rate of cooling will again depend on the target material and the initial distribution of heat during impact. After cooling, or in regions unaffected by the hydrothermal system, colonization will occur by non-thermophilic organisms. This successional transition has clearly occurred in the shocked rocks of the Haughton impact structure in the present-day.

The subsequent changes that occur in the crater amount to a long-term phase of post-impact succession and climax in which microbial ecosystems become established that correspond to the gross geological changes occurring in and around the crater. For example, the formation of the crater cavity can lead to the formation of an intra-crater lake as observed in many craters today, with their associated biota (Ashton, 1999; Schoeman and Ashton, 1982; Ashton and Shoeman, 1983; Ashton and Schoeman, 1988; Bouchard, 1989; Gronlund et al., 1990; Maltais and Vincent, 1997; Cremer and Wagner, 2003; Wani et al., 2006; Surakasi et al., 2007; Joshi et al., 2008; Antony et al., 2014). There may later be a phase of colonization of intra-crater lake sediments if the crater rim is breached and a lake is drained (Hickey et al., 1988). Successional changes will occur as new organisms are blown into the crater and colonise the various available substrates. In the case of impact events in aquatic habitats, successional changes will be associated first with local disruption of the overlying water column, and possibly sediments, such as seems to have been the case at Mjølnir, Norway (Smelror and Dypvik, 2006). These lake successional events are not unique to impact craters *per se*, but the crater cavity, the confining effects of the rim, the potential for the residual influence of the hydrothermal system *etc.* underscore the fact that the geological and thermal changes caused by a given impact will influence the nature of successional changes and the biota that occur in that location.

Shocked lithologies will change the habitats for microorganisms, as has already been reviewed, throughout all of these phases. The communities that colonize impact-metamorphosed rocks observed in craters today are, in most cases, unlikely to be the same as the communities that did so immediately after the impact. There is no new crater existing today in which immediate post-successional microbial communities can be examined. The communities either taking advantage of shocked rocks, or excluded from deleteriously altered lithologies, will be influenced by the regional climate in which the impact occurs, and in the early stages after impact, possibly the impact hydrothermal system and intra-crater water bodies. However, in the case of the rocks at Haughton, for instance, the bulk chemistry and physical features of the rock have not been substantially altered by non-impact processes. Therefore, the colonization of the rocks and the effects of impact on them can be considered to provide a faithful insight into the effects of impact on microbial recolonization processes.

A late phase of ecological assimilation in which erosion of the crater causes the ecology to become indistinguishable from the outlying ecology was suggested (Cockell and Lee, 2002). However, the observations from the Chesapeake structure show that this phase may never actually be achieved until the crater is completely eroded and its entire geologic manifestation erased, such as, for example, by subduction.

Perspectives for astrobiology
The data provide insights into the influence of impacts on habitats for life during the early history of Earth, when impacts were more frequent than today (Abramov and Mojzsis, 2009). The Chesapeake structure yields evidence that the deep fracturing of rocks in the subsurface, or the emplacement of fractured rocks into the bottom of a crater cavity, can provide improved habitats for life underground. Although in the subsurface, nutrients are generally more of a limitation to life than pore space (Wellsbury et al., 1997), impact fracturing improves the flow of nutrients and redox couples for microbial life. During the early history of Earth, the planet was subjected to sterilizing impact pulses on its surface (Maher and Stevenson, 1988; Sleep et al., 1989; Abramov and Mojzsis, 2009). The data from the Chesapeake structure show that impacts would have created deep refugia (Sleep and Zahnle, 1998) in which life would have been protected from the destructive effects of subsequent impacts.

Impact fractured and vesicularised rocks could also have provided protection for photosynthetic organisms from the more intense UV radiation environment on the early Earth when the planet lacked a significant ozone shield (and speculatively on any anoxic planet with impact-shocked substrates on its surface). Samples of impact-shocked gneiss from Haughton impact structure containing the cyanobacterium, *Chroococcidiopsis*, were flown to the International Space Station and exposed for 22 months on the outside of the station to a simulated early earth UV flux using the extraterrestrial spectrum and cut-off filters. It was empirically demonstrated that shocked rocks could have provided a refugium for life under the worst-case flux assumed for the early Earth (Bryce et al., 2014).

The study of the microbiology of impact craters also reveals insights into the potential influence of impacts on habitability elsewhere. Transient hydrothermal systems represent an obvious potential habitat for life in these contexts (Newsom, 1980; Osinski et al., 2001; Rathburn and Squyres, 2002; Koeberl and Reimold, 2004; Abramov and Kring, 2005; Squyres et al., 2008; Hode et al., 2009; Osinski et al., 2013), although they may be less abundant and in some cases more short-lived than volcanic hydrothermal systems (Pope et al., 2006). On Mars, impact craters may harbour deep subsurface locations where fractured rocks would enhance water flow in aquifers and geochemical turnover, possibly in combination with deep hydrothermal systems. In particular, future robotic and human exploration efforts on Mars might focus on the search for geological interfaces within craters where fluid flow between lithological units has enhanced the availability of energy and nutrients. These locations would be promising regions for deep drilling efforts to assess the geological conditions and potential habitability of that planet in its early history. In summary, the formation of hydrothermal systems, intra-crater lakes, fractured and permeabilised target rocks and geological interfaces all show the high potential of impact craters as sites to test the hypothesis of life on Mars and explore the formation of habitable conditions.

Summary

Asteroid and comet impacts exert lasting and important effects on microbial patterns of colonization in and around their craters. Future observations should improve our understanding of how physical and chemical changes in differing impact lithologies influence the abundance, diversity and distribution of microorganisms today and in the past and how geochemical changes in and around impact craters influence the availability of nutrients and redox couples for microbial communities. By coupling surface and subsurface microbiological studies of craters, the influence of impacts on global microbial processes will be better understood and the implications for life on the more violent early Earth can be examined. As impact events are a universal process, the study of their effects on a microbiota will yield insights into the potential for life in the face of impact bombardment elsewhere and where we might search for life and investigate the conditions for habitability in locations such as Mars. Paleobiological studies will advance our understanding of the early stages of the interactions of microbial life and impact cratering; for example, the study of ancient intra-crater hydrothermal systems, lake sediments, and impact metamorphosed rocks and their preserved biota will yield better insights into how to search for signatures of fossilized life on Earth and elsewhere.

References

Abramov, O., and Kring, D.A. (2004). Numerical modelling of an impact-induced hydrothermal system at the Sudbury crater. Journal of Geophysical Research 109, E10007. doi 10.1029/2003JE002213.

Abramov, O., and Kring, D.A. (2005). Impact-induced hydrothermal activity on early Mars. Journal of Geophysical Research 110, E12S09.

Abramov, O., and Kring, D.A. (2007). Numerical modeling of impact-induced hydrothermal activity at the Chicxulub crater. Meteoritics and Planetary Science 42, 93-112 (2007).

Abramov, O., and Mojzsis, S.J. (2009). Microbial habitability of the Hadean Earth during the late heavy bombardment. Nature 459, 419-422.

Adushkin, V.V., and Nemchinov, I.V. (1994). Consequences of impacts of cosmic bodies on the surface of the Earth. in Hazards due to Asteroids and Comets (ed., Gehrels, T.), 721-778 (University of Arizona Press, Arizona).

Alvarez, L.W., Alvarez, W., Asaro, F., and Michel, H.V. (1980). Extraterrestrial cause for the Cretaceous-Tertiary extinction – Experimental results and theoretical interpretation. Science 208, 1095-1108.

Ames, D.E., Golightly, J.P., Lightfoot, P.C., and Gibson, H.L. (2002). Vitric compositions in the Onaping Formation and their relationship to the Sudbury Igneous Complex, Sudbury structure. Economic Geology and the Bulletin of the Society for Economic Geologists 97, 1541-1562.

Antony, C.P., Shimpi, G.G., Cockell, C.S., Patole, M.S., and Shouche, Y.S. (2014). Molecular characterisation of prokaryotic communities associated with Lonar Crater basalts. Geomicrobiology Journal 31, 519-528.

Ashton, P.J., and Schoeman, F.R. (1983). Limnological studies on the Pretoria Salt Pan, a hypersaline maar lake. 1. Morphometric, physical and chemical features. Hydrobiologia 99, 61-73.

Ashton, P.J., and Schoeman, F.R. (1988). Thermal stratification and the stability of meromixis in the Pretoria Salt Pan, South Africa. Hydrobiologia 158, 253-265.

Ashton, P.J. (1999). Limnology of the Pretoria Saltpan Crater-lake. In Tswaing. Investigations into the Origin, Age and Paleoenvironments of The Pretoria Saltpan. (ed. Partridge, T.C.) 72-90 (Council for Geoscience. Geological Survey of South Africa. Memoir 85. Pretoria).

Blackhurst, R.L., Genge, M.J., Kearsley, A.T., and Grady, M.M. (2005). Cryptoendolith alteration of Antarctic sandstone substrates: pioneers or opportunists? Journal of Geophysical Research 110, E12S24. doi number: 10.1029/2005JE002463.

Bouchard, M. A. (1989). L'histoire naturelle du Cratere du Nouveau-Quebec. Collection Environment et Géologie, v. 7 (Départment de Géologie, Université de Montréal).

Bowden, S.A., and Parnell, J. (2007). Intracrystalline lipids within sulfates from the Haughton impact structure – implications for survival of lipids on Mars. Icarus 187, 422-429.

Bown, P. (2005). Selective calcareous nannoplankton survivorship at the Cretaceous–Tertiary boundary. Geology 33, 653–656.

Bryce, C., Horneck, G., Rabbow, E., Edwards, H.G.M., and Cockell, C.S. (2015). Impact shocked rocks as protective habitats on an anoxic early Earth. Intern. Journ. Astrobiology 14, 115-122.

Büdel B., Weber B., Kühl M., Pfanz H., Sültemeyer D., and Wessels D. (2004). Reshaping of sandstone surfaces by cryptoendolithic cyanobacteria: bioalkalization causes chemical weathering in arid landscapes. Geobiology 2, 261-268.

Chapelle, F.H., Zelibor, J.L., Grimes, D.J., and Knobel, L.L. (1987). Bacteria in deep Coastal Plain sediments of Maryland: A possible source of CO_2 to groundwater. Water Research 23, 1625-1632.

Christakis, C.A., Polymenakou, P.N., Mandalakis, M., Nomikou, P., Kristoffersen, J.B., Lampridou, D., Kotoulas, G., and Magoulas, A. (2017). Microbial community differentiation between active and inactive sulfide chimneys of the Kolumbo submarine volcano, Hellenic Volcanic Arc. Extremophiles 22, 13-27.

Cockell, C.S., and Lee, P. (2002). The biology of impact craters - a review. Biological Reviews 77, 279-310.

Cockell, C.S., and Stokes, M.D. (2004). Widespread colonization by polar hypoliths. Nature 431, 414.

Cockell, C.S., and Stokes, M.D. (2006). Hypolithic colonization of opaque rocks in the Arctic and Antarctic polar desert. Arctic, Antarctic and Alpine Research 38, 335-342.

Cockell, C.S., and Osinski, G.R. (2007). Impact-induced impoverishment and transformation of a sandstone habitat for lithophytic microorganisms. Meteoritics and Planetary Science 42, 1985-1993.

Cockell, C.S., Osinski, G.R., and Lee, P. (2003). The impact crater as a habitat: effects of impact processing of target materials. Astrobiology 3, 181-191.

Cockell, C.S., Lee, P.C., Schuerger, A.C., Hidalgo, L., Jones, J.A., and Stokes, M.D. (2001). Microbiology and vegetation of micro-oases and polar desert, Haughton impact crater, Devon Island, Nunavut, Canada. Arctic, Antarctic and Alpine Research 33, 306-318.

Cockell, C.S., Lee, P., Osinski, G., Horneck, G., and Broady, P. (2002). Impact-induced microbial endolithic habitats. Meteoritics and Planetary Science 37, 1287-1298.

Cockell, C.S., Osinski, G.R., and Lee, P. (2003).The impact crater as a habitat : effects of impact-processing of target materials. Astrobiology 3, 181-191.

Cockell, C.S., Lee, P., Broady, P., Lim, D.S.S., Osinski, G.R., Parnell, J., Koeberl, C., Pesonen, L., and Salminen, J. (2005). Effects of asteroid and comet impacts on habitats for lithophytic organisms - A synthesis. Meteoritics and Planetary Science 40, 1901-1914.

Cockell, C.S., Kennerley, N., Lindstrom, M., Watson, J., Ragnarsdottir, V., Sturkell, E., Ott, S., and Tindle, A.G. (2007). Geomicrobiology of a weathering crust from an impact crater and a hypothesis for its formation. Geomicrobiology Journal 24, 425-440.

Cockell, C.S., Gronstal, A.L., Voytek, M.A., Kirshtein, J.D., Finster, K., Sanford, W.E., Glamoclija, M., Gohn, G.S., Powars, D.S., and Wright Horton, J. Jr. (2009). Microbial abundance in the deep subsurface of the Chesapeake Bay impact crater: Relationship to lithology and impact processes. Geological Society of America Special Papers 458, 941-950.

Cockell, C.S., Osinski, G.R., Banerjee, N.R., Howard, K.T., Gilmour, I., and Watson, J.S. (2010). The microbe–mineral environment and gypsum neogenesis in a weathered polar evaporite. Geobiology 8, 293-308.

Colwell, F.S., Onstott, T.C., Delwiche, M.E., Chandler, D., Fredrickson, J.K., Yao, Q.-J., McKinley, J.P., Boone, D.R., Griffiths, R., Phelps, T.J., Ringelberg, D., White, D.C., LaFreniere, L., Balkwill, D., Lehman, R.M., Konisky, J., and Long, P.E. (1997). Microorganisms from deep, high temperature sandstones: constraints on microbial colonization. FEMS Microbiology Reviews 20, 425-435.

Cremer, H., and Wagner, B. (2003). The diatom flora in the ultra-oligotrophic Lake El'gygytgyn, Chukotka. Polar Biology 26, 105-114.

D'Hondt, S, Jørgensen, B.B., Miller, D.J., Batzke, A., Blake, R., Cragg, B.A., Cypionka, H., Dickens, G.R., Ferdelman, T., Hinrichs, K.U., Holm, N.G., Mitterer, R., Spivack, A., Wang, G., Bekins, B., Engelen, B., Ford, K., Gettemy, G., Rutherford, S.D., Sass, H., Skilbeck, C.G., Aiello, I.W., Guèrin, G., House, C.H., Inagaki, F., Meister, P,. Naehr, T., Niitsuma, S., Parkes, R.J., Schippers, A., Smith, D.C., Teske, A., Wiegel, J., Padilla, C.N., and Acosta, J.L. (2004). Distributions of microbial activities in deep subseafloor sediments. Science 306, 2216-2221.

French, B.M. (2004). The importance of being cratered: The new role of meteorite impact as a normal geological process. Meteoritics and Planetary Science 39, 169-197.

Friedmann, E.I. (1982). Endolithic microorganisms in the Antarctic cold desert. Science 215, 1045-1053.

Glamoclija, M., Steele, A., Fries, M., Schieber, J., Voytek, M.A., and Cockell, C.S. (2009). Association of anatase (TiO2) and microbes: Unusual fossilization effect or a potential biosignature? Geological Society of America Special Papers 458, 965-975.

Gohn, G., Koeberl, C., Miller, K.G., Reimold, U., Browning, J.C., Cockell, C.S., Horton, J.W., Kenkman, T., Kulpecz, A.A., Powars, D.S., Sanford, W.E., and Voytek, M.A. (2008). Deep drilling into the Chesapeake Bay impact structure. Science 320, 1740-1745.

Gronlund, T., Lortie, G., Guilbault, J.P., Bouchard, M.A., and Saanisto, M. (1990). Diatoms and arcellaceans from lac du Cratere du Nouveau-Quebec, Ungava, Quebec, Canada. Canadian Journal of Botany 68, 1187-1200.

Gronstal, A.L., Voytek, M.A., Kirshtein, J.D., von der Heyde, N.M., Lowit, M.D., and Cockell, C.S. (2009). Contamination assessment in microbiological sampling of the Eyreville core, Chesapeake Bay impact structure. Geological Society of America Special Papers 458, 951-964.

Gurov, Ye P., and Gurova, Ye P. (1983). Laws of distribution of faults around a meteor crater; example of Elgygytgyn Crater. Doklady Akademii Nauk SSSR 269(5), 1150-1153.

Henkel, H. (1992). Geophysical aspects of meteorite impact craters in eroded shield environment, with special emphasis on electric resistivity. Tectonophysics 216, 63-89.

Hickey, L.J., Johnson, K.R., and Dawson, M.R. (1988). The stratigraphy, sedimentology, and fossils of the Haughton formation - a post-impact crater-fill, Devon Island, NWT, Canada. Meteoritics 23, 221-231.

Hode, T., Cady, S.L., von Dalwigk, I., and Kristiansson, P. (2009). Evidence of ancient microbial life in an impact structure and its implications for astrobiology – A case study. In From Fossils to Astrobiology (eds, J. Seckbach, J. and Walsh, M.), 249–273 (Springer).

Horsfield, B., Kieft, T. L., Amann, H., Franks, S. G., Kallmeyer, J., Mangelsdorf, K., Parkes, R. J., Wagner, D., Wilkes, H., and Zink, K.-G. (2007). The Geobiosphere. in Continental Scientific Drilling (eds., Harms, U., Koeberl, C., Zoback, M.D.) 163-211 (Springer, Heidelberg).

Horton, J.W., Powars, D.S., and Gohn, G.S. (2005). Studies of the Chesapeake Bay Impact Structure – The USGS-NASA Langley corehole, Hampton, Virginia, and related coreholes and geophysical surveys. (USGS Professional Paper #1688, US Geological Survey, Reston, Virginia).

Horton, J.W., Gibson, R.L., Reimold, W.U, Wittman, A., Gohn, G.S., and Edwards, L.E. (2009). Geologic columns for the ICDP-USGS Eyreville B core, Chesapeake Bay impact structure: Impactites and crystalline rocks, 1766 to 1096 m depth. Geological Society of America Special Papers, 458, https://doi.org/10.1130/2009.2458(02).

Izawa, M.R.M., Banerjee, N.R., Osinski, G.R., Flemming, R.L., Parnell J., and Cockell, C.S. (2011). Weathering of post-impact hydrothermal deposits from the Haughton Impact Structure: implications for microbial colonization and biosignature preservation. Astrobiology 11, 537-550.

Joshi, A.A., Kanekar, P.P., Kelkar, A.S., Shouche, Y.S., Vani, A.A., Borgave, S.B., and Sarnaik, S.S. (2008). Cultivable bacterial diversity of alkaline Lonar Lake, India. Microbial Ecology 55, 163-172.

Kallmeyer, J., Pockalny, R., Adhikari, R.R., Smith, D.C., and D'Hondt, S. (2012) Global distribution of microbial abundance and biomass in subseafloor sediment. Proceedings of the National Academy of Sciences 109, 16213-16216.

Kieffer, S.W. (1971). Shock metamorphism of the Coconino sandstone at Meteor crater, Arizona. Journal of Geophysical Research 76, 5449-5473.

Kieffer, S.W., Phakey, P.P. and Christie, J.M. (1976). Shock processes in porous quartzite: transmission electron microscope observations and theory. Contributions to Mineralogy and Petrology 59, 41-93.

Koeberl, C., Poag, C.W., Reimold, W.U., and Brandt, D. (1996). Impact origin of the Chesapeake Bay structure, and source of the North American tektites. Science 271, 1263-1266.

Koeberl, C., and Reimold, W.U. (2004). Post-impact hydrothermal activity in meteorite impact craters and potential opportunities for life. Bioastronomy 2002: Life among the Stars, 213, 299-304.

Kring, D.A. (1997). Air blast produced by the Meteor Crater impact event and a reconstruction of the affected environment. Meteoritics and Planetary Science 32, 517-530.

Kring, D.A. (2003). Environmental consequences of impact cratering events as a function of ambient conditions on Earth. Astrobiology 3, 133-152.

Kumar, P.S. (2005) Structural effects of meteorite impact on basalt: evidence from Lonar Crater, India. Journal of Geophysical Research 110, B12402, doi number: 10.1029/2005JB003662.

Kumar, P.S., and Kring, D.A. (2008). Impact fracturing and structural modification of sedimentary rocks at Meteor Crater, Arizona. Journal of Geophysical Research 113, E09009, doi: 10.1029/2008JE003115.

Leroux, H. (2005). Weathering features in shocked quartz from the Ries impact crater: Germany. Journal of Geophysical Research 40, 1347-1352.

Lindgren, P., Parnell, J., Bowden, S., Taylor, C., Osinski, G.R., and Lee, P. (2009). Preservation of biological markers in clasts within impact melt breccias from the Haughton Impact Structure, Devon Island. Astrobiology 9, 391-400.

Magnabosco, C., Lin, L.-H., Bomberg, M., Ghiorse, W., Stan-Lotter, H., Pedersen, K., Kieft, T.L., van Heerden, E., and Onstott, T.C. (2018), The biomass an biodiversity of the continental subsurface. Nature Geoscience 11, 707-717.

Maher, K.A., and Stevenson, D.J. (1988). Impact frustration of the origin of life. Nature 331, 612-614.

Melosh, H.J. (1989). Impact Cratering: A Geologic Process. (Oxford University Press, Oxford).

Malinconico, M.L., Horton, J.W., and Sanford, W.E (2009). Postimpact heat conduction and compaction-driven fluid flow in the Chesapeake Bay impact structure based on downhole vitrinite reflectance data, ICDP-USGS Eyreville deep core holes and Cape Charles test holes. Geological Society of America Special Papers 458, https://doi.org/10.1130/2009.2458(38).

Maltais, M.J., and Vincent, W.F. (1997). Periphyton community structure and dynamics in a subarctic lake. Canadian Journal of Botany 75, 1556-1569.

Metzler, A., Ostertag, R., Redeker, H.J., and Stoffler, D. (1988). Composition of the crystalline basement and shock metamorphism of crystalline and sedimentary target rocks at the Haughton-impact-crater, Devon Island, Canada. Meteoritics 23, 197-207.

Naumov, M.V. (2005). Principal features of impact-generated hydrothermal circulation systems: mineralogical and geochemical evidence. Geofluids 5, 165-184.

Newson, H.E. (1980). Hydrothermal alteration of impact melt sheets with implications for Mars. Icarus 44, 207-216.

Omelon, C.R., Pollard, W.H., and Ferris, F.G. (2006). Chemical and ultrastructural characterization of high arctic cryptoendolithic habitats. Geomicrobiology Journal 23, 189-200.

Osinski, G.R. (2008). Impact metamorphism of $CaCO_3$-bearing sandstones at the Haughton structure, Canada Meteoritics and Planetary Science 42, 1945-1960.

Osinski, G.R., Spray, J.G., and Lee, P. (2001). Impact-induced hydrothermal activity within the Haughton impact structure: generation of a transient, warm, wet oasis. Meteoritics and Planetary Science 36, 731-745.

Osinski, G.R., Lee, P., Parnell, J., Spray, J.G., and Baron, M.T. (2005a). A case study of impact-induced hydrothermal activity: The Haughton impact structure, Devon Island, Canadian high arctic. Meteoritics and Planetary Science 40, 1859-1877.

Osinski, G.R., Spray, J.G., and Lee, P. (2005b). Impactites of the Haughton impact structure, Devon Island, Canadian High Arctic. Meteoritics and Planetary Science 40, 1789-1812.

Osinski, G. R., Lee, P., Spray, J.G., Parnell, J., Lim, D.S.S., Bunch, T.E., Cockell, C.S., and Glass, B. (2005c). Geological overview and cratering model of the Haughton impact structure, Devon Island, Canadian High Arctic. Meteoritics and Planetary Science 40, 1759-1776.

Osinski, G.R., Tomabene, L.L., Banerjee, N.R., Cockell, C.S., Flemming, R., Izawa, M.R.M., McCutcheon, J., Parnell, J., Preston, L.J., Pickersgill, A.E., Pontefract, A., Sapers, H.M., and Southam, G. (2013). Impact-generated hydrothermal systems on earth and Mars. Icarus 224, 347-363.

Parkes, R.J., Cragg, B.A., Bale, S.J., Getliff, J.M., Goodman, K., Rochelle, P.A., Fry, J.C. Weightman, A.J., and Harvey, S.M. (1994). Deep bacterial biosphere in Pacific Ocean sediments. Nature 371, 410-413.

Parnell, J., Lee, P., Cockell, C.S., and Osinski, G.R. (2004). Microbial colonization in impact-generated hydrothermal sulphate deposits, Haughton impact structure, and implications for sulphates on Mars. International Journal of Astrobiology 3, 247-256.

Parnell, J., Boyce, A., Thackrey, S.N., Muirhead, D.K., Lindgren, P., Mason, C., Taylor, C.W., Still, J.W., Bowden, S., Osinski, G.R., and Lee, P. (2010) Sulfur isotope signatures for rapid colonization of an impact crater by thermophilic microbes. Geology 38, 271-274.

Parnell, J., Osinski, G., Lee, P., Green, P., and Baron, M. (2005). Thermal alteration of organic matter in an impact crater, and the duration of post-impact heating. Geology 33, 373-376.

Patwardhan, S., Foustoukos, D.I., Giovannelli, D., Yücel, M., and Vetriani, C. (2018). Ecological succession of sulfur-oxidizing Epsilon- and Gammaproteobacteria during colonization of a shallow-water gas vent. Frontiers in Microbiologyy 9; doi.org/10.3389/fmicb.2018.02970.

Pesonen, L.J., Elo, S, Lehtinen, M., Jokinen, T., Puranen, R., and Kivekäs, L. (1999). Lake Karikkoselkä impact structure, central Finland: New geophysical and petrographic results. In Large Meteorite Impacts and Planetary Evolution II (eds., B.O. Dressler, and V.L. Sharpton) 131-147 Geological Society of America Special Paper, Boulder, Colorado, 339).

Pilkington, M., and Grieve, R.A.F. (1992). The geophysical signatures of terrestrial impact craters. Reviews of Geophysics 30, 161-181.

Plado, J., Pesonen, L.J., Koeberl, C., and Elo, S. (2000). The Bosumtwi meteorite impact structure, Ghana: A magnetic model. Meteoritics and Planetary Science 35, 723-732.

Poag, C.W. Koeberl, C., and Reimold, W.U. (2004). The Chesapeake Bay crater – geology and geophysics of a late Eocene submarine impact structure. (Impact Studies Series, Springer, Heidelberg).

Pontefract, A., Osinski, G.R., Lindgren, P., Parnell, J., Cockell, C.S., and Southam, G. (2012). The effects of meteorite impacts on the availability of bioessential elements for endolithic organisms. Meteoritics and Planetary Science 47, 1681-1691.

Pontefract, A., Osinski, G.R., Cockell, C.S., Moore, C.A., Moores, J.E., and Southam, G. (2014). Impact-generated endolithic habitat within crystalline rocks of the Haughton Impact Structure, Devon Island, Canada. Astrobiology 14, 522-533.

Pontefract, A., Osinski, G.R., Cockell, C.S., Souitham, G., McCausland, P.J.A., Umoh, J., and Holdsworth, D.W. (2016). Microbial diversity of impact-generated habitats. Astrobiology 16, 775-786.

Pope, K.O., Kieffer, S.W., and Ames, D.E. (2006). Impact melt sheet formation on Mars and its implication for hydrothermal systems and exobiology. Icarus 183, 1-9.

Pospichal, J.J. (1996). Calcareous nannofossils and clastic sediments at the Cretaceous–Tertiary boundary, Northeastern Mexico. Geology 24, 255–258.

Powars, D.S., and Bruce, T.S. (1999) The effects of the Chesapeake Bay impact crater on the geological framework and correlation of hydrogeologic units of the lower York-James Peninsula, Virginia. (USGS Professional Paper #1612, US Geological Survey, Reston, Virginia).

Rathburn, J.A., and Squyres, S.W. (2002). Hydrothermal systems associated with Martian impact craters. Icarus 157, 362-372.

Reimold, W.U., and Gibson, R.L. (1996). Geology and evolution of the Vredefort Impact Structure, South Africa. Journal of African Earth Sciences 23, 125-162.

Riding, R. (1979). Origin and diagenesis of lacustrine algal bioherms at the margin of the Ries crater, Upper Miocene, Southern Germany. Sedimentology 26, 645-680.

Salminen, J. (2004). Petrophysics and paleomagnetism of Jänsijärvi impact structure. MSc thesis, 155 pp.

Sanford, W.E. (2003). Heat flow and brine generation following the Chesapeake Bay bolide impact. Journal of Geochemical Exploration 78/79, 243-247.

Sanford, W.E., Voytek, M.A., Powars, D.S., Jones, B.F., Cozzarelli, I.M., Cockell, C.S., and Eganhouse, R.P. (2009). Pore-water chemistry from the ICDP-USGS core hole in the Chesapeake Bay impact structure—Implications for paleohydrology, microbial habitat, and water resources. Geological Society of America Special Papers 458, 867-890.

Sapers, H.M., Osinski, G.R., Banerjee, N.R., and Preston, L.J. (2014). Enigmatic tubular features in impact glass. Geology 42, 471-474.

Schoeman, F.R., and Ashton, P.J. (1982). The diatom flora of the Pretoria Salt Pan, Transvaal, Republic of South Africa. Bacillaria 5, 63-99.

Schulte, P., Alegret, L., Arenillas, I., Arz, J.A., Barton, P.J., Bown, P.R., Bralower, T.J., Christeson, G.L., Claeys, P., Cockell, C.S., Collins, G.S., Deutsch, A., Goldin, T.J., Goto, K., Grajales-Nishimura, J.M., Grieve, R.A.F., Gulick, S.P.S., Johnson, K.R., Kiessling, W., Koeberl, C., Kring, D.A., MacLeod, K.G., Matsui, T., Melosh, J., Montanari, A., Morgan, J.V., Neal, C.R., Nichols, D.J., Norris, R.D., Pierazzo, E., Ravizza, G., Rebolledo-Vieyra, M., Reimold, W.U., Robin, E., Salge, T., Speijer, R.P., Sweet, A.R., Urrutia-Fucugauchi, J., Vajda, V., Whalen, M.T., and Willumsen, P.S. (2010). The Chicxulub asteroid impact and mass extinction at the Cretaceous-Paleogene boundary. Science 327, 1214-1218.

Singleton, A.C., Osinski, G.R., McCausland, P.J.A., and Moser, D.E. (2011). Shock-induced changes in density and porosity in shock-metamorphosed crystalline rocks, Haughton impact structure, Canada. Meteoritics and Planetary Science 46, 1774-1786.

Sleep, N.H., Zahnle, K.J., Kasting, J.F., and Morowitz, H.J. (1989). Annihilation of ecosystems by large asteroid and comet impacts on the early Earth. Nature 342, 139-142.

Sleep, N.H., and Zahnle, K.J. (1998). Refugia from asteroid impacts on early Mars and the early Earth. Journal of Geophysical Research 103, 28529-28544.

Smelror, M., and Dypvik, H. (2006). The sweet aftermath: environmental changes and biotic restoration following the Marine Mjølnir impact (Volgian-Ryazanian Boundary, Barents Shelf). in Biological Processes Associated with Impact Events (eds., Cockell, C.S., Koeberl, C., Gilmour, I) 143-178 (Impact Studies Series, Springer, Heidelberg).

Squyres, S.W., Arvidson, R.E., Ruff, S., Gellert, R., Morris, R.V., Ming, D.W., Crumpler, L., Farmer, J.D., Marais, D.J., Yen, A., McLennan, S.M., Calvin, W., Bell,. JF. Clark, B.C., Wang, A., McCoy, T.J., Schmidt, M.E., and de Souza, P.A. (2008). Detection of silica-rich deposits on Mars. Science 320, 1063-1067.

Staudigel, H., Chastain, R.A., Yayanos, A., and Boucier, W. (1995). Biologically mediated dissolution of glass. Chemical Geology 126, 147-154.

Staudigel, H., Yayanos, A., Chastain, R., Davies, G., Th Verdurmen, E.A., Schiffman, P., Boucier, R., and de Baar, H. (1998). Biologically mediated dissolution of volcanic glass in seawater. Earth and Planetary Science Letters 164, 233-244.

Stuart, J.C., and Binzel R.P. (2004). Bias-corrected population, size distribution, and impact hazard for the near-Earth objects. Icarus 170, 295-311.

Surakasi, V.P., Wani, A.A., Shouche, Y.S., and Ranade, D.R. (2007). Phylogenetic analysis of methanogenic enrichment cultures obtained from Lonar Lake in India: Isolation of Methanocalculus sp. and Mathanoculleus sp. Microbial Ecology 54, 697-704.

Sylvan, J.B., Toner, B.M., and Edwards, K.J. (2012). Life and death of deep-sea vents: Bacterial diversity and ecosystem succession on inactive hydrothermal sulfides. mBio 3, e00279-11.

Takai, K., Mormile, M.R., McKinley, J.P., Brockman, F.J., Holben, W.E., Kovacik, W.P. Jr, and Fredrickson, J.K. (2008). Shifts in archaeal communities associated with lithological and geochemical variations in subsurface Cretaceous rock. Proceedings of the National Academy of Sciences 105, 10949–10954.

Thorseth, I.H., Furnes, H., and Tumyr, O., (1991). A textural and chemical study of Icelandic palagonite of varied composition and its bearing on the mechanism of the glass-palagonite transformation, Geochimica Cosmochimica Acta 55, 731-749.

Toon, O.W., Zahnle, K., Morrison, D., Turco, R.P., and Covey, C. (1997). Environmental perturbations caused by the impacts of asteroids and comets. Reviews in Geophysics 35, 41-78.

Torsvik, T., Furnes, H., Muehlenbachs, K., Thorseth, I.H., and Tumyr, O. (1998). Evidence for microbial activity at the glass-alteration interface in oceanic basalts. Earth and Planetary Science Letters 162, 165-176.

Versh, E., Kirsimae, K., Joeleht, A., and Plado, J. (2005). Cooling of the Kardla impact crater: I. The mineral paragenetic sequence observation. Meteoritics and Planetary Science 40, 3-19.

Wani, A.A., Surakasi, V.P., Siddharth, J., Raghavan, R.G., Patole, M.S., Ranade, D., and Shouche, Y.S. (2006). Molecular analyses of microbial diversity associated with the Lonar soda lake in India: An impact crater in a basalt area. Research in Microbiology 10, 928-937.

Weber, B., Wessels, D.C.J., and Büdel, B. (1996). Biology and ecology of cryptoendolithic cyanobacteria of a sandstone outcrop in the Northern Province, South Africa. Algological Studies 83, 565-579.

Wellsbury, P., Goodman, K., Barth, T., Cragg, B.A., Barnes, S.P., and Parkes, J. (1997). Deep marine biosphere fuelled by increasing organic matter availability during burial and heating. Nature 388, 573-576.

Wessels, D.C.J., and Büdel, B. (1995). Epilithic and cryptoendolithic cyanobacteria of Clarens sandstone cliffs in the Golden Gate Highlands National Park, South Africa. Botanica Acta 108, 220-226.

Whitman, W.B., Coleman, D.C., and Wiebe, W.J. (1998). Prokaryotes: the unseen Proceedings of the National Academy of Sciences 95, 6578-6583.

Young, K.E., van Soest, M.C., Hodges, K.V., Watson, E.B., Adams, B.A., and Lee, P. (2013). Impact thermochronology and the age of Haughton impact structure, Canada. Geophysical Research Letters 40, 3836-3840.

Chapter 5

Impact of Simulated Martian Conditions on (Facultatively) Anaerobic Bacterial Strains from Different Mars Analogue Sites

Kristina Beblo-Vranesevic[1*], Maria Bohmeier[1], Sven Schleumer[1,2], Elke Rabbow[1], Alexandra K. Perras[3,4], Christine Moissl-Eichinger[3,7], Petra Schwendner[5,6], Charles S. Cockell[5], Pauline Vannier[8], Viggo T. Marteinsson[8,9], Euan P. Monaghan[10], Andreas Riedo[10], Pascale Ehrenfreund[10,11], Laura Garcia-Descalzo[12], Felipe Gómez[12], Moustafa Malki[13], Ricardo Amils[13], Frédéric Gaboyer[14], Keyron Hickman-Lewis[14], Frances Westall[14], Patricia Cabezas[15], Nicolas Walter[15] and Petra Rettberg[1]

[1]Institute of Aerospace Medicine, Radiation Biology Department, German Aerospace Center (DLR), Cologne, Germany
[2]Hochschule Niederrhein, Wirtschaftingenieurswesen, Krefeld, Germany
[3]Department of Internal Medicine, Medical University of Graz, Graz, Austria
[4]Department of Microbiology and Archaea, University of Regensburg, Regensburg, Germany
[5]UK Center for Astrobiology, School of Physics and Astronomy, University of Edinburgh, Edinburgh, UK
[6]Department of Plant Pathology, Space Life Sciences Laboratory, University of Florida, FL, USA
[7]BioTechMed Graz, Graz, Austria
[8]MATIS - Prokaria, Reykjavík, Iceland
[9]Faculty of Food Science and Nutrition, University of Iceland, Reykjavík, Iceland
[10]Leiden Observatory, Universiteit Leiden, Leiden, Netherland
[11]Space Policy Institute, George Washington University, Washington DC, USA
[12]Instituto Nacional de Técnica Aeroespacial - Centro de Astrobiología (INTA-CAB), Madrid, Spain
[13]Centro de Biología Molecular Severo Ochoa, Universidad Autónoma de Madrid (UAM), Madrid, Spain
[14]Centre de Biophysique Moléculaire, Centre National de la Recherche Scientifique (CNRS), Orléans, France
[15]European Science Foundation (ESF), Strasbourg, France

*kristina.beblo@dlr.de

DOI: https://doi.org/10.21775/9781912530304.05

Abstract

Five bacterial (facultatively) anaerobic strains, namely *Buttiauxella* sp. MASE-IM-9, *Clostridium* sp. MASE-IM-4, *Halanaerobium* sp. MASE-BB-1, *Trichococcus* sp. MASE-IM-5, and *Yersinia intermedia* MASE-LG-1 isolated from different extreme natural environments were subjected to Mars relevant environmental stress factors in the laboratory under controlled conditions. These stress factors encompassed low water activity, oxidizing compounds, and ionizing radiation. Stress tests were performed under permanently anoxic conditions. The survival rate after addition of sodium perchlorate (Na-perchlorate) was found to be species-specific. The intercomparison of the five microorganisms revealed that *Clostridium* sp. MASE-IM-4 was the most sensitive strain (D_{10}-value (15 min, $NaClO_4$) = 0.6 M). The most tolerant microorganism was *Trichococcus* sp. MASE-IM-5 with a calculated D_{10}-value (15 min, $NaClO_4$) of 1.9 M. Cultivation in the presence of Na-perchlorate in Martian relevant concentrations up to 1 wt% led to the observation of chains of cells in all strains. Exposure to Na-perchlorate led to a lowering of the survival rate after desiccation. Consecutive exposure to desiccating conditions and ionizing radiation led to additive effects. Moreover, in a desiccated state, an enhanced radiation tolerance could be observed for the strains *Clostridium* sp. MASE-IM-4 and *Trichococcus* sp. MASE-IM-5. These data show that anaerobic microorganisms from Mars analogue environments can resist a variety of Martian-simulated stresses either individually or in combination. However, responses were species-specific and some Mars-simulated extremes killed certain organisms. Thus, although Martian stresses would be expected to act differentially on microorganisms, none of the expected extremes tested here and found on Mars prevent the growth of anaerobic microorganisms.

Introduction

Mars has been a favored target in the search of extinct or extant life beyond the Earth. Various articles have been published discussing the habitability of early and present-day Mars (e.g. Tosca et al., 2008; Westall et al., 2013; Cockell et al., 2016; Eigenbrode et al., 2018). The present-day Martian surface is characterized by the absence of liquid water. If there is liquid water in the near-surface environment it likely occurs temporarily as brines (Orosei et al., 2018), *i.e.* with high concentrations of different salts, including perchlorates. In addition, the Martian surface is exposed to a high radiation flux in form of ionizing radiation and solar UV-radiation due to a thin anoxic atmosphere consisting of mainly CO_2, and by the lack of a planetary magnetic field (Horneck, 2000, Jakosky et al., 2001, Martin-Torres et al., 2015, McEwen et al., 2011, Hassler et al., 2014, Matthiä et al., 2016, Schubert et al., 2000, Gu et al., 2018).

Another potentially harmful environmental factor on the Martian surface is the ubiquitous presence of oxidizing compounds, especially perchlorates, which might have a strong impact on habitability. The Phoenix lander detected significant concentrations up to 0.6 weight percent (wt%) of perchlorate ions at the landing site in the northern polar regions (Hecht et al., 2009). The MSL mission showed that perchlorates are present presumably on the entire surface of Mars (Archer et al., 2013). At distinct places different types of perchlorates have been detected. For example, a mixture of sodium perchlorate (Na-perchlorate) and magnesium perchlorate (Mg-perchlorate) were inferred in the Palikir and Hale crater. At Horowitz crater Na-perchlorate has been suggested and at Gale Crater calcium-perchlorate was inferred (Ojha et al., 2015; Glavin et al., 2013). The detailed formation mechanism of perchlorates is still not fully understood. Two different formation mechanisms of Martian perchlorates have been suggested. One hypothesis suggests that the perchlorates were produced on the surface whereby Martian surface minerals catalyze the photochemical oxidation of chlorides to perchlorates (Schuttlefield et al., 2011; Kim et al., 2013). It was shown that in chloride-containing Martian soil simulants, perchlorates are produced in the presence of ultraviolet light (Carrier and Kounaves, 2015). Another formation mechanism might be through the reaction of atmospheric oxidants probably on dust particles in the arid environment on Mars (Catling et al., 2010). Perchlorates, as hygroscopic substances, bind water from the atmosphere and contribute to the formation of brines with high concentrations of different dissolved salts, including chlorides, sulfates, and perchlorates. In some regions these brines are thought to remain liquid even at the low temperatures prevailing on the surface of Mars (Gough et al., 2011; Toner and Catling, 2016; Kounaves et al., 2014; Fox-Powell et al., 2016; Martín-Torres et al., 2015).

The working program of the European Community's Seventh Framework Program project MASE (Mars analogues for space exploration; grant agreement n° 607297) included sampling from terrestrial Mars analogue sites to obtain new (facultatively) anaerobic model microorganisms adapted to extreme conditions (Cockell et al., 2017). In this study, some of these microorganisms were exposed to a representative subset of environmental conditions as they occur on present-day Mars. They were perchlorates at different concentrations, absence of water, ionizing radiation, and a Martian atmosphere pressure. As these are some of the most prominent stress factors in a Martian environment, the biological effects on the microorganisms where examined when exposed to individual stress factors and in combination.

Material and Methods
Strains and culture conditions
In order to get an impression of the natural distribution of tolerances to simulated Martian conditions, only the wild type organisms were used. These strains were obtained in the context of the MASE project from various extreme environments which were considered Mars analogues sites (Cockell et al., 2017). The following microorganisms were investigated: *Buttiauxella* sp. MASE-IM-9 (DSM 105071), *Clostridium* sp. MASE-IM-4 (DSM 105631), *Halanaerobium* sp. MASE-BB-1 (DSM

Table 1. Strains, origin and cultivation conditions. [a] at the applied cultivation conditions, no spores of *Clostridium* sp. MASE-IM-4 were detectable.

Strain	Origin	Medium	Supplements in MASE medium (wt%)	Gas phase (vol%)	Temp. (°C)
Buttiauxella sp. MASE-IM-9	Islinger Mühlbach, Germany	MASE-II / TSA	0.1% Yeast extract	80% N_2, 20% CO_2	30
Clostridium sp. MASE-IM-4[a]	Islinger Mühlbach, Germany	MASE-II -$FeCl_2$ / TSA	0.01% Dimethylamine 0.001% $FeCl_2$	15% H_2, 25% CO_2, 60% N_2	30
Halanaerobium sp. MASE-BB-1	Boulby Mine, Great Britain	HACE. No growth on solid surfaces.	0.1% Yeast extract	15% H_2, 25% CO_2, 60% N_2	45
Trichococcus sp. MASE-IM-5	Islinger Mühlbach, Germany	MASE-II -$FeCl_2$ / TSA	0.01% Na_2SO_4 0.01% $C_6H_5Na_3O_7$ x 2 H_2O 0.02% KNO_3	15% H_2, 25% CO_2, 60% N_2	30
Yersinia intermedia MASE-LG-1	Lake Grænavatn, Iceland	MASE-I / TSA	0.01% KNO_3 0.01% C-Org-Mix	80% N_2, 20% CO_2	30

105537), *Trichococcus* sp. IM-5 (DSM 105632), and *Yersinia intermedia* MASE-LG-1 (DSM 102845). Medium compositions and strain-specific anoxic cultivation conditions are summarized in Table 1 and described in detail by Cockell et al. (2017). The incubation was carried out at the indicated cultivation temperature and cultures were shaken at 50 rpm.

Individual and combined stress tests
All stress tests were performed in anoxic MASE/HACE medium (Table 1). The influence of individual and combined stress factors on the selected MASE isolates were examined (Table 2).

Table 2. Overview of performed individual and combined stress tests with the MASE-strains.

Individual stress	Conditions
Na-perchlorate exposure	15 minutes, 4 hours, 24 hours 96 hours; ≤ 4 M
Growth in the presence of Na-perchlorate	0.5 wt% / 1 wt%, 24 hours
Combined Stress	**Conditions**
Na-perchlorate addition and desiccation	0.5 wt% / 1 wt%; 24 hours
Desiccation and X-rays	24 hours; ≤ 3000 Gy
Desiccation and Mars atmosphere	24 hours; ≤ 1 month

For the investigation of the effect of Na-perchlorate as individual stress factor the cells were exposed to different concentrations of Na-perchlorate up to 4 M including Martian relevant concentrations of 36 mM (~ 0.5 wt%) and 71 mM (~ 1.0 wt%). Na-perchlorate was added to an overnight culture grown at strain-specific optimal conditions. After anoxic incubation up to 96 hours at room temperature, the exposure was terminated by sudden dilution (1:400), followed by the determination of the survival rate by the most probable number technique (MPN) or a plating assay.

In addition to these exposure experiments, the ability of the model organisms to actively grow and multiply in the presence of perchlorates was tested. Martian relevant concentrations of Na-perchlorate (0.5 wt% or 1.0 wt%) were added before inoculation and cultivation was conducted at optimal growth conditions in the presence of Na-perchlorate.

For experiments using desiccation as an individual stress factor as well as in combination with other stresses, *i.e.* desiccation and exposure to radiation or exposure to Martian atmosphere, (see also Table 2), the cells were cultivated under optimal growth conditions until stationary growth phase was reached (~24 h). Desiccation experiments were performed as described earlier (Beblo et al., 2009). One milliliter of culture (equivalent to ~10^7 cells) was spread evenly on sterile glass slides or quartz discs and dried under anoxic conditions within an anaerobic chamber (Coy Laboratory Products Inc.; [O_2] < 5 ppm, relative humidity 13 ± 0.5 vol%). Perchlorate treated cells were desiccated in an identical way.

Exposure to ionizing radiation, in the form of X-rays, was performed as described in earlier studies with the X-ray source Gulmay RS 225A (Gulmay Medical Ltd.) at 200 kV and 15 mA (Beblo et al., 2011). The cells were irradiated at a distance of 19.5 cm below the X-ray source with 20 Gy/min ± 5 Gy/min up to 3000 Gy. The dose rate was measured with a UNIDOS dosimeter (PTW Freiburg GmbH). All irradiation experiments were performed at room temperature.

The combination of anoxic desiccation and ionizing irradiation and the combination of anoxic desiccation and exposure to a simulated Martian atmosphere (a Mars-like gas composition (2.7 vol% N_2, 1.6 vol% Ar, 0.15 vol% O_2 in CO_2) at a pressure of 10^3 Pa), were carried out using the previously described TRex-Box (Beblo-Vranesevic et al., 2017a).

If not stated otherwise all exposures and treatments were performed under anoxic conditions at room temperature.

Determination of survival rate
Growth and morphology of the unfixed native cells was observed by phase-contrast light microscopy (ZeissR AxiolmagerTM M2) with 400× or 1000× magnification.

Determination of the survival rate and enumeration of cultivable cells was achieved by the MPN-technique via dilution series with ten-fold dilution steps in anoxic MASE/HACE medium (Franson, 1985). The cells were incubated at their optimal growth temperature for up to four weeks. Alternatively, the survival rate was determined by applying a plating assay on tryptic soy agar (TSA) plates under anoxic conditions for *Buttiauxella* sp. MASE-IM-9, *Clostridium* sp. MASE.IM-4, *Trichococcus* sp. MASE-IM-5, *Yersinia intermedia* MASE-LG-1 (see also Table 1). The plates were incubated in an anaerobic chamber at room temperature up to one week. All experiments were repeated independently at least three times to represent biological replicas.

Results
Survival of MASE-strains after exposure to perchlorates as individual stress factor

To get an overview of the tolerance of the MASE-strains towards Na-perchlorate, survival curves after 15 minutes exposure and the corresponding calculation of the D_{10}-values were performed as shown in Figure 1 and Table 3. All tested strains survived the treatment with Na-perchlorate as an individual stress factor for 15 minutes at room temperature.

Table 3. Calculated D_{10}-values of MASE-strains after addition of Na-perchlorate for 15 minutes at room temperature.
* data from Beblo-Vranesevic et al. 2017b.

Strain	D_{10}-value [M] NaClO$_4$
Clostridium sp. MASE-IM-4	0.6
Buttiauxella sp. MASE-IM-9	1.2
Halanaerobium sp. MASE-BB-1	1.5
Yersinia intermedia MASE-LG-1	1.7
Trichococcus sp. MASE-IM-5	1.9
Escherichia coli	1.3*
Bacillus subtilis (vegetative cells)	1.9*
Deinococcus radiodurans	2.7*

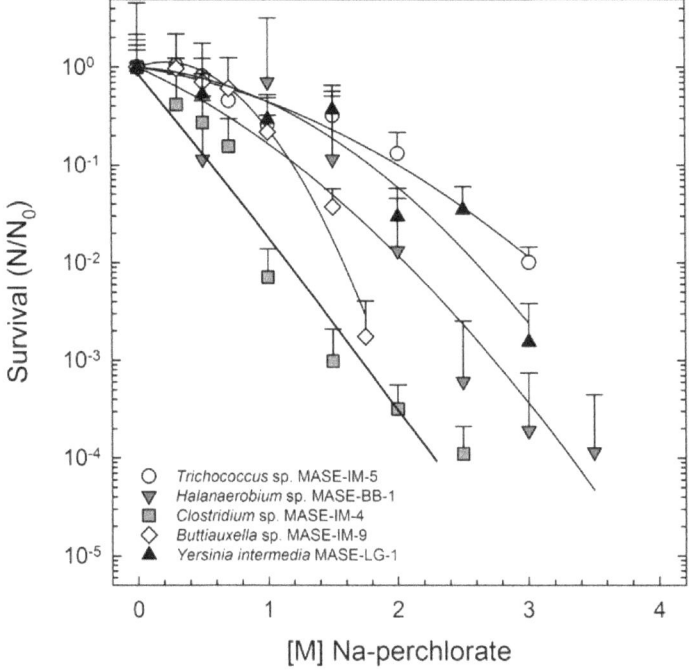

Figure 1. Survival after 15 minutes anoxic exposure to Na-perchlorate. N_0: viable non-treated cells; N: viable cells after exposure to Na-perchlorate for 15 minutes (n=3). Error bars are representing the standard deviation.

The MASE-strains were ranked in terms of their tolerance to Na-perchlorate after an exposure for 15 minutes in the following descending order from tolerant to sensitive: *Trichococcus* sp. MASE-IM-5 > *Yersinia intermedia* MASE-LG-1 > *Halanaerobium* sp. MASE-BB-1 > *Buttiauxella* sp. MASE-IM-9 > *Clostridium* sp. MASE-IM-4. This is also visible in the calculated D_{10}-values of the investigated MASE-strains ranking from 0.6 M to 1.9 M (15 min, $NaClO_4$) (Table 3).

It could be shown that exposure to lower concentrations of Na-perchlorate (36 mM = 0.5 wt% and 71 mM = 1.0 wt%), equivalent to concentrations occurring on Mars, for up to 96 hours at room temperature under anoxic conditions did not lead to an alteration in survivability of all five MASE-strains. The number of living cells remained the same or it was elevated due to further growth at room temperature for the tested time period of up to 96 hours (Figure 2).

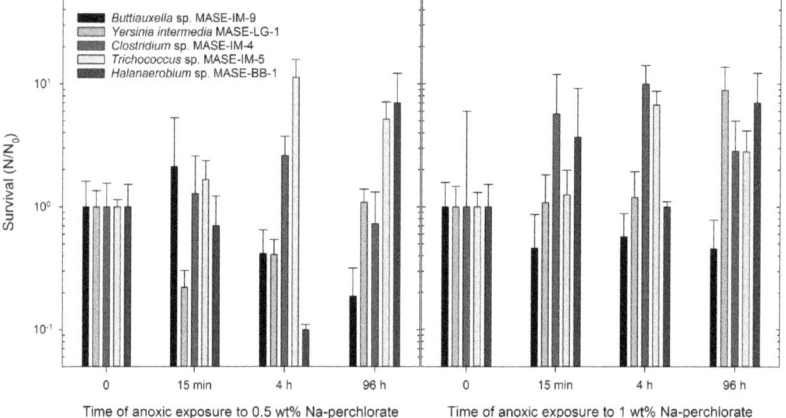

Figure 2. Response of organisms to Na-perchlorate. MASE-strain cultures (*Buttiauxella* sp. MASE-IM-9, *Yersinia intermedia* MASE-LG-1, *Clostridium* sp. MASE-IM-4, *Trichococcus* sp. MASE-IM-5, *Halanaerobium* sp. MASE-BB-1) were incubated in the presence of Na-perchlorate under anoxic conditions at room temperature. N_0: Viable cells without added perchlorate, N: Viable cells after storage in the presence of perchlorates (0.5 wt% or 1.0 wt%). Recovery was performed under standard cultivation conditions without perchlorate (n=3). Error bars are representing the standard deviation.

Growth in the presence of perchlorate
Despite the fact that all MASE-strains were growing mainly as single cells under standard cultivation conditions, microscopically observations revealed that all strains showed extensive cell agglomerations (*i.e.* cell-cell-connections) with increasing Na-perchlorate concentrations if perchlorate was present during cultivation (Figure 3). *Buttiauxella* sp. MASE-IM-9, *Clostridium* sp. IM-5, *Halanaerobium* sp. MASE-BB-1, and *Yersinia intermedia* MASE-LG-1 showed chain formation of increasing length in the presence of Na-perchlorate. The longest cell-chains were observed for *Halanaerobium* sp. MASE-BB-1 and *Buttiauxella* sp. MASE-IM-9. *Yersinia intermedia* MASE-LG-1 and *Clostridium* sp. MASE-IM-4 showed additionally an elongation of the cells itself if they were growing in chains. *Trichococcus* sp. MASE-IM-5, normally growing as diplo-coccus or three to four cells attached together, formed chains like pearls on a string.

Perchlorates and desiccation – combined stresses
The exposure to Martian concentrations of Na-perchlorate and subsequent desiccation showed additive effects of both stresses with regard to survivability. Na-perchlorate had a negative influence on the desiccation tolerance of all tested organisms (Figure 4). In the presence of Na-perchlorate, the survival rate was lowered up to five orders of magnitude compared to desiccation for 24 hours without perchlorates. The lowest

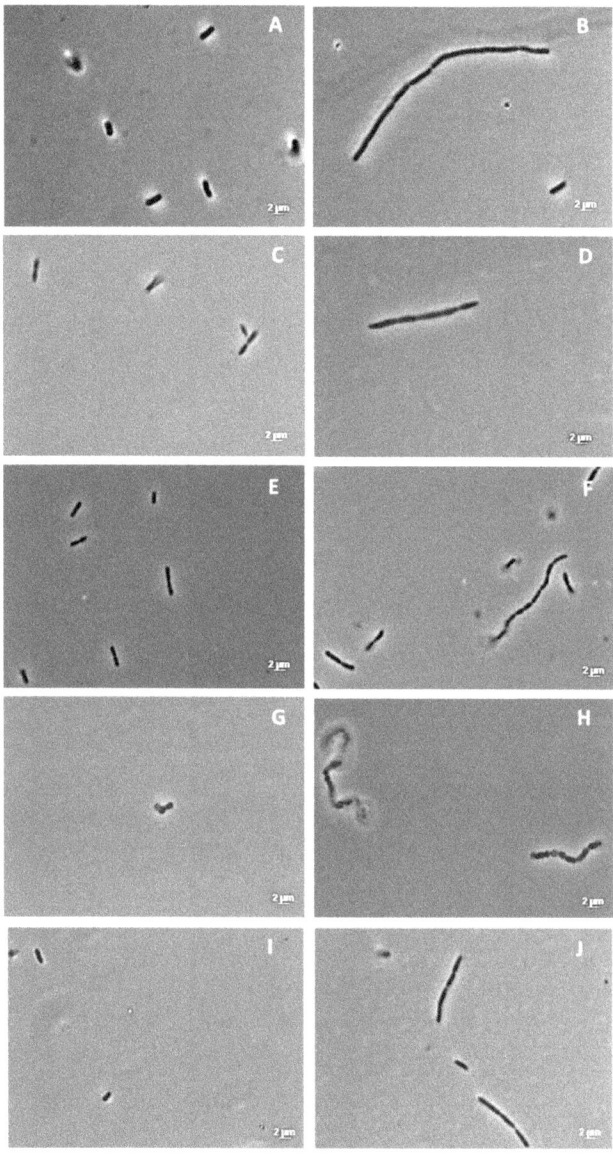

Figure 3. Light microscopy image of *Buttiauxella* sp. MASE-IM-9 (A, B), *Clostridium* sp. MASE-IM-4 (C, D), *Halanaerobium* sp. MASE-BB-1 (E, F), *Trichococcus* sp. MASE-IM-5 (G, H), *Yersinia intermedia* MASE-LG-1 (I, J) grown under standard cultivation conditions (first row, A, C, E, G, I) and cultivated in the presence of 1 %wt Na-perchlorate (second row, B, D, F, H, J).

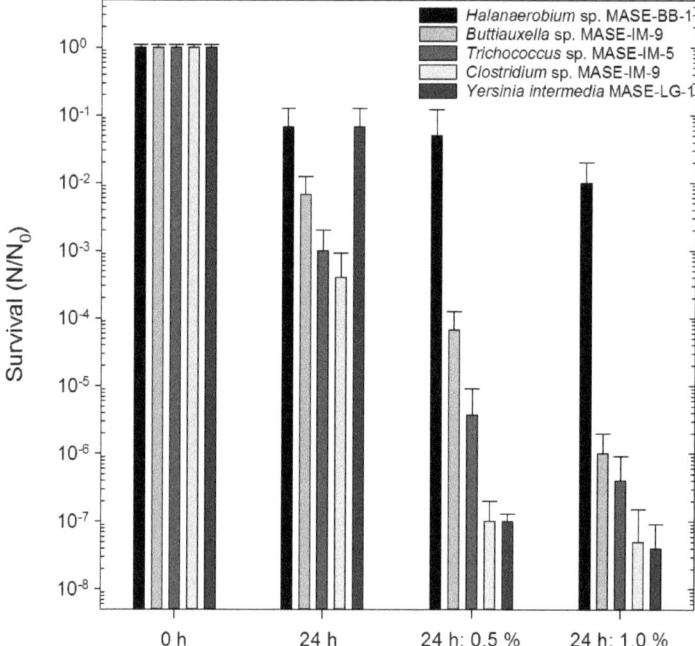

Time of desiccation [hours] and amount of Na-perchlorate [wt%]

Figure 4. Influence of perchlorates on desiccation tolerance of MASE-strains (*Halanaerobium* sp. MASE-BB-1, *Buttiauxella* sp. MASE-IM-9, *Trichococcus* sp. MASE-IM-5, *Clostridium* sp. MASE-IM-4, *Yersinia intermedia* MASE-LG-1) under anoxic conditions. N_0: Viable cells without desiccation, N: Viable cells after desiccation in the absence or presence of perchlorates (0.5 wt% or 1.0 wt%). Recovery was performed under standard cultivation conditions without perchlorate (n=3). Error bars are representing the standard deviation.

effect was observed for *Halanaerobium* sp. MASE-BB-1, the highest effect was visible for *Yersinia intermedia* MASE-LG-1. In all cases, the reduction in the survival after desiccation did not reveal a substantial difference for both Mars relevant concentrations of 0.5 wt% and 1.0 wt% Na-perchlorate.

The combination of desiccation and ionizing radiation, led to additive effects of both stresses in all model organisms (Figure 5). Moreover, differential reactions to the stresses were observed: *Clostridium* sp. MASE-IM-4 and *Trichococcus* sp. MASE-IM-5 showed enhanced radiation tolerance under stress in a dried state and both strains were able to multiply after application of ionizing radiation (3 kGy) in a desiccated state. In contrast, *Buttiauxella* sp. MASE-IM-9 and *Trichococcus* sp. MASE-IM-5 only showed additive effects after irradiation in a dried state.

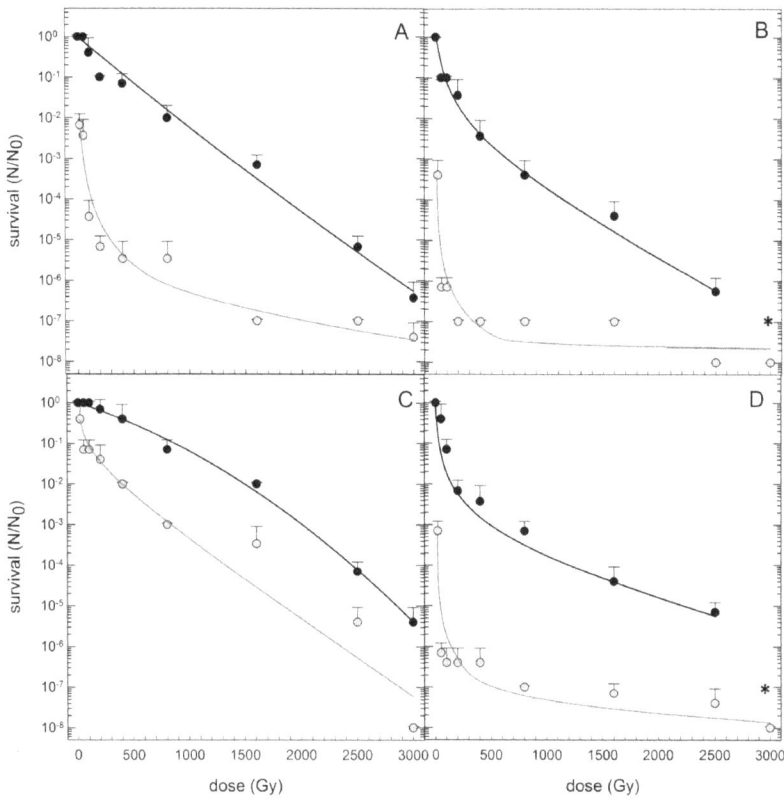

Figure 5. Survival of the MASE isolates (A: *Buttiauxella* sp. MASE-IM-9, B: *Clostridium* sp. MASE-IM-4, C: *Halanaerobium* sp. MASE-BB-1, D: *Trichococcus* sp. MASE-IM-5) after anoxic irradiation up to 3 kGy in liquid medium (black lines) and after the combination of anoxic desiccation (24 h) and subsequent exposure to ionizing radiation (up to 3 kGy) under anoxic conditions (grey lines). *: no viable cells detected. N_0: viable non-desiccated non-irradiated cells; N: viable cells after irradiation or after combined desiccation and irradiation (n=3). Error bars are representing the standard deviation.

An additional decrease of the survival of about one to two orders of magnitude was observed after the cells were exposed to desiccating conditions in combination with Martian atmosphere and pressure (2.7 vol% N_2, 1.6 vol% Ar, 0.15 vol% O_2 in CO_2 at a pressure of 10^3 Pa) (Figure 6). One exception was *Clostridium* sp. MASE-IM-4: the exposure to the Martian atmosphere enhanced the survival after desiccation after four weeks of storage (Figure 6B). In all other organisms, namely *Buttiauxella* sp. MASE-IM-9, *Halanaerobium* sp. MASE-BB-1, and *Trichococcus* sp. MASE-IM-5, additive effects of desiccation and the application of simulated Martian atmosphere were observed.

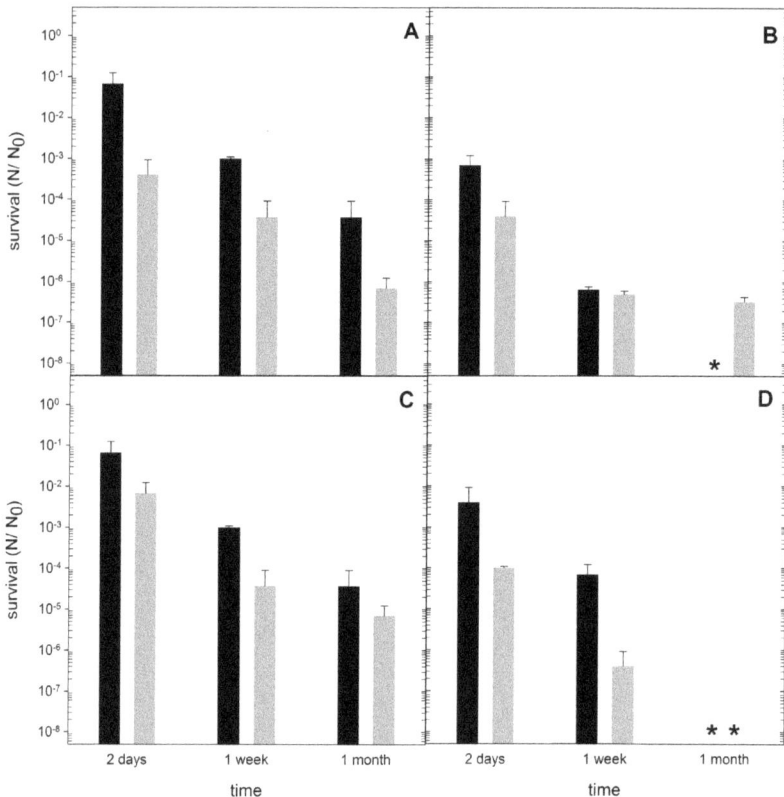

Figure 6. Survival of the MASE-strains (A: *Buttiauxella* sp. MASE-IM-9, B: *Clostridium* sp. MASE-IM-4, C: *Halanaerobium* sp. MASE-BB-1, D: *Trichococcus* sp. MASE-IM-5) after exposure to anoxic desiccation and Martian atmosphere. N_0: viable cells without desiccation and without exposure Martian atmosphere, N: viable cells after desiccation and exposure Martian atmosphere (n=3). Error bars are representing the standard deviation. *: no viable cells detected. Black: Cells were desiccated under anoxic conditions. Grey: Cells were desiccated under anoxic conditions and exposed to Martian atmosphere (Mars gas at a pressure of 10^{-3} Pa).

Discussion

In this study, we investigated the effects of individual and combined Mars-relevant stresses on (facultatively) anaerobic microorganisms. We focused on physical and chemical stress factors that are known to be prominent on the Martian surface and in the near-surface environment. They include the presence of perchlorates, desiccation, ionizing radiation and Martian atmospheric conditions. Each of the individual stress factors tested, *i.e.*, desiccation, radiation, and perchlorates led to a reduction in survival rate of the organisms. For the investigated organisms it is not known whether they are able to metabolize perchlorates. Perchlorate concentrations similar to those studied here are not thought to exist in their natural environment. Therefore, it is not surprising that the presence of perchlorates played a major role in cell stress. Even short-term exposure of 15 minutes led to cell damaging effects. Comparison of the D_{10}-values with data from previous studies shows that the MASE-strains have a slightly lower tolerance to perchlorates than other model organisms, such as *Escherichia coli* and vegetative *Bacillus subtilis* cells (Beblo-Vranesevic et al., 2017b).

In this study the tolerance of different microorganisms to Martian relevant concentrations of perchlorates was tested (0.5 wt% and 1.0 wt% Na-perchlorate). The literature reports different tolerance levels amongst Bacteria and Archaea which do not metabolize perchlorates: one example of a sensitive organism is the acidophilic iron sulfur bacterium *Acidithiobacillus ferrooxidans*, possibly able to grow under Mars-like geochemical conditions but not able to multiply in the presence of 0.022 M (~ 0.5 wt%) Mg-perchlorate (Bauermeister, 2012; Bauermeister et al., 2014). *Halobacterium* sp. NRC-1 a model halophilic archaeon cannot grow in the presence of 0.04 M Na-perchlorate (Laye and DasSarma, 2018). Different methanogenic archaea (three *Methanobacterium* strains and two *Methanosarcina* strains) are also negatively influenced in their growth behavior at low concentrations (up to 0.01 M) of Na-perchlorate (Shcherbakova et al., 2015). It has been reported that several halotolerant strains show only slight alterations in their growth pattern in the presence of perchlorates: nearly all of the halotolerant isolates grew in the presence of 0.05 M Mg-perchlorate (Al Soudi et al., 2017). Comparable results were shown for halophilic bacteria, such as *Alkalibacillus* and *Halomonas* (Matsubara et al., 2017). Unfortunately, the absolute tolerance levels of organisms capable of metabolizing perchlorates, such as *Dechloromonas hortensis* and *Aeropyrum pernix*, are absent from the literature. These organisms were treated with perchlorate concentrations between 10 mM and 50 mM (Wolterink et al., 2005; Liebensteiner et al., 2015).

We also observed morphological changes, including filaments and cell-chain formation associated with the presence of Na-perchlorate during growth. Such modification of size and shape has been observed earlier as a response to changes in the environmental conditions or to stress factors. Filamentation is one observed shape alteration that can be influenced by several factors such as nutrient deprivation, oxidative stress, DNA damage, exposure to antibiotics and temperatures shifts (Young, 2006; Justice et al., 2008; Perfumo et al., 2014). The observation of chain formation at 0.5 wt% Na-perchlorate in *Buttiauxella* sp. MASE-IM-9, *Halanaerobium* sp. MASE-BB-1, *Clostridium* sp. MASE-IM-4, and *Yersinia intermedia* MASE-LG-1 and cluster-like structures in *Trichococcus* sp. MASE-IM-5 shows that Mars relevant concentrations of perchlorates are capable of causing microbial morphological anomalies. Morphological changes due to Na-perchlorate are also described for halophilic, thermophilic, and methanogenic microorganisms. Grown at the highest tolerated perchlorate concentrations, *Halobacterium salinarum*, *Haloferax mediterranei*, and *Haloarcula marismortui* were unusually swollen and deformed. For instance, *Halomonas elongata* cells appeared normal up to 0.2 M Na-perchlorate, but in a medium with 0.4 M Na-perchlorate, the cells had a thin and wrinkled appearance (Oren et al., 2014). The thermophilic bacterium *Hydrogenothermus marinus* tends to grow in chains in the presence of 0.3 M Na-perchlorate (Beblo-Vranesevic et al., 2017b). The methanogenic strain, *Methanobacterium arcticum,* shows other morphological changes and builds cyst-like cells in the presence of Mg-perchlorate (Shcherbakova et al., 2015). The reason for the morphological changes within the MASE-strains is not known. It is possible that the chain formation provides a survival advantage during cell damaging conditions such as has been hypothesized for biofilms (Cvitkovitch, 2004).

For experiments with combined stress factors with desiccation and radiation, desiccation and simulated Martian atmosphere, desiccation and perchlorates mainly additive effects could be shown. The negative influence of perchlorates on the cellular metabolism is also visible in the combination of perchlorate exposure and desiccation. Even low concentrations of perchlorates led to a reduction in survivability up to four orders of magnitude compared to desiccated cells without perchlorates. Interestingly, if *Yersinia intermedia* MASE-LG-1 cells were irradiated in the presence of different perchlorates in Martian relevant concentrations no influence on the survival after exposure to ionizing radiation was detected (Beblo-Vranesevic et al., 2017a). The combination of exposure to perchlorates and UV irradiation led to additive and even synergistic bactericidal effect for vegetative *Bacillus subtilis* cells (Wadsworth and Cockell, 2017).

An enhanced radiation tolerance was visible when the *Clostridium* sp. MASE-IM-4 and *Trichococcus* sp. MASE-IM-5 were exposed to ionizing radiation in a dried form. This effect is also reported for some halophilic archaea (Leuko et al., 2015) and can be explained by the low abundance or absence of intra- and extracellular water within the desiccated samples. Consequently, radiolysis cannot occur due to exposure to ionizing radiation. A lower concentration of reactive oxygen species can be assumed. Absorption effects of the ingredients of the medium can be excluded since ionizing radiation penetrates through the (in-) organic residues of the medium. The exposure of dried cells to Martian atmosphere had no effect on the survival; i.e. once the cells survived the first desiccation step, they were stable with respect to additional desiccation in Martian atmosphere. These data are in accordance to data from literature: if *Deinococcus radiodurans* is exposed to Martian atmosphere (7 days) the survival rate was reduced less than one order of magnitude (Pogoda de la Vega et al., 2007).

Conclusion
The selected organisms obtained from extreme anoxic analogue environments on Earth were shown to possess not only a high tolerance against the environmental stresses that occur in their normal habitat, but also exhibit a substantial tolerance to individual and combined Mars relevant stress factors, *i.e.* desiccation, Martian atmosphere and pressure, ionizing radiation, and the presence of perchlorates. The observed effects of combined stress factors were found to be additive in the case of *Buttiauxella* sp. MASE-IM-9 and *Halanaerobium* sp. MASE-BB-1, and synergistic in the case of *Clostridium* sp. MASE-IM-4 and *Trichococcus* sp. MASE-IM-5. The desiccation step seems to give a relative advantage to cells to cope with other stress factors which provides constraints for the search for live on Mars. Moreover, these MASE strains were even able to grow in the presence of Martian relevant concentrations of Na-perchlorate under anoxic conditions. Our data show that survival in Martian environments, *i.e.* Martian brines, is in principle possible for some organisms. This work advances our understanding of the limits of survival in Mars-relevant conditions.

References
Al Soudi, AF., Farhat, O., Chen, F., Clark, BC., and Schneegurt, MA. (2017). Bacterial growth tolerance to concentrations of chlorate and perchlorate salts relevant to Mars. Int. J. Astrobiol. 16, 229-235.
Archer, PD. Jr., Sutter, B., Ming, DW., McKay, CP., Navarro-Gonzalez, R., Franz, HB., McAdam A., Mahaffy PR., and the MSL Science Team. (2013). Possible detection of perchlorates by evolved gas analysis of rocknest soils: global implications. In: Proceedings of the 44th Lunar and Planetary Science Conference Abstracts (Houston, TX: Lunar and Planetary Institute).
Bauermeister, A. (2012). Characterization of stress tolerance and metabolic capabilities of acidophilic iron-sulfur-transforming bacteria and their relevance to Mars. Doctoral dissertation, University of Duisburg-Essen, Duisburg.
Bauermeister, A., Rettberg, P., and Flemming, HC. (2014). Growth of the acidophilic iron–sulfur bacterium *Acidithiobacillus ferrooxidans* under Mars like geochemical conditions. Planet. Space Sci. 98, 205-215.
Beblo, K., Rabbow, E., Rachel, R., Huber, H., and Rettberg, P. (2009). Tolerance of thermophilic and hyperthermophilic microorganisms to desiccation. Extremophiles. 13, 521-531.
Beblo, K., Douki, T., Schmalz, G., Rachel, R., Wirth, R., Huber, H., Reitz, G., and Rettberg, P. (2011). Survival of thermophilic and hyperthermophilic microorganisms after exposure to UV-C, ionizing radiation and desiccation. Arch. Microbiol. 193, 797-809.
Beblo-Vranesevic, K., Bohmeier, M., Perras, AK., Petra Schwendner, P., Rabbow, E., Moissl-Eichinger, C., Charles S. Cockell, CS., Pukall, R., Vannier, P., Marteinsson, VT., Monaghan, EP., Ehrenfreund, P., Garcia-Descalzo, L., Gómez, F., Malki, M., Amils, R., Gaboyer, F., Westall, F.,

Cabezas, P., Walter, N., and Rettberg, P. (2017a). The responses of an anaerobic microorganism, *Yersinia intermedia* MASE-LG-1 to individual and combined simulated Martian stresses. PloS One. 12, e0185178.

Beblo-Vranesevic, K., Huber, H., and Rettberg, P. (2017b) High tolerance of *Hydrogenothermus marinus* to sodium perchlorate. Front.Microbiol. 8, 1369.

Carrier, BL., and Kounaves, SP. (2015). The origins of perchlorate in the Martian soil. Geophys. Res. Lett. 42, 3739-3745.

Catling, DC., Claire, MW., Zahnle, KJ., Quinn, RC., Clark, BC., Hecht, MH., and Kounaves, S. (2010). Atmospheric origins of perchlorate on Mars and in the Atacama. J. Geophys. Res. Planet. 115, E00E11.

Cockell, CS., Bush, T., Bryce, C., Direito, S., Fox-Powell, M., Harrison, JP., Lammer, H., Landenmark, H., Martin-Torres, J., Nicholson, N., Noack, L., O'Malley-James, J., Payler, SJ., Rushby, A., Samuels, T., Schwendner, P., Wadsworth, J., and Zorzano, MP. (2016). Habitability: a review. Astrobiology. 16, 89-117.

Cockell, CS., Schwendner, P., Perras, A., Rettberg, P., Beblo-Vranesevic, K., Bohmeier, M., Rabbow, E., Moissl-Eichinger, C., Wink, L., Marteinsson, V., Vannier, P., Gomez, F., Garcia-Descalzo, L., Ehrenfreund, P., Monaghan, E.P., Westall, F., Gaboyer, F., Amils, R., Malki, M., Pukall, R., Cabezas, P. and Walter, N. (2017). Anaerobic microorganisms in astrobiological analogue environments: from field site to culture collection. Int. J. Astrobiol. 1-15.

Cvitkovitch, D. (2004). Genetic exchange in bioflms. In: Ghannoum, M., O'Toole, G. (ed). Microbial Bioflms. Washington, DC, ASM Press, 192-205.

Dartnell, LR., Desorgher, L., Ward, JM., and Coates, AJ. (2007). Modelling the surface and subsurface martian radiation environment: implications for astrobiology. Geophys. Res. Let. 34.

Eigenbrode, JL., Summons, RE., Steele, A., Freissinet, C., Millan, M., Navarro-González, R., Sutter, B., McAdam, AC., Franz, HB., Glavin, DP., Archer, PD., Mahaffy, PR., Conrad PG., Hurowitz JA., Grotzinger JP., Gupta S., Ming DW., Sumner DY., Szopa C., Malespin C., Buch A., and Coll, P. (2018). Organic matter preserved in 3-billion-year-old mudstones at Gale crater, Mars. Science. 360, 1096-1101.

Fox-Powell, MG., Hallsworth, JE., Cousins, CR., and Cockell, CS. (2016). Ionic strength is a barrier to the habitability of Mars. Astrobiology. 16, 427.-442.

Franson, MAH. (ed.). (1985). Standard methods for the examination of water and wastewater. In: American Public Health Association, 16th edition, Washington DC.

Glavin, DP., Freissinet, C., Miller, KE., Eigenbrode, JL., Brunner, AE., Buch, A., Sutter, B., Archer, PD. Jr., Atreya, SK., Brinckerhoff, WB., Cabane, M., Coll, P., Conrad, PG., Coscia, D., Dworkin, JP., Franz, HB., Grotzinger, JP., Leshin, LA., Martin, MG., McKay, C., Ming, DW., Navarro-González, R., Pavlov, A., Steele, A., Summons, RE., Szopa, C., Teinturier, S., and Mahaffy, PR. (2013). Evidence for perchlorates and the origin of chlorinated hydrocarbons detected by SAM at the Rocknest aeolian deposit in Gale Crater. J. Geophys. Res. Planets. 118, 1955-1973.

Gough, RV., Chevrier, VF., Baustian, KJ., Wise, ME., and Tolbert, MA. (2011). Laboratory studies of perchlorate phase transitions: support for metastable aqueous perchlorate solutions on Mars. Earth Planet. Sci. Lett. 312, 371-377.

Gu, W., Li, Y., Tang, M., Jia, X., Ding, X., Bi, X., and Wang, X. (2017).Water uptake and hygroscopicity of perchlorates and implications for the existence of liquid water in some hyperarid environments. RSC Advances. 7, 46866-46873.

Hassler, DM., Zeitlin, C., Wimmer-Schweingruber, RF., Ehresmann, B., Rafkin, S., Eigenbrode, JL., Brinza, DE., Weigle, G., Böttcher, S., Böhm, E., Burmeister, S., Guo, J., Köhler, J., Martin, C., Reitz, G., Cucinotta, FA., Kim, M-H., Grinspoon, D., Bullock, MA., Posner, A., Gómez-Elvira, J., Vasavada, A., Grotzinger, JP., and MSL Science Team. (2014). Mars' surface radiation environment measured with the Mars Science Laboratory's Curiosity rover. Science. 343, 1244797.

Hecht, MH., Kounaves, SP., Quinn, RC., West, SJ., Young, SM., Ming, DW., Catling, DC., Clark, BC., Boynton, WV., Hoffman, J., DeFlores, LP., Gospodinova, K., ,Kapit, J., and Smith, PH. (2009). Detection of perchlorate and the soluble chemistry of Martian soil at the Phoenix lander site. Science. 325, 64-67.

Horneck, G. (2000). The microbial world and the case for Mars. Planet. Space Sci. 48, 1053-1063.

Jakosky, BM., and Phillips, RJ. (2001). Mars' volatile and climate history. Nature. 412, 237-244.

Justice, SS., Hunstad, DA., Cegelski, L., and Hultgren, SJ. (2008). Morphological plasticity as a bacterial survival strategy. Nat. Rev. Microbiol. 6, 162-168.

Kim, YS., Wo, KP., Maity, S., Atreya, SK., and Kaiser, RI. (2013). Radiation induced formation of chlorine oxides and their potential role in the origin of Martian perchlorates. J. Am. Chem. Soc. 135, 4910-4913.

Kounaves, SP., Chaniotakis, NA., Chevrier, VF., Carrier, BL., Folds, KE., Hansen, VM., McElhoney, KM., O'Neil, GD., and Weber, AW. (2014). Identification of the perchlorate parent salts at the Phoenix Mars landing site and possible implications. Icarus. 232, 226-231.

Laye,VJ., and DasSarma, S. (2018). An Antarctic Extreme Halophile and Its Polyextremophilic Enzyme: Effects of Perchlorate Salts. Astrobiology. 18, 412-418.

Leuko, S., Domingos, C., Parpart, A., Reitz, G., and Rettberg, P. (2015). The survival and resistance of *Halobacterium salinarum* NRC-1, *Halococcus hamelinensis*, and *Halococcus morrhuae* to simulated outer space solar radiation. Astrobiology. 15, 987-997.

Liebensteiner, MG., Pinkse, MW., Nijsse, B., Verhaert, PD., Tsesmetzis, N., Stams, AJ., Lomans, BP. (2015). Perchlorate and chlorate reduction by the Crenarchaeon *Aeropyrum pernix* and two thermophilic Firmicutes. *Environ.* Microbiol. Rep. 7, 936-945.

Matsubara, T., Fujishima, K., Saltikov, CW., Nakamura, S., and Rothschild, LJ. (2017). Earth analogues for past and future life on Mars: isolation of perchlorate resistant halophiles from Big Soda Lake. Int. J. Astrobiol. 16, 218-228.

Matthiä, D., Ehresmann, B., Lohf, H., Köhler, J., Zeitlin, C., Appel, J., Sato, T., Slaba, T., Martin, C., Berger, T., Boehm, E., Boettcher, S., Brinza, DE., Burmeister, S., Guo, J., Hassler, DM,, Posner, A., Rafkin, SCR., Reitz, G., Wilson, JW., and Wimmer-Schweingruber RF. (2016). The Martian surface radiation environment–a comparison of models and MSL/RAD measurements. J. Space Weather Space Clim. 6, A13.

Martín-Torres, FJ., Zorzano, MP., Valentín-Serrano, P., Harri, AM., Genzer, M., Kemppinen, O., Rivera-Valentin, EG., Jun, I., Wray, J., Madsen, MB., Goetz, W., McEwen, AS., Hardgrove, C., Renno, N., Chevrier, VF., Mischna, M., Navarro-González, R., Martínez-Frías, J., Conrad, P., McConnochie, T., Cockell, C., Berger, G., Vasavada, AR., Sumner, D., and Vaniman, D. (2015). Transient liquid water and water activity at Gale crater on Mars. Nat. Geosci. 8, 357-361.

McEwen, AS., Ojha, L., Dundas, CM., Mattson, SS., Byrne, S., Wray, JJ., Cull SC., Murchie SL., Thomas N., and Gulick VC. (2011). Seasonal flows on warm Martian slopes. Science. 333, 740-743.

Ojha, L., Wilhelm, MB., Murchie, SL., McEwen, AS., Wray, JJ., Hanley, J., Massé M., and Chojnacki M. (2015). Spectral evidence for hydrated salts in recurring slope lineae on Mars. Nat. Geosci. 8, 829-832.

Oren, A., Bardavid, RE., and Mana, L. (2014). Perchlorate and halophilic prokaryotes: implications for possible halophilic life on Mars. Extremophiles. 18, 75-80.

Orosei, R., Lauro, SE., Pettinelli E., Cicchetti, A., Coradini, M., Cosciotti, B., Di Paolo BF., Flamini, E., Mattei, E., Pajola, M., Soldovieri, F., Cartacci, M., Cassenti, F., Frigeri, A., Giuppi, S., Martufi, R., Masdea, A., Mitri, G., Nenna, C., Noschese, R., Restano, M., Seu, R. (2018) Radar evidence of subglacial liquid water on Mars. Science. eaar7268.

Perfumo, A., Elsaesser, A., Littmann, S., Foster, RA., Kuypers, MM., Cockell, CS., and Kminek, G. (2014). Epifluorescence, SEM, TEM and nanoSIMS image analysis of the cold phenotype of *Clostridium psychrophilum* at subzero temperatures. FEMS Microbiol. Ecol. 90, 869-882.

Pogoda de La Vega, U., Rettberg, P., and Reitz, G. (2007). Simulation of the environmental climate conditions on Martian surface and its effect on *Deinococcus radiodurans*. Adv. Space Res. 40, 1672-1677.

Schubert, G., Russell, CT., and Moore, WB. (2000). Geophysics: Timing of the Martian dynamo. Nature. 408, 666.

Schuttlefield, JD., Sambur, JB., Gelwicks, M., Eggleston, CM., and Parkinson, BA. (2011). Photooxidation of chloride by oxide minerals: implications for perchlorate on Mars. J. Am. Chem. Soc. 133, 17521-17523.

Shcherbakova, V., Oshurkova, V., and Yoshimura, Y. (2015). The effects of perchlorates on the permafrost methanogens: implication for autotrophic life on Mars. Microorganisms. 3, 518-534.

Toner, JD., and Catling, DC. (2016). Water activities of $NaClO_4$, $Ca(ClO_4)_2$, and $Mg(ClO_4)_2$ brines from experimental heat capacities: water activity > 0.6 below 200K. Geochim. Cosmochim. Acta. 181, 164-174.

Tosca, NJ., Knoll, AH., and McLennan, SM. (2008). Water activity and the challenge for life on early Mars. Science. 320, 1204-1207.

Wadsworth, J., and Cockell, CS. (2017). Perchlorates on Mars enhance the bacteriocidal effects of UV light. Scientific reports. 7, 4662.

Westall, F., Loizeau, D., Foucher, F., Bost, N., Betrand, M., Vago, J., and Kminek, G. (2013). Habitability on Mars from a microbial point of view. Astrobiology. 13, 887-897.

Wolterink, A., Kim, S., Muusse, M., Kim, IS., Roholl, PJ., van Ginkel, CG., Stams, AJ., Kengen, SW. (2005). *Dechloromonas hortensis* sp. nov. and strain ASK-1, two novel (per)chlorate-reducing bacteria, and taxonomic description of strain GR-1. Int. J. Syst. Evol. Microbiol. 55, 2063-2068.

Young, KD. (2006). The selective value of bacterial shape. Microbiol. Mol. Biol. Rev. 70, 660-703.

Chapter 6

Exploring Deep-Sea Brines as Potential Terrestrial Analogues of Oceans in the Icy Moons of the Outer Solar System

André Antunes[1*], Karen Olsson-Francis[2] and Terry J. McGenity[3]

[1]State Key Laboratory of Lunar and Planetary Sciences, Macau University of Science and Technology (MUST), Taipa, Macau SAR, China
[2]School of Environment, Earth and Ecosystem Sciences, The Open University, Milton Keynes MK7 6AA, UK
[3]School of Life Sciences, University of Essex, Colchester CO4 3SQ, UK

*aglantunes@must.edu.mo

DOI: https://doi.org/10.21775/9781912530304.06

Abstract

Several icy moons of the outer solar system have been receiving considerable attention and are currently seen as major targets for astrobiological research and the search for life beyond our planet. Despite the limited amount of data on the oceans of these moons, we expect them to be composed of brines with variable chemistry, some degree of hydrothermal input, and be under high pressure conditions. The combination of these different conditions significantly limits the number of extreme locations, which can be used as terrestrial analogues. Here we propose the use of deep-sea brines as potential terrestrial analogues to the oceans in the outer solar system. We provide an overview of what is currently known about the conditions on the icy moons of the outer solar system and their oceans as well as on deep-sea brines of the Red Sea and the Mediterranean and their microbiology. We also identify several threads of future research, which would be particularly useful in the context of future exploration of these extra-terrestrial oceans.

The icy moons of the Outer Solar System

Icy moons are natural satellites that are characterised by a surface that is composed predominantly of ice, which may contain a sub-surface ocean, and possibly a rocky core of silicate or metallic rocks. In the outer Solar System, icy moons orbit the gaseous planets, Jupiter and Saturn. The icy moons of Jupiter — Io, Europa, Ganymede and Callisto — were discovered by Galileo Galilei in 1610 and are frequently referred to as the Galilean moons of Jupiter (Showman and Malhotra, 1999); whilst Saturn is orbited by numerous icy moons, which include Enceladus, Mimas, Tethys, Dione, and Titan.

The icy moons are of extensive interest for future exploration due to their potential sub-surface oceans. For example, Europa (Kivelson et al., 2000), Ganymede (Kivelson et al., 2002), Callisto (Khurana et al., 1998), and Enceladus (Spencer and Nimmo, 2013) present evidence of a briny ocean under an icy surface. As our understanding of these icy moons evolves so does interest from an astrobiological point of view. The following sections will discuss two of these icy moons, Enceladus and Europa, in more detail.

Enceladus

Surface
Enceladus is covered with a thick icy shell that varies in depth dependant on latitude. Based on data from Cassini, modelling has predicted that the ice shell is approximately 18 to 22 km thick on average, but less than 5 km at the South Pole region (Cadek et al., 2016). The composition is believed to be almost pure water with spectroscopy data suggesting that it also contains small amounts of carbon dioxide, hydrogen peroxide, light organics, and perhaps ammonium (Brown et al., 2006, Emery et al., 2005). The temperature at the surface at low latitude varies from approximately -223°C at night to -193°C during the day (Howett et al., 2010). The surface has a particularly high albedo (Howett et al., 2010) and contains several regions of cratered and smooth terrain, which may denote differences in their respective age (Squyres et al., 1983a).

In the South Polar Region unique surface features called "tiger stripes" are observed, which are central to present-day geological activity (Spencer and Nimmo, 2013; Spencer et al., 2006). The stripes are the source of thermal activity and associated plume emissions venting from the surface (Squyres et al., 1983a), which were observed by the Cassini flyby mission. The presence of these plumes suggests the existence of a sub-surface ocean consisting of liquid water (Waite et al., 2017) and analysis of the icy grains in the plumes imply that they are rich in sodium (Postberg et al., 2011). The salt-composition is similar to that expected for water that has equilibrated with Enceladus's presumed silicate core (Zolotov, 2007). The plumes may be caused by hydrothermal activity at the ocean floor (Hsu et al., 2015; Waite et al., 2017), which is supported by data from the Cassini mission that shows tidal dissipation in the rocky core.

Sub-surface ocean

Contrary to previous assumptions, the ocean of Enceladus is believed to have a global distribution rather than being restricted to polar regions (Patthoff and Kattenhorn, 2011). The energy required to maintain liquid water most likely originates from the dissipation of tidal energy from the friction of the sub-surface ocean with the internal silicate interior (Nimmo et al., 2007).

Information regarding the composition of the ocean has been obtained from plume analysis, which showed that particle emissions are dominated by water ice and is rich in sodium and potassium salts (0.5 to 2 % by mass), but also contains sodium bicarbonate/ sodium carbonate (Postberg et al., 2011). The ocean is expected to contain low concentrations of ammonia, methane, carbon dioxide and molecular hydrogen (Waite et al., 2017). These elements are potentially produced as a result of geochemical reactions occurring at the interface of the chondrite-like core and the ocean at temperatures below 100°C (Waite et al., 2017). The circulation of the water would drive the chemical evolution of both the rock material and the ocean water, producing a chemical gradient, which could be used by microorganisms to generate energy (Barge and White, 2017). The presence of silica nanoparticles in the plumes is also evidence for hydrothermal reactions occurring in the interior, which may be the source of molecular hydrogen (Sekine et al., 2015).

Based on the chemistry of the plumes, the pH of the ocean is predicted to be alkaline, with current estimates placing it between 8.5 and 13 (e.g. Postberg et al., 2009; Hsu et al., 2015). These values have been calculated by either modelling the equilibrium reactions at the water-rock interface or by the concentrations of salts in the plumes (Zolotov, 2007; Postberg et al., 2009). The temperature varies within the sub-surface oceans. The estimated temperature at the ocean-ice interface is approximately -0.15°C (Glein et al., 2015); whereas SiO_2 nanoparticles in the plumes suggest a minimal localised temperature of 90°C at the water-rock interface. Gravimetric measurements have suggested that that rock-ocean and ice-oceans interfaces are at depths of 50 km (~ 5.3 MPa) and 35-40 km (3.6-4.2 MPa) (less et al., 2014). However, values have been estimated to be as high at 10 MPa at the rock-ocean interface (Zolotov, 2007) and 8 MPa at the ice-ocean interface (Hsu et al., 2015). It should be noted that these pressures are lower than that on Earth due to the gravitational force associated with the icy moons.

Europa

Surface

Europa is covered by an icy shell, with a highly debated thickness that is estimated to range between a few km to over 30 km (Billings and Kattenhorn, 2005; Quick and Marsh, 2015). The surface of Europa is dominated by ridges and chaotic terrain, which is thought to be relatively young, or still geologically active (Squyres et al., 1983b; Carr et al., 1998). These regions indicate locations where resurfacing of material from the sub-surface ocean may have occurred (Hand and Carlson, 2015). The surface is a distinct yellow-brown colour, which has been postulated to result from sulfur chemistry, either from an exogenous or endogenous source (Geissler et al., 1998; Carlson et al., 2009). However, Hand and Carlson suggested that the discolouration might be due to sodium chloride from the sub-surface, which yields a yellow-brown colour when exposed to the radiation conditions present at the surface of Europa (Hand and Carlson, 2015). Observations of the thermal emissions by the Galileo mission showed low latitude diurnal brightness temperatures between -194°C in the Polar Regions and approximately -143°C at the equatorial regions (Spencer et al., 1999). Images from the Hubble Space Telescope suggest the presence of water-vapour plumes, appearing to rise 200 km above the disk of Europa's solid body, which would indicate transport of material from the interior ocean to the surface (e.g. Ross and Schubert, 1987; Roth et al., 2014; Sparks et al., 2016; Jia et al., 2018). Furthermore, tidal stress has been suggested as playing a role in the opening and closing of fractures at the surface to allow water vapour to be released from the sub-surface ocean (Roth et al., 2014).

Sub-surface ocean
Magnetometer measurements suggest that Europa has a global-scale ocean (Ross and Schubert, 1987; Kivelson et al., 2000). The formation of this liquid ocean is thought to be due to tidal heating, tidal flexing, and/or to radioactive decay of the silicate core (Ross and Schubert, 1987; Han and Showman, 2010). Based on the expected salinity of the ocean brine the temperature will be at a minimum, -13°C (Zolotov and Kargel 2009), with the maximum temperature occurring at the ocean floor where hydrothermal activity may occur (~90°C) (Kargel et al., 2000; Zolotov and Kargel, 2009). Therefore, temperatures that are significantly higher than freezing should only exist in porous channels below the seafloor, or close to hydrothermally active regions (Vance and Goodman, 2009). At the ocean floor the pressure has been modelled to be approximately 110 MPa (Hand et al., 2009).

Due to lack of direct measurements of the sub-surface ocean, the chemical composition remains unknown. However, initial experiments and theoretical simulations strongly suggest that the composition of the ice surface is a direct result of the ocean material from below and could thus be used to infer its composition (Kargel et al., 2000). Yet, more recently the chemical species present at the surface (e.g. sulfates) were suggested to be a result of radiation (Hand and Carlson, 2015). Therefore, geochemical models have been used to predict the composition of the salty brine, which results in widely varying compositions. For example, modelling of the initial chondritic composition leads to a magnesium and sulfur-rich ocean (Kargel et al., 2000); whereas modelling of the water-to-rock cycling at the silicate seafloor leads to a chloride-rich ocean (Glein and Shock, 2010). The chemical composition of the ocean would influence the pH, which limits our ability to make conclusive predictions in this regard. For example, a brine dominated by sulfate would be acidic and a brine rich in chloride would be basic or neutral (Zolotov, 2007; 2009). Due to surface irradiation and tidal forces associated with Europa it has been postulated that a metastable dynamic state would occur, which could support microbial metabolism. For example, the water-rock interaction at the ocean floor would produce electron-rich energy sources, while oxidants would be produced due to radiolysis at the ice surface (Hand et al., 2004; Russell et al., 2017).

Icy moon analogues
Terrestrial analogue environments exhibit conditions that are similar to those of planetary bodies and moons in the Solar System (Martins et al., 2017). Historically, the majority of terrestrial analogue sites have been selected for studying Mars. More recently, however, attention has focused on the icy moons with a number of studies using sulfate lakes (e.g. Prieto-Ballesteros et al., 2003), cold-springs (e.g. Gleeson et al., 2012) and sub-glacial lakes as analogues for the sub-surface oceans (Garcia-Lopez and Cid, 2017). For example, the hypersaline subglacial lake beneath the Devon ice cap in the Canadian Arctic has been discovered recently and considered as an analogue for the sub-surface ocean of Europa (Rutishauser et al., 2018). However, to fully understand the processes that occur, and the potential life, in the sub-surface oceans of the icy moons an ideal analogue would be terrestrial deep seas. These environments are among the last earthly frontiers for discovery. Many areas of the ocean floor remain inadequately mapped, and new geomorphological features, and several new life forms have been isolated. The discovery and exploration of deep-hypersaline anoxic basins is particularly relevant from an astrobiological perspective, as it shows striking parallels with the oceans of the icy moons of the Outer Solar System.

While detailed information on the exact physical-chemical conditions present in the exooceans of the icy moons of the outer Solar system is still quite limited, we now know that these are brine oceans and likely associated with localized hydrothermal input. Salinity estimates vary but most seem to point to a somewhat lower value than the Earth's oceans, although these have been reported as having a high degree of uncertainty of two orders of magnitude (Lunine, 2017). Furthermore, several authors propose mechanisms that would lead to expected localized significant increases in salinity. Indeed, increase in salinity might arise from brine exclusion during freezing at the interface between the ice crust and the underlying ocean (equivalent to the formation of brine inclusions on marine ice on Earth) or interactions between seawater and rocks at the bottom of these exooceans (Hendrix et al., 2019). Furthermore, strong double diffusive convection processes have been predicted for Europa (Vance and

Goodman, 2009; Elsenousy et al., 2015), leading us to expect density gradients driven by temperature and salinity. According to some of the possible scenarios proposed by these authors, Europa might have a stratified ocean and precipitation from rising plumes could cause "snowing" effects with precipitation of salt, which would increase the salinity closer to the bottom and might even lead to thick layers of salt deposits.

Deep-hypersaline anoxic basins provide an interesting proxy for several of these conditions and their wide range of physical-chemical conditions allow us to partly circumvent the wide variability of predictions on the physical-chemical settings in these exooceans. The following section will focus on these unusual deep-sea extreme environments and their microbes, with the discussion being centred on the Red Sea and the Mediterranean.

Deep-hypersaline anoxic basins (DHABs)

Origin and main characteristics
DHABs are very unusual environments, which combine a unique range of environmental extremes and are regarded as one of the most extreme environments on our planet (e.g. Antunes, 2017; Merlino et al., 2018; Antunes et al., 2019). They were accidentally discovered in the axial region of the Red Sea after the collection of unexpectedly warm and salty deep-sea water during an expedition of the RV Albatross in 1947/48 (Bruneau et al., 1953). Further studies led to the identification and description of the first DHAB in the Red Sea (Miller, 1964) and wider surveys identified several additional ones scattered across the central axis of the Red Sea. Similar environments were later detected in the Mediterranean Sea (Cita et al., 1985) and in the Gulf of Mexico (Shokes et al., 1977), with on-going exploration regularly increasing the number of identified DHABs (e.g. Ehrhardt et al., 2005; Yakimov et al., 2013, 2015).

All of these DHABs seem to be associated with tectonic activity in areas of the globe with deeper saline strata formed *via* evaporation of ancient seas. Such tectonic activity promotes the formation of topographical depressions and exposes these evaporite strata to dissolution and leaching events and formation of brines that accumulate in several of these depressions as a result of their significantly higher density. The sheltered location of the brines together with their higher densities, and very weak deep-sea currents, contribute to very limited mixing with overlying seawater and lead to stable brine bodies with very sharp brine-seawater interfaces, which display drastic transitions in physical-chemical conditions often occurring within the span of a few meters (De Lange et al., 1990a; Eder et al., 2001; van der Wielen et al., 2005; Borin et al., 2009; Antunes et al., 2011d).

The brine-seawater interfaces are very complex environments, with significant shifts in salinity, O_2 concentration, pH, and temperature over a relatively small vertical scale. Such transitions provide a variety of environmental niches and ideal conditions for the establishment of local redox cycles, which involve biotic input. Furthermore, the density difference occurring at the brine-seawater interface leads to increased particle load by trapping sinking particles and mineral precipitates produced in local redox reactions. The trapping of organic matter combines with the redox gradient to create a variety of electron donor-acceptor couplings that are favourable for microbial generation of energy, making this compartment of the brines the one with the highest activity.

There is extensive discussion about the extent to which these DHABs were formed by leaching of tectonically exhumed halite or by release of relic brines trapped within the evaporites (Vengosh et al., 1998; Camerlenghi, 1990; Cita, 2006), or supercritical heating at great depth of seawater resulting in hydrothermally formed salts and brines (Hovland et al., 2018), with the diverse sources of the brines resulting in distinct chemistries (Table 1).

The primary difference between the DHABs of the Red Sea and the eastern Mediterranean Sea is that those in the Mediterranean are generally cooler (~15°C), which reflects the inherently lower temperature of the deep Eastern Mediterranean (~13.5°C; Tsimplis and Baker, 2000) compared with that of the Red Sea (~21.5°C; Yao and Hoteit, 2018), as well as a lesser contribution by hydrothermal fluids in the Mediterranean (Cita, 2006). They are also the deepest DHABs at more than 3000 mbsl (Cita, 2006). Overall, the Red Sea DHABs are, perhaps, geochemically less diverse than those of the Mediterranean (based on dissolved ion concentrations) but have more variable temperatures and generally higher concentrations of heavy metals (Antunes et al., 2011d).

Red Sea DHABs
The DHABs of the Red Sea were formed as a result of the tectonic split of the Arabian and African tectonic plates and re-dissolution of evaporites from the Miocene (mostly composed of halite and anhydrite).

The Red Sea has the highest number of known DHABs, which include: Albatross, Atlantis II, Chain, Conrad, Discovery, Erba, Kebrit, Nereus, Oceanographer, Port Sudan, Shaban, Shagara, Suakin, Valdivia, and Wando Basin (Bruneau et al., 1953; Backer and Schoell, 1972; Pautot et al., 1984; Cochran et al., 1986). Based on their geochemical properties (Table 1), a recent cluster analysis by Schmidt et al. (2015) split the DHABs of the Red Sea into two groups, which could be correlated with the local sedimentary and tectonic setting of the individual deeps.

Red Sea Type 1 Brines – Oceanographer and Kebrit DHABs
Type 1 brines have high salinity, low pH, low temperature, and low trace-metal concentrations which are probably linked with their high H_2S content (e.g. Weber and Gurskii 1982; Schmidt et al., 2003). They are located

Table 1. Overview of major physical-chemical characteristics of the DHABs of the Red Sea and the Mediterranean Sea.

	Na^+	Cl^-	Mg^{2+}	K^+	Ca^{2+}	SO_4^{2-}	pH	T (°C)
Red Sea								
Albatross	4344.88	4608.75	48.84	55.96	111.98	14.89	6.8	23.9
Atlantis II	4677.73	5189.97	35.42	86.88	148.71	10.74	5.21	68.2
Chain	4800	5200	36.8	66.8	143.3	8.5	6.48	53.2
Conrad	4295.43	4619.67	101.01	67.7	24.63	48.77	6.1	23.0
Discovery	4635.53	5021.75	36.78	84.86	144.87	10.66	6.2	44.8
Erba	2621.99	2678.14	71.43	30.92	29.57	47.15	7.4	26.1
Kebrit	4805.74	5135.67	121.13	36.7	56.56	20.08	5.5	23.3
Nereus	3556.62	4131.47	64.06	77.42	230.8	11.25	7.43	30.2
Oceanographer	4353.37	4716.86	280.64	77.09	86.21	10.32	5.6	24.9
Port Sudan	3784.68	3855.22	70.89	46.5	35.63	47.77	6.43	36.2
Shaban (N)	4784.51	4900.52	97.96	49.29	19.74	51.57	6.0	25.3
Suakin	2416.77	2612.84	62.87	35.42	57.14	36.05	7.85	24.6
Valdivia	4100	4600	95.3	52.0	25.0	72.0	6.21	29.8
Mediterranean Sea								
Bannock	4200	5378	643.9	126.3	16.3	135.2	6.5	15.1
Discovery	840.17	10154.29	5142.97	89.52	1	110.35	4.5	14.5
Hephaestus	93	9120	4720	28	2	203	5.0	15.3
Kryos	1236.32	9054.24	4402.39	84.4	1	322.71	5.4	14.5
L'Atalante	4654.24	5302.8	658.3	368.31	7.49	333.13	NA	14.3
Medee	4178	5259	788	471	2.8	201	6.7	15.4
Thetis	4760	5300	604	230	9	265	NA	15.1
Tyro	5300	5350	71.1	19.2	35.4	52.7	6.75	14.2
Urania	3505	3730	315	122	31.6	107	6.8	18.3

All chemical composition data provided in mM. NA: Not Available. Data compiled from Antunes et al., 2011d; De Lange et al., 1990b; La Cono et al., 2011; 2019; Merlino et al., 2018; Sass et al., 2001; Schmidt et al., 2015; Van der Wielen et al., 2005; Yakimov et al., 2013; 2015; and references cited therein.

slightly off-axis and controlled by evaporite dissolution rather than contributions from sediment alteration with hydrothermal fluid derived from water/volcanic rock interaction, which appear to be of minor importance.

Red Sea Type 2 Brines – Suakin, Port Sudan, Erba, Albatross, Discovery, Atlantis II, Nereus, Shaban, and Conrad DHABs

These brines are influenced by hydrothermal activity and variable contributions from volcanic/ magmatic rock alteration, which explain e.g. the high concentration of Mn, Fe, and Zn and low concentrations of Mg and sulfate (Seyfried, 1987). This hydrothermal influence is strongest in Atlantis II, which has a multi-layered brine and is also the DHAB with highest recorded temperature.

Mediterranean Sea DHABs

The DHABs of the Mediterranean share clear similarities with those in the Red Sea, although they occur under a different tectonic setting (convergent plate boundary). The source of the brine is underlying evaporites (deposited in the Mediterranean basin during the Messinian Salinity Crisis 5.97 to 5.32 million years ago), mostly from the acme evaporitic stage (5.60 to 5.55 MYA), which has thick beds of halite with associated minerals (Roveri et al., 2004). However, hydrothermally-derived salt and brine has also been proposed as an alternative source (Hovland et al., 2018).

The Mediterranean DHABs are (in order of discovery): Tyro (Jongsma et al., 1983; De Lange et al., 1990b), Bannock (Scientific staff of Cruise Bannock 1984-12, 1985; De Lange et al., 1990b), Urania (Medriff Consortium, 1995), L'Atalante (Medriff Consortium, 1995), Discovery (Medriff Consortium, 1995), Thetis (La Cono et al., 2011), Medee (Yakimov et al., 2013), Kryos (Yakimov et al., 2015), and Hephaestus (La Cono et al., 2019). Their chemical differences (outlined in Table 1) allow us to split them into three groups.

Mediterranean Type 1 Brines – Urania DHAB

Urania DHAB is >100 m deep, filled with NaCl-rich brine, similar to type 2 brines but with lower salinity (about six times seawater salinity). Additional features include a second stable but less pronounced chemocline (in addition to the brine-seawater interface), as well as very high sulfide concentrations of up to 16 mM (van der Wielen et al., 2005; Borin et al., 2009). This makes Urania brine among the most sulfidic water body on Earth.

An intriguing feature of the deeper western part of the horse-shoe shaped Urania DHAB is the high-temperature bubbling mud vent (45°C; 10% salinity) beneath the relatively cold Urania brine (16°C; 27% salinity) resulting in a reverse halocline (Yakimov et al., 2007a).

Mediterranean Type 2 Brines – Bannock, L'Atalante, Medee, Thetis DHABs

Mediterranean Type 2 Brines are dominated by Na^+ and Cl^- ions, have an ionic composition that largely reflects concentrated seawater, and have a salinity that is about eight times higher than that of seawater, which differentiates them from Mediterranean Type 1 Brines. The proximity of the different basins seems to have little effect on chemical composition. For example, L'Atalante basin (Type 2) is close to Urania basin (Type 1) and Discovery basin (Type 3), whereas it is more distant from other Type-2 brines (Merlino et al., 2018).

Mediterranean Type 3 Brines – Discovery, Hephaestus and Kryos DHABs
DHABs within this group are perhaps the most intriguing of all owing to the exceptionally high $MgCl_2$ concentration of their brines (~5 M), which are presumably derived from bischofite ($MgCl_2.6H_2O$) that forms in the very latest stages of seawater evaporitic cycles (see La Cono et al., 2019). This high concentration of the divalent ion Mg^{2+} coupled with Cl^- results in brines that are below the currently accepted water activity (A_w) limit of life, which has recently decreased from the long-established value of 0.605 (Stevenson et al., 2015) to 0.585 (Stevenson et al., 2016), as well as brines that are extremely destabilising towards biological macromolecules (i.e. chaotropic) (Hallsworth et al., 2007). The presence of life in such hostile brines would alter our perception of life's limits on Earth and elsewhere.

Life in Deep Hypersaline Anoxic Basins

Microbial isolates
The Red Sea DHABs were originally declared sterile in the 1960s (Watson and Waterbury, 1969) and left mostly unexplored for several decades. Later studies, particularly based on molecular-based methodologies, have shown them to be teeming with life.

Work in the DHABs of the Red Sea has proven particularly fruitful in the isolation of microbes representing novel higher taxa. These include *Flexistipes sinusarabici* (the first representative of the phylum Deferribacteres; Fiala et al., 1990; Garrity et al., 2001), *Salinisphaera shabanensis* (representing the new family Salinisphaeraceae within the Gammaproteobacteria; Antunes et al., 2003; Vetriani et al., 2014), and *Haloplasma contractile* (the first representative of the order Haloplasmatales and likely representing a novel bacterial phylum; Antunes et al., 2008a). Further fully described taxa include several novel species and a new genus, and have been often associated with whole-genome sequencing (e.g. Antunes et al., 2008b; 2011b; Zhang et al., 2017a, 2017b; 2017c; see Table 2 and Table 3 for a full overview). Several additional studies reported the isolation of further microbial strains, unfortunately never fully described but still providing us with further insights into the cultivated microbial diversity of these locations. (e.g. Eder et al., 2001; Sagar et al., 2013a, 2013b; Zhang et al., 2016a, 2016b; see Table 2 and Table 3 for a full overview).

The exploration of the DHABs of the Mediterranean has led to the description of one new species: the archaeon *Natrinema salaciae* (Albuquerque et al., 2012). Studies on the Mediterranean have frequently focused on cultivation-independent approaches or have reported on the isolation of microbial strains that have not been fully described. Enrichment campaigns by Daffonchio et al. (2006) isolated fermentative halophiles such as *Halanaerobiales*, a *Halothiobacillus* that aerobically oxidised thiosulfate with CO_2 as sole carbon source over a NaCl range of 0.5–23%,

Table 2. Proteobacteria isolated from DHABs of the Red Sea and the Mediterranean Sea.

Class Family	Genus/Species/Strain	Origin			Description	Genome	Ref
Alphaproteobacteria							
Rhodobacteraceae	strain GMDJE10F1	Med	Bannock	BSI	-	-	1
	Ponticoccus marisrubri	RS	Erba	BSI	new species	+	2
	Ruegeria marisrubri	RS	Erba	BSI	new species	+	3
	Ruegeria profundi	RS	Erba	BSI	new species	+	3
	Roseobacter sp.	Med	Urania	BSI	-	-	4
	Sediminimonas sp.	RS	Kebrit	BSI	-	-	5
	Sulfitobacter sp.	RS	Kebrit	BSI	-	-	5
Rhodospirillaceae	ND	Med	Urania	BSI	-	-	4
Erythrobateraceae	*Erythrobacter* sp.	Med	Urania	BSI	-	-	4
Betaproteobacteria							
Nitrosomonadaceae	*Nitrosovibrio* sp.	Med	Bannock	Sed	-	-	6
Gammaproteobacteria							
Alcanivoracaceae	*Alcanivorax* sp.	Med	Bannock	BSI	-	-	1
	Fundibacter sp.	Med	Urania	BSI	-	-	4
Alteromonadaceae	*Alteromonas* spp.	Med	Bannock	BSI	-	-	1
			Urania	BSI	-	-	4
			L'Atalante	BSI	-	-	7
			Discovery	BSI	-	-	7
	Marinobacter spp.	Med	Bannock	BSI	-	-	1
			Urania	BSI	-	-	4
		RS	Erba	BSI	-	-	5
		RS	Nereus	BSI	-	-	8
	Marinobacter salsuginis	RS	Shaban	BSI	new species	-	9
Colwelliaceaea	Isolate S11L1B	RS	Shaban	BSI	-	-	10
Enterobacteriaceae	*Salmonella* sp.	Med	Bannock	BSI	-	-	1
Halomonadaceae	*Halomonas* spp.	Med	Bannock	BSI	-	-	1, 7
			Urania	BSI	-	-	4, 7
		RS	Erba	BSI	-	-	8
			Kebrit	BSI	-	-	5, 8
			Nereus	BSI	-	-	5, 8
	Chromohalobacter sp.	RS	Atlantis II	BSI	-	-	5
			Discovery	BSI	-	-	5, 8
			Kebrit	Sed	-	-	5
Halothiobacillaceae	*Halothiobacillus* spp.	Med	Bannock	BSI	-	-	1
			Urania	BSI	-	-	7
			Urania	Sed	-	-	11

Idiomarinaceae	Idiomarina spp.	RS	Discovery	BSI	-	-	5
		RS	Erba	BSI	-	-	5, 8
		RS	Kebrit	BSI	-	-	5
		RS	Nereus	BSI	-	-	5
		RS	Shaban	BSI	-	-	12
		Med	Bannock	BSI	-	-	1
Oceanospirillaceae	Oceanospirillum sp.	Med	Urania	BSI	-	-	4
Pseudoalteromonadaceae	Pseudoalteromonas sp.	Med	Bannock	BSI	-	-	1
Pseudomonadaceae	Pseudomonas spp.	RS	Suakin	Sed	-	-	13
		Med	Bannock	BSI	-	-	1
			Urania	BSI	-	-	4, 14
			Bannock	Sed	-	-	6
Salinisphaeraceae	Salinisphaera shabanensis	RS	Shaban	BSI	new order	+	15
Vibrionaceae	Vibrio sp.	Med	Bannock	BSI	-	-	1
Deltaproteobacteria							
Desulfovibrionaceae	Desulfovibrio sp.	RS	Atlantis II	BSI	-	-	16
Cystobacteraceae	ND	Med	Urania	BSI	-	-	4
Epsilonproteobacteria							
Campylobacteraceae	Sulfurospirillum sp.	Med	Bannock	BSI	-	-	1

Med, Mediterranean; RS, Red Sea; BSI, Brine-seawater interface; Sed, Sediment.; ND: Not determined; Data from: 1, Daffonchio et al., 2006; 2, Zhang et al., 2017a; 3, Zhang et al., 2017b; 4, Sass et al., 2001; 5, Sagar et al., 2013b; 6, Rodondi et al., 1996; 7, Borin et al., 2008; 8, Sagar et al., 2013a; 9, Antunes et al., 2007; 10, Eder et al., 2002; 11, Sorokin et al., 2006; 12, A. Antunes, M. Taborda and M.S. da Costa (unpublished); 13, Heitzer and Ottow, 1976; 14, Brusa et al., 2001; 15, Antunes et al., 2003; 16, Trüper, 1969.

as well as members of the *Epsilonprotobacteria* and the *Bacteroidetes*. A later study by Sass et al. (2008) reports on a large collection of over 80 strains obtained from several Mediterranean DHABs and including *Bacillus*-like isolates, and a few *Halomonas* and *Alteromonas*. More recent, noteworthy isolation and description results include: i) a strain of *Halanaeroarchaeum sulfurireducens*, an unusual sulfur-reducing and acetate-utilising haloarchaeon obtained from Medee DHAB, ii) a strain closely related to the Red Sea DHAB *Halorhabdus tiamatea* (Werner et al., 2014), and iii) the description of a three-component microbial consortium from Thetis DHAB (consisting of *Halobacteroides*, *Methanohalophilus*, and *Halanaerobium*), which linked anaerobic glycine betaine degradation with methanogenesis (La Cono et al., 2015).

Molecular-based studies
The original misconception of DHABs as sterile environments was drastically revised after the pioneering 16S rRNA gene phylogenetic studies by Eder et al., (1999, 2001, 2002). These studies uncovered thriving microbial communities and detected several bacterial and archaeal sequence groups which were completely new to Science, which were later

Table 3. Non-proteobacterial prokaryotes isolated from DHABs of the Red Sea and the Mediterranean Sea.

Class Family	Genus/Species/Strain	Origin			Description	Genome	Ref
Actinobacteria							
Microbacteriaceae	Microbacterium sp.	Med	Urania	BSI	-	-	1
Micrococcaceae	Arthrobacter sp.	Med	Urania	BSI	-	-	1
Bacilli							
Staphylococcaceae	Staphylococcus sp.	Med	Bannock	Sed	-	-	2
	Staphylococcus sp.	RS	Kebrit	Sed			3
Bacillaceae	Bacillus spp.	Med	L'Atalante	Sed	-	-	4
			Bannock	Sed	-	-	4
				BSI	-	-	5, 6
			Urania	Sed	-	-	4
			Discovery	BSI	-	-	5
				Sed	-	-	4
	Halobacillus spp.	RS	Kebrit	Sed	-	-	7
		Med	L'Atalante	BSI	-	-	5
			Urania	Sed	-	-	4
	Pontibacillus sp.	Med	L'Atalante	Sed	-	-	4
	Pontibacillus sp.	Med	Urania	Sed	-	-	4
	Thalassobacillus sp.	Med	Bannock	Sed	-	-	4
	Virgibacillus sp.	Med	Bannock	Sed	-	-	4
		RS	Nereus	B	-	-	3
Clostridia							
Halanaerobiaceae	Halanaerobium spp.	RS	Kebrit	BSI	-	-	8
			Shaban	BSI	-	-	9
		Med	Bannock	BSI	-	-	6
Cytophagia							
Cytophagaceae	Cytophaga sp.	Med	Bannock	BSI	-	-	6
Deferribacteres							
Deferribacteraceae	Flexistipes sinusarabici	RS	Atlantis II	BSI	new phylum	+	10
Firmicutes/ Mollicutes							
Haloplasmataceae	Haloplasma contractile	RS	Shaban	BSedI	new order	+	11
Flavobacteria							
Flavobacteriaceae	ND	Med	Urania	BSI	-	-	1
	Zunongwangia sp.	RS	Atlantis II	BSI	-	-	3
Halobacteria							
Halobacteriaceae	Halorhabdus tiamatea	RS	Shaban	BSedI	new species	+	12
Haloferacaceae	Halanaeroarchaeum sp.	Med	Medee	B	-	+	13
	Haloferax marisrubri	RS	Discovery	BSI	new species	+	14
	Haloferax profundi	RS	Discovery	BSI	new species	+	14
	Haloprofundus marisrubri	RS	Discovery	BSI	new genus	+	15
Natrialbaceaea	Natrinema salaciae	Med	Medee	B	new species	+	16

Med, Mediterranean; RS, Red Sea; BSI, Brine-seawater interface; B, Brine; BSedI, Brine-sediment interface; Sed, Sediment; ND: Not determined; Data from: 1, Sass et al., 2001; 2, Rodondi et al., 1996; 3, Sagar et al., 2013b; 4, Sass et al., 2008; 5, Borin et al., 2008; 6, Daffonchio et al., 2006; 7, Sagar et al., 2013a; 8, Eder et al., 2001; 9, Eder et al., 2002; 10, Fiala et al., 1990; 11, Antunes et al., 2008a; 12, Antunes et al., 2008b; 13, Messina et al., 2016; 14, Zhang et al., submitted; 15, Zhang et al., 2017c; 16, Albuquerque et al., 2012.

further expanded by similar studies in the Mediterranean DHABs (e.g. van der Wielen et al., 2005; Daffonchio et al., 2006). Some of these new sequence groups seemed to be specific to individual brines, while others (e.g., KB1, MSBL1) were detected in multiple DHABs both in the Red Sea and the Mediterranean (reviewed in Antunes et al., 2011d). Such a widespread distribution pointed to a high degree of adaptation of these new groups to life in such unique settings.

More recent years generated a considerable amount of data linked with further molecular-based studies, whole-genome sequencing, single-cell genomics, and metagenomics. These studies have confirmed the vertical stratification of microbial and viral communities across the brine-seawater interface and offered important insights into the taxonomic and physiological diversity of the brine-seawater interface and the sediments of DHABs (e.g. Antunes et al., 2015; Bougouffa et al., 2013). Significant contributions were made in the clarification of microbial groups involved in sulfate reduction, methanotrophy, methanogenesis and in several aspects of the Carbon, and the Nitrogen cycle (e.g. Yakimov et al., 2007b; La Cono et al., 2011; Siam et al., 2012; Abdallah et al., 2014; Pachiadaki et al., 2014; Guan et al., 2015; Ngugi et al., 2015, 2016; Sorokin et al., 2016).

Whole-genomic sequencing projects have provided some much-needed data on isolated strains from several of the Red Sea DHABs (Table 2 and Table 3), while single-cell genomics efforts based both on Red Sea and Mediterranean samples provided relevant insights into non-cultivated strains from the bacterial KB1 group and the euryarchaeal MSBL1 and SA1 groups (Yakimov et al., 2013; La Cono et al., 2015; Mwirichia et al., 2016; Nigro et al., 2016; Ngugi and Stingl, 2018), which helped to clarify their physiology and might assist in their cultivation (see sections below).

Further insights into the Life in DHABs
Life in the DHABs is increasingly seen as more complex than anticipated, extending past its prokaryotic communities, with first insights available both on their viral and eukaryote communities.

Metagenomic-based analysis on samples from the Red Sea DHABs gave us some clues on viral diversity and community structure, and revealed that viral communities are stratified along the brine-seawater interface (Antunes et al., 2015), similarly to what had been previously reported for prokaryotes.

In the Mediterranean DHAB, sediments were shown to have viral abundances similar to oxic deep-sea sediments (Danovaro et al., 2005), and had high levels of viral infection, resulting in the release of large amounts of prokaryote DNA (Corinaldesi et al., 2014). Viral lysis, rather than grazing by eukaryotes, was proposed to be the main top-down control of prokaryotes in DHAB sediments, as well as a major means of recycling nutrients (Corinaldesi et al., 2014).

The uncovering of eukaryotic life associated with DHABs started with the molecular-based detection of an incredibly diverse and novel range of protists (mostly composed of dinoflagellates, ciliates and other alveolates, as well as fungi) present in Bannock and Discovery, in the Mediterranean Sea (Edgcomb et al., 2009). Danovaro et al. (2010) reported three novel species of the animal phylum *Loricifera* in L'Atalante DHAB sediment, proposing that they were active and thus the first described metazoans seemingly completing their life cycle in permanently anoxic conditions

(Danovaro et al., 2016). Our knowledge of eukaryotes associated with DHABs has further escalated with the discovery of macrofauna in a few DHABs in the Red Sea. These findings include: the description of a new genus of bivalve (Oliver et al., 2015), the detection of enriched zooplankton and discovery of fish apparently feeding on the thick particle load present at the brine-seawater interface (Kaartvedt et al., 2016; Vestheim and Kaartvedt, 2016). This rich macrofauna associated with some of the investigated DHABs is in contrast with the extremely poor benthos usually reported for the Red Sea and has been suggested to be linked with increased microbial load and activity in the brine-seawater interfaces (Vestheim and Kaartvedt, 2016).

Despite these recent advances, DHAB research into these two fields is still in its infancy. Further efforts are essential to understand the likely grazing effects and complex trophic interactions between the viral, prokaryotic, and the eukaryotic components of these ecosystems.

Microbiology and geochemistry of the DHABs
The link between local geochemistry of the brines and their microbiota often predated the first data from microbial studies on several DHABs. Indeed, geochemical data hinted at biogenic sulfate reduction in sulfide-forming processes as well as biotic methane oxidation occurring at the brine-seawater interface (Blum and Puchelt, 1991; Faber et al., 1998). Complex redox cycles involving iron and manganese were also postulated to occur in the brine-seawater interface of several DHABs of the Red Sea (Stoffers et al., 1998).

In recent years we have accumulated increasing support for the existence of microbes associated with sulfate reduction, acetogenesis, methanogenesis, autotrophic, and heterotrophic activity, as well as different steps of the nitrogen cycle, including AnAmmOx by Planctomyces (Borin et al., 2009; 2013; Speth et al., 2017) and ammonia oxidation by Thaumarchaeota (Ngugi et al., 2015). These processes have been recently reviewed by Merlino et al. (2018) and Antunes et al. (2019), and are outside the direct scope of this review. The topic of methanogenesis in DHABs has been reviewed by McGenity and Sorokin (2018), but will be discussed briefly in the following section, as methane and other volatile compounds could be detected remotely to inform on the possibility of life in icy moons, as is currently being done with the NOMAD detector on the Mars Trace Gas Orbiter (Vandaele et al., 2015).

Methanogenesis in the DHABs
Early investigators of the Red Sea DHABs, such as the Kebrit and Shaban deeps, suspected the presence of methanogens based on archaeal biomarkers (Michaelis et al., 1990) and euryarchaeal 16S rRNA gene amplicons (Eder et al., 2002), both of which could derive from non-methanogenic Archaea. There is also evidence for methylotrophic

methanogenesis in Mediterrenean DHABs due to the common, but inconsistent, detection or isolation of *Methanohalophilus* sequences (Hallsworth et al., 2007; Yakimov et al., 2007b; Yakimov et al., 2015; La Cono et al., 2015; McGenity and Sorokin, 2018), as well as demonstration of the osmolyte and compatible solute, glycine betaine, fuelling methane production *via* fermentation to methylamines (Yakimov et al., 2013; La Cono et al., 2015). In the Mediterranean DHABs methane production rates (µM CH_4 d^{-1}) of 85.8 (Urania), 16.9 (l'Atalante), 4.2 (Bannock) and 2.6 (Discovery) were detected (van der Wielen et al., 2005; Borin et al., 2009). The most abundant uncultivated archaeal clones (termed MSBL-1) in most of these four basins (van der Wielen et al., 2005), the Urania hydrothermal mud vent (Yakimov et al., 2007a), Thetis DHAB (La Cono et al., 2011), Medee DHAB (Yakimov et al., 2013) and Kryos DHAB (Yakimov et al., 2015), are phylogenetically related to methanogens. This observation, coupled with coincident methane production and/or detection of *mcrA* transcripts, together with the occasional absence of known methanogenic taxa (Yakimov et al., 2013), led to the suggestion that Candidate Division MSBL-1 Archaea may be methanogens. However, this supposition is not supported by a single-cell genomics analysis of MSBL-1, where the core methanogenesis genes were not detected, and pathway analysis suggest a mixotrophic lifestyle, fermenting glucose or, in the absence of organic carbon, fixing CO_2 (Mwirichia et al., 2016).

So, which microbes are carrying out methanogenesis (in addition to *Methanohalophilus* when present)? Several studies from the Red Sea now report the presence of methanogenesis and methanogens in the brines and the brine-seawater interface (Antunes et al., 2011d). Guan et al. (2015), for example, detected *mcrA* genes in several of the interfaces, finding typical halophilic methylotrophic genera *Methanohalophilus* and *Methanococcoides*, as well as phylotypes similar to *Methanomassiliicoccales*, which were also detected in Kryos DHAB (Yakimov et al., 2015). *Methanomassiliicoccales* is a methanogenic order in the *Thermoplasmata* that can reduce methylamines and methanol with hydrogen ("methyl reduction"; Borrel et al., 2014), although, currently, only non-halophilic members of this order have been characterized. Arguably the most interesting group of halophilic methanogens is the Methanonatronarchaeia (SA1), which were isolated from anoxic hypersaline environments and shown to carry out methanogenesis by methyl reduction as with the *Methanomassiliicoccales* (Sorokin et al., 2017). The methyl-reduction pathway is distinguished from methylotrophic methanogenesis, as the C_1 methylated compounds are used only as electron acceptors while H_2 is the external electron donor (Sorokin et al., 2017). Comparison of sequences from Archaea in the DHABs revealed that the uncultivated Candidate Division Shaban Archaea (SA1), which is commonly detected in Red Sea and Mediterranean DHABs (Eder et al., 2002; Yakimov et al., 2013; Merlino et al., 2018; McGenity and Sorokin, 2018), is closely related to Methanonatronarchaeia. Thus, this type of methanogenesis may be quantitatively important in the DHABs. Ngugi and Stingl (2018), however, added a note of caution when they carried out single-cell genomic analysis on SA1 cells, and revealed no core methanogenesis genes. However, there is a lot of diversity in the SA1 group, and the genome-sequenced lineage had only 93% identity in 16S rRNA sequence to Methanonatronarchaeia (Ngugi and Stingl, 2018). Further cultivation-dependent and -independent explorations of SA1 group Archaea are required to understand the metabolic diversity within this lineage.

In the context of detecting methane from subterranean hypersaline oceans on icy moons it is important to consider that a high proportion may be microbially oxidised before entering the atmosphere. There is evidence for anaerobic oxidation of methane (AOM) in the DHABs, primarily via detection of sequences from uncultured Anaerobic Methane oxidizers Group 1 (ANME-1) (van der Wielen et al., 2005; Daffonchio et al., 2006; Lloyd et al., 2006; Yakimov et al., 2007a, b; La Cono et al., 2011; Pachiadaki et al., 2014). Indeed ANME-1 seem to be the main group carrying out AOM in a wide range of anoxic hypersaline environments (McGenity and Sorokin, 2018). The methane-oxygen counter gradients in DHAB brine-seawater interfaces are conducive to aerobic methane oxidation, and carbon isotope analyses suggest its occurrence in Atlantis II, Discovery and Kebrit DHABs from the Red Sea (Faber et al., 1998; Schmidt et al., 2003), and Type-I methanotrophs (and a small proportion of Type-II methanotrophs) were detected in the brine-seawater interfaces of the aforementioned DHABs (Abdallah et al., 2014).

Further discussions on the feasibility of specific types of microbial metabolism are obviously hampered by the limited amount of actual data on the physical-chemical conditions on these exooceans and is outside the scope of this paper. Nonetheless, it is worth noting that a few authors have been looking into the possibility of microbial activity occurring under predicted exoocean conditions. A relevant example is the study by Taubner et al. (2018), where methane production by the methanogenic archaeal species *Methanothermococcus okinawensis* under putative Enceladus-like conditions was tested and confirmed.

Future research

The limits of life
Studying the limits of life is one of the main pillars of Astrobiology. The DHABs provide an opportunity to explore the impact of multiple extremes on microbes, especially the interactions of high pressure and high salinity, including both kosmotropic and chaotropic salts. It is well known that microbial activities are influenced by high pressure (Tamburini et al., 2013) and high salinity (Oren, 2011), but synergistic and antagonistic effects on microbes of multiple environmental stressors, despite being the norm, requires much more in-depth investigation in order to predict biological

processes on Earth and elsewhere (Harrison et al., 2013; Robinson and Mikucki, 2018). The classic experiments by Yayanos (1986) investigated the interaction of temperature and pressure on the growth rate of isolates, while Kaye and Baross (2004) showed improved growth of *Halomonas* strains under high-pressure at intermediate, compared with low, salinities. However, there have been few systematic studies investigating polyextremophily, and only modest mechanistic understanding has been obtained, e.g. compatible solutes have been proposed to protect cellular macromolecules from both high salt and pressure stress (Yancey, 2004). In order to understand the interacting effects of those multiple extremes that are most relevant to DHABs and icy-moon brines, a much broader range of microbes and communities must be investigated, especially those that are native to the DHABs. To date, no microbes from the DHABs have been isolated under high hydrostatic pressure, despite evidence from metagenomic expression of esterases, which exhibited optimal function at high hydrostatic pressure and salinity, indicating that they probably derived from piezophiles (Ferrer et al., 2005).

It is known that chaotrope-induced cellular damage can be offset (Hallsworth et al., 2007) and growth enhanced (Zajc et al., 2014; Lima de Alves et al., 2015) by adding kosmotropic solutes to chaotropic solutions/media, which reduces the water activity but counterbalances the chaotropicity, thus demonstrating that chaotropic-solute-induced destabilisation of macromolecules is a key growth-limiting factor. Despite their hostile chemistry, there is some evidence of microbial activity in both Discovery (van der Wielen et al., 2005) and Kryos DHABs (Yakimov et al., 2015; Steinle et al., 2018), but no evidence to date of life in the youngest (700-year-old) $MgCl_2$-dominated Hephaestus DHAB beyond 2.97 M $MgCl_2$ (La Cono et al., 2019). Therefore, it will be important to understand which microbes may be performing these activities and how they protect themselves against extremely chaotropic brines. Equally, an improved understanding of the compensatory effects of chaotropic and kosmotropic salts (Hallsworth et al., 2007), and differential microbial adaptations to both types of salts, are needed to predict where life will occur, particularly given the uncertainty of the ionic compositions of icy-moon brines (Marion et al., 2003). Ultimately, the ionic composition (balance of kosmotropes and chaotropes) as well as the overall ionic strength of brines elsewhere in the solar system, together with the physical conditions (pressure, temperature etc.) and microbial adaptations to counter these effects, will be key to determining brine habitability (Hallsworth et al., 2007; Fox-Powell et al., 2016). To this end the Mediterranean type-3 DHABs, rich in $MgCl_2$, provide ideal environments to explore the water activity and chaotropicity limits of life, and potentially to find obligate chaophiles.

Source of microbial strains for exposure experiments
The use of exposure experiments and testing the resilience of several microbial strains to stressful conditions (equivalent to those found outside

our planet) has been a strong focus of recent research as highlighted by several dedicated large-scale research projects (e.g. BIOMEX-Biology and Mars experiment, MEXEM- Mars exposed extremophile mixture, IceXpose-Icy exposure of microorganisms) and discussed by Martins et al. (2017). These experiments are a vital source of data for discussions in Astrobiology, namely on the feasibility of Life outside our planet and on Planetary Protection.

The testing of microbes obtained from extreme terrestrial analogues has been suggested as particularly advantageous as they were hypothesized to have highly effective cellular and molecular adaptations and repair mechanisms which would allow them to better withstand conditions present in their extra-terrestrial counterparts (currently being tested as part of the MEXEM project). In this context, microbes isolated from DHABs (Table 2 and Table 3) constitute a rich source of interesting new targets for exposure experiments and should be prioritized in future studies. The availability of microbial strains representing novel higher-ranked taxonomic groups would significantly extend the range of tested taxa and constitute a relevant source of new data. Furthermore, DHABs' strains belonging to previously known taxa would provide an interesting target for comparative studies which would: a) pair them with closely-related strains isolated from different environments and b) look into potential differences in their resilience when exposed to stress. The existence of genomic data for many of the strains isolated from DHABs (Table 2 and Table 3) would further facilitate a more in-depth analysis.

Long-term viability and preservation: microbes and biomolecules
As is frequently the case when surveying previously uncharted extreme environments, the microbial exploration of DHABs has uncovered several novel phylogenetic groups. Surprisingly, several of these novel groups seem to be DHAB-specific and have a very wide-spread distribution, being consistently detected in brines of the Red Sea and the Mediterranean (e.g. KB1, MSBL1, and SA2 phylogenetic groups), despite the isolation imposed by their geographical location and by the nature of the brines. Previous studies on haloarchaea, the most extreme of halophiles, pointed to their apparent capability of global dispersal across isolated coastal hypersaline environments (Clark et al., 2017), but the mode of dispersal is most likely via birds and wind, where they may be protected inside halite crystals (Clark et al., 2017; Kemp et al., 2018), especially given that many haloarchaea lyse in seawater (Torreblanca et al., 1986). No such vectors or protection would be available for transporting haloarchaea to the deep sea (though our understanding of dispersal mechanisms is limited). Dispersal of other DHAB-specific novel groups may be facilitated by overlying seawater. Haloclines do not represent an absolute barrier for penetration by motile microbes in the micrometre range (Doostmohammadi et al., 2012), but any DHAB-adapted microbe would presumably have to be actively moving while in the deep-sea water (or induced to reactivate from a dormant state

upon reaching the halocline), and then able to overcome the rapid increase in viscosity and salinity as it returns to its preferred habitat. Nevertheless, to date, DHAB-specific taxa have not been detected in seawater.

An alternative explanation for this apparent paradox is microbes from these groups could be trapped in the salt crystals and/or brine inclusions, present in the possibly linked salt deposits under the Red Sea and Eastern Mediterranean, and released during the formation of the brines where they would thrive (Antunes et al., 2011d). Such long-term preservation and continued viability of microbes in ancient salt deposits is in line with several previous studies (e.g. Grant et al., 1998; McGenity et al., 2000; Mormile et al., 2003). Indeed, evidence is mounting in support of survival over millions of years (Røy et al., 2012), specifically in halite (McGenity et al., 2000; Schubert et al., 2010; Gramain et al., 2011; Jaakkola et al., 2016), and the interconnection between the geosphere and the biosphere, specifically in the context of the DHABs (biosphere) and the Messinian evaporites (geosphere) was discussed by McGenity et al. (2008). Collection and analysis of sediment cores of the evaporite layers would allow us to confirm the presence of such microbial groups and help clarify the process involved in the microbial colonization of the DHABs. Such studies would be very relevant for the future exploration of the icy moons of the outer Solar System.

An equally relevant and under-studied aspect regarding long-term preservation is the effect of such brines in the stability of bio-molecules. Previous reports pointed to the preserving effects of samples from the DHABs of the Mediterranean on DNA (Borin et al., 2008). One should note that this preservation would be affected by the type of salts present in the brine. In this regard, one particularly interesting location is the $MgCl_2$-saturated Discovery lake. Chaotropic $MgCl_2$ destabilises macromolecules and would therefore significantly restrict any microbial/enzyme activity (Hallsworth et al., 2007). Therefore, Discovery DHAB, together with the other known $MgCl_2$-rich brines, Lake Kryos and Hephaestus, would provide excellent locations for preserving biomolecules (Hallsworth et al., 2007; Sass et al., 2008).

The wide-range of chemical conditions present in the DHABs provide an ideal testing ground for looking into the effects of brine on microbes and bio-molecules, which would be easily linked to the astrobiological study of biosignatures under high-salinity conditions and lead to technical improvements. As an example of such an improvement, the discussion on this brine-preserving effect and detrimental impact on the reliability of results obtained from standard molecular-based analysis led to a switch to alternative RNA-based approaches in several subsequent studies.

Technology development and testing: Fine-scale sampling and laboratorial replication of complex environments

One of the major benefits of working with terrestrial analogues is the opportunity to use such sites as a basis for developing, testing and fine-tuning different technologies and equipment in preparation for future space missions. Relevant examples of this approach include e.g. the use of Boulby mine in the MINAR- Mining and Analogue Research project (e.g. Payler et al., 2017; Cockell et al., 2018) or the current testing of prototype ice-melting probes in glaciers and regions with ice-sheets across the globe (e.g. Dachwald et al., 2014; Funke and Horneck, 2018). The exploration of deep-sea terrestrial locations and access to water bodies below thick ice layers are seen as vital for the future exploration of Mars and the icy moons of the outer Solar System so such inputs are an invaluable source of information.

The complex nature of DHABs and the drastic transitions observed at their brine-seawater interfaces also provide remarkable technical challenges that are relevant for Astrobiology. We believe that they also provide unique opportunities and ideal settings to develop and study new equipment for fine-scale sampling which would be relevant for exploring equivalent settings in the oceans of the icy moons of the Outer Solar System. The development of MODUS (Mobile Docker for Underwater Sciences), a Remotely Operated Vehicle designed to improve precision of sampling technologies and facilitate finer-scale sampling in such challenging environments was tested *in situ* in the DHABs of the Mediterranean (Malinverno et al., 2006) and was seen as a very relevant advance. The attempt to deploy the recently developed Microbial Sampler-Submersible Incubation Device (MS-SID) at the brine-seawater interface of Urania DHAB is also a welcome advance, notwithstanding the technical issues that affected its use (Pachiadaki et al., 2016). Despite these improvements, there is a clear need for further technical developments and improvements in this field.

Another technical issue to keep in mind is our limited capability to successfully replicate gradient-rich environments in the lab, a bottleneck which significantly hampers research in this field. Recent advances making use of gel-stabilised gradient plates (More et al., submitted) or the use of simulation chambers (e.g. Herschy et al., 2014) for the simulation of chemical gradients in liquid media, should help to fill this technical gap and are welcome additions to our current range of tools.

Acknowledgements

AA is funded by the Science and Technology Development Fund, Macau SAR. TJM acknowledges support from the European Cooperation in Science and Technology (COST) Action CA15103, "Uncovering the Mediterranean Salt Giant (MEDSALT)". KOF acknowledges support from the Science Technology Facilities Council, UK

References

Abdallah, R.Z., Adel, M., Ouf, A., Sayed, A., Ghazy, M.A., Alam, I., Essack, M., Lafi, F.F., Bajic, V.B., El-Dorry, H., and Siam, R. (2014). Aerobic methanotrophic communities at the Red Sea brine-seawater interface. Front. Microbiol. 5, 487. https://doi.org/10.3389/fmicb.2014.00487

Albuquerque, L., Taborda, M., La Cono, V., Yakimov, M., and da Costa, M.S. (2012). *Natrinema salaciae* sp. nov., a halophilic archaeon isolated from the deep, hypersaline anoxic Lake Medee in the Eastern Mediterranean Sea. Syst. Appl. Microbiol. 35, 368-373. https://doi.org/10.1016/j.syapm.2012.06.005

Antunes, A. (2017). Extreme Red Sea: life in the deep-sea anoxic brine lakes. In Human Interaction with the Environment in the Red Sea, D.A. Agius, E. Khalil, E. Scerri, and A. Williams, eds. (Leiden, Netherlands: Brill), pp. 30-47. https://doi.org/10.1163/9789004330825_004

Antunes, A., Eder, W., Fareleira, P., Santos, H., and Huber, R. (2003). *Salinisphaera shabanensis* gen. nov., sp. nov., a novel, moderately halophilic bacterium from the brine–seawater interface of the Shaban Deep, Red Sea. Extremophiles 7, 29-34. https://doi.org/10.1128/JB.05459-11

Antunes, A., França, L., Rainey, F.A., Huber, R., Nobre, M.F., Edwards, K.J., and da Costa, M.S. (2007). *Marinobacter salsuginis* sp. nov., isolated from the brine–seawater interface of the Shaban Deep, Red Sea. Int. J. Syst. Evol. Microbiol. 57, 1035-1040. https://doi.org/10.1099/ijs.0.64862-0

Antunes, A., Rainey, F.A., Wanner, G., Taborda, M., Pätzold, J., Nobre, M.F., da Costa, M.S., and Huber, R. (2008a). A new lineage of halophilic, wall-less, contractile bacteria from a brine-filled deep of the Red Sea. J. Bacteriol. 190, 3580-3587. https://dx.doi.org/10.1128%2FJB.01860-07

Antunes, A., Taborda, M., Huber, R., Moissl, C., Nobre, M.F., and da Costa, M.S. (2008b). *Halorhabdus tiamatea* sp. nov., a non-pigmented, extremely halophilic archaeon from a deep-sea, hypersaline anoxic basin of the Red Sea, and emended description of the genus *Halorhabdus*. Int. J. Syst. Evol. Microbiol. 58, 215-220. https://dx.doi.org/10.1099/ijs.0.65316-0

Antunes, A., Alam, I., Bajic, V.B., and Stingl, U. (2011a). Genome sequence of *Haloplasma contractile*, an unusual contractile bacterium from a deep-sea anoxic brine lake. J Bacteriol. 193, 4551-4552. https://doi.org/10.1128/JB.05461-11

Antunes, A., Alam, I., Bajic, V.B., and Stingl, U. (2011b). Genome sequence of *Halorhabdus tiamatea*, the first archaeon isolated from a deep-sea anoxic brine lake. J. Bacteriol. 193, 4553-4554. https://doi.org/10.1128/JB.05462-11

Antunes, A., Alam, I., Bajic, V.B., and Stingl, U. (2011c). Genome sequence of *Salinisphaera shabanensis*, a gammaproteobacterium from the harsh, variable environment of the brine-seawater interface of the Shaban Deep in the Red Sea. J. Bacteriol. 193, 4555-4556. https://doi.org/10.1128/JB. 05459-11

Antunes, A., Ngugi, D.K., and Stingl, U. (2011d). Microbiology of the Red Sea (and other) deep-sea anoxic brine lakes. Environ. Microbiol. Rep. 3(4), 416-433. https://doi.org/10.1111/j.1758-2229.2011.00264.x

Antunes, A., Alam, I., Simões, M.F., Daniels, C., Ferreira, A.J., Siam, R., El-Dorry, H., and Bajic, V.B. (2015). First insights into the viral communities of the deep-sea anoxic brines of the Red Sea. Genomics Proteomics Bioinformatics 13, 304-309. https://doi.org/10.1016/j.gpb.2015.06.004

Antunes, A., Kaartvedt, S., and Schmidt, M. (2019). Geochemistry and life at the interfaces of brine-filled deeps in the Red Sea. In Oceanographic and Biological Aspects of the Red Sea, N.M.A. Rasul and I.C.F. Stewart, eds. (Cham, Switzerland: Springer Nature), pp. 185-194.

Backer, H. and Schoell, M. (1972). New deeps with brines and metalliferous sediments in the Red Sea. Nat. Phys. Sci. 240, 153-158. https://doi.org/10.1038/physci240153a0

Barge, L.M., and White, L.M. (2017). Experimentally Testing Hydrothermal Vent Origin of Life on Enceladus and Other Icy/Ocean Worlds. Astrobiology 17, 820-833. https://doi.org/10.1089/ast.2016.1633

Billings, S.E., and Kattenhorn, S.A. (2005). The great thickness debate: Ice shell thickness models for Europa and comparisons with estimates based on flexure at ridges. Icarus 177, 397-412. https://doi.org/10.1016/j.icarus.2005.03.013

Blum, N. and Puchelt, H. (1991). Sedimentary-hosted polymetallic massive sulfide deposits of the Kebrit and Shaban Deeps, Red Sea. Miner. Depos. 26, 217-227. https://doi.org/10.1007/BF00209261

Borin, S., Crotti, E., Mapelli, F., Tamagnini, I., Corselli, C. and Daffonchio, D. (2008). DNA is preserved and maintains transforming potential after contact with brines of the deep anoxic hypersaline lakes of the Eastern Mediterranean Sea. Saline Systems. 4, 10. https://doi.org/10.1186/1746-1448-4-10

Borin, S., Brusetti, L., Mapelli, F., D'Auria, G., Brusa, T., Marzorati, M., Rizzi, A., Yakimov, M., Marty, D., De Lange, G.J., Van der Wielen, Bolhuis, P.H., McGenity, T.J., Polymenakou, P.N., Malinverno, E., Giuliano, L., Corselli, C. and Daffonchio, D. (2009). Sulfur cycling and methanogenesis primarily drive microbial colonization of the highly sulfidic Urania deep hypersaline basin. Proc. Natl. Acad. Sci. U. S. A. 106, 9151-9156. https://doi.org/10.1073/pnas.0811984106

Borin, S., Mapelli, F., Rolli, E., Song, B., Tobias, C., Schmid, M.C., De Lange, G.J., Reichart, G.J., Schouten, S., Jetten, M. and Daffonchio, D. (2013). Anammox bacterial populations in deep marine hypersaline gradient systems. Extremophiles 17, 289-299. https://doi.org/10.1007/s00792-013-0516-x

Borrel, G., Parisot, N., Harris, H.M.B., Peyretaillade, E., Gaci, N., Tottey, W., Bardot, O., Raymann, K., Gribaldo, S., Peyret, P., O'Toole, P.W., and Brugère, J.-F. (2014). Comparative genomics highlights the unique biology of Methanomassiliicoccales, a Thermoplasmatales-related seventh order of methanogenic archaea that encodes pyrrolysine. BMC genomics. 15(1), 679. https://doi.org/10.1186/1471-2164-15-679

Bougouffa, S., Yang, J.K., Lee, O.O., Wang, Y., Batang, Z., Al-Suwailem, A., and Qian, P.Y. (2013). Distinctive microbial community structure in highly stratified deep-sea brine water columns. Appl. Environ. Microbiol. 79, 3425-3437. http://dx.doi.org/10.1128/AEM.00254-13

Brown, R.H., Clark, R.N., Buratti, B.J., Cruikshank, D.P., Barnes, J.W., Mastrapa, R.M.E., Bauer, J., Newman, S., Momary, T., Baines, K.H., Bellucci, G., Capaccioni, F., Cerroni, P., Combes, M., Coradini, A., Drossart, P., Formisano, V., Jaumann, R., Langevin, Y., Matson, D.L., Mccord, T.B., Nelson, R.M., Nicholson, P.D., Sicardy, B., and Sotin, C. (2006). Composition and physical properties of Enceladus' surface. Science. 311, 1425-1428. http://dx.doi.org/10.1126/science.1121031

Bruneau, L., Jerlov, N.G., and Koczy, F.F. (1953). Physical and chemical methods (+ Appendix), In Reports of the Swedish deep-sea expedition, vol. III. Physics and chemistry, no. 4, N.G. Jerlov, ed. (Goteborg, Sweden: Elanders), pp. 99-112.

Brusa, T., Borin, S., Ferrari, F., Sorlini, C., Corselli, C., and Daffonchio, D. (2001). Aromatic hydrocarbon degradation patterns and catechol 2,3-dioxygenase genes in microbial cultures from deep anoxic hypersaline lakes in the eastern Mediterranean sea. Microbiol. Res. 156, 49-57.

Cadek, O., Tobie, G., Van Hoolst, T., Masse, M., Choblet, G., Lefevre, A., Mitri, G., Baland, R.M., Behounkova, M., Bourgeois, O., and Trinh, A. (2016). Enceladus's internal ocean and ice shell constrained from Cassini gravity, shape, and libration data. Geophys. Res. Lett. 43, 5653-5660. https://doi.org/10.1002/2016GL068634

Camerlenghi, A. (1990) Anoxic basins of the eastern Mediterranean: geological framework. Mar. Chem. 31, 1-19. https://doi.org/10.1016/0304-4203(90)90028-B

Carlson, R.W., Calvin, W.M., Dalton III, J.B., Hansen, G.B., Hudson, R.L., Johnson, R.E., Mccord, T.B., and Moore, M.H. (2009). Europa's surface composition. In Europa, R.T. Pappalardo, W.B. McKinnon, and K.K. Khurana, eds. (Tucson, AZ, USA: University of Arizona Press), pp. 283-327.

Carr, M.H., Belton, M.J.S., Chapman, C.R., Davies, A.S., Geissler, P., Greenberg, R., Mcewen, A.S., Tufts, B.R., Greeley, R., Sullivan, R., Head, J.W., Pappalardo, R.T., Klaasen, K.P., Johnson, T.V., Kaufman, J., Senske, D., Moore, J., Neukum, G., Schubert, G., Burns, J.A., Thomas, P., and Veverka, J. (1998). Evidence for a subsurface ocean on Europa. Nature. 391, 363-365. https://doi.org/10.1038/34857

Cita, M.B., Kastens, K.A., McCoy, F.W., Aghib, F., Cambi, A., Camerlenghi, A., Corselli, C., Erba, E., Giambastiani, M., Herbert, T., Leoni, C., Malinverno, P., Nosetto, A., and Parisi, E. (1985). Gypsum precipitation from cold brines in an anoxic basin in the eastern Mediterranean. Nature. 314, 152-154. https://doi.org/10.1038/314152a0

Cita, M.B. (2006). Exhumation of Messinian evaporites in the deep-sea and creation of deep anoxic brine-filled collapsed basins. Sediment. Geol. 188, 357-378. https://doi.org/10.1016/j.sedgeo.2006.03.013

Clark, D.R., Mathieu, M., Mourot, L., Dufossé, L., Underwood, G.J.C., Dumbrell, A.J., and McGenity, T.J. (2017). Biogeography at the limits of life: Do extremophilic microbial communities show biogeographic regionalisation? Glob. Ecol. Biogeogr. 26, 1435-1446. https://doi.org/10.1111/geb.12670

Cochran, J.R., Martinez, F., Steckler, M.S., and Hobart, M.A. (1986). Conrad Deep: A new Northern Red Sea Deep. Origin and implications for continental rifting. Earth Planet. Sci. Lett. 78, 18-32. https://doi.org/10.1016/0012-821X(86)90169-X

Cockell, C.S., Holt, J., Campbell, J., Groseman, H., Josset, J.L., Bontognali, T.R., Phelps, A., Hakobyan, L., Kuretn, L., Beattie, A., Blank, J., Bonaccorsi, R., McKay, C., Shirvastava, A., Stoker, C., Willson, D., McLaughlin, S., Payler, S., Stevens, A., Wadsworth, J., Bessone, L., Maurer, M., Sauro, F., Martin-Torres, J., Zorzano, M.P., Bhardwaj, A., Soria-Salinas, A., Mathanlal, T., Nazarious, M.I., Ramachandran, A.V., Vaishampayan, P., Guan, L., Perl, S.M., Telling, J., Boothroyd, I.M., Tyson, O., Realff, J., Rowbottom, J., Lauernt. B., Gunn, M., Shah, S., Singh, S., Paling, S., Edwards, T., Yeoman, L., Meehan, E., Toth, C., Scovell, P., and Suckling, B. (2018). Subsurface scientific exploration of extraterrestrial environments (MINAR 5): analogue science, technology and education in the Boulby Mine, UK. Int. J. Astrobiol. 1-26. https://doi.org/10.1017/S1473550418000186

Corinaldesi, C., Tangherlini, M., Luna, G.M., and Dell'Anno, A. (2014). Extracellular DNA can preserve the genetic signatures of present and past viral infection events in deep hypersaline anoxic basins. Proc. R. Soc. Lond. B: Biol. Sci. 281, 20133299. https://doi.org/10.1098/rspb.2013.3299

Dachwald, B., Mikucki, J., Tulaczyk, S., Digel, I., Espe, C., Feldmann, M., Francke, G., Kowalski, J., and Xu, C. (2014). IceMole: a maneuverable probe for clean in situ analysis and sampling of subsurface ice and subglacial aquatic ecosystems. Ann. Glaciol. 55, 14-22.
https://doi.org/10.3189/2014AoG65A004

Daffonchio, D., Borin, S., Brusa, T., Brusetti, L., van der Wielen, P.W.J.J., Bolhuis, H., D'Auria, G., Yakimov, M., Giuliano, L., Tamburini, C., Marty, D., McGenity, T.J., Hallsworth, J.E., Sass, A.M., Timmis, K.N., Tselepides, A., de Lange, G.J., Huebner, A., Thomson, J., Varnavas, S.P., Gasparoni, F., Gerber, H.W., Malinverno, E., Corselli, C., and Biodeep Scientific Party (2006). Stratified prokaryote network in the oxic- anoxic transition of a deep-sea halocline. Nature 440, 203-207. https://doi.org/10.1038/nature04418

Danovaro, R., Corinaldesi, C., Dell'Anno, A., Fabiano, M., and Corselli, C. (2005). Viruses, prokaryotes and DNA in the sediments of a deep-hypersaline anoxic basin (DHAB) of the Mediterranean Sea. Environ. Microbiol. 7, 586-592. https://doi.org/10.1111/j.1462-2920.2005.00727.x

Danovaro, R., Dell'Anno, A., Pusceddu, A., Gambi, C., Heiner, I., and Kristensen, R.M. (2010). The first metazoa living in permanently anoxic conditions. BMC Biol. 8, 30. https://doi.org/10.1186/1741-7007-8-30

Danovaro, R., Gambi, C., Dell'Anno, A., Corinaldesi, C., Pusceddu, A., Neves, R.C., and Kristensen, R.M. (2016). The challenge of proving the existence of metazoan life in permanently anoxic deep-sea sediments. BMC Biol. 14, 43. https://doi.org/10.1186/s12915-016-0263-4

De Lange, G.J., Catalano, G., Klinkhammer, G.P., and Luther III, G.W. (1990a). The interface between oxic seawater and the anoxic Bannock brine; its sharpness and the consequences for the redox-related cycling of Mn and Ba. Mar. Chem. 31, 205-217. https://doi.org/10.1016/0304-4203(90)90039-F

De Lange, G.J., Middelburg, J.J., Van der Weijden, C.H., Catalano, G., Luther Iii, G.W., Hydes, D.J., Woittiez, J.R.W., and Klinkhammer, G.P. (1990b) Composition of anoxic hypersaline brines in the Tyro and Bannock Basins, Eastern Mediterranean. Mar. Chem. 31, 63-88. https://doi.org/10.1016/0304-4203(90)90031-7

Doostmohammadi, A., Stocker, R., and Ardekani, A.M. (2012). Low-Reynolds-number swimming at pycnoclines. Proc. Natl. Acad. Sci. USA 109, 3856–3861. https://doi.org/10.1073/pnas.1116210109

Eder, W., Ludwig, W., and Huber, R. (1999). Novel 16S rRNA gene sequences retrieved from highly saline brine sediments of Kebrit Deep, Red Sea. Arch. Microbiol. 172, 213-218. https://doi.org/10.1007/s0020300507

Eder, W., Jahnke, L. L., Schmidt, M., and Huber, R. (2001). Microbial diversity of the brine-seawater interface of the Kebrit Deep, Red Sea, studied via 16S rRNA gene sequences and cultivation methods. Appl. Environ. Microbiol. 67, 3077-3085. https://doi.org/10.1128/AEM.67.7.3077-3085.2001

Eder, W., Schmidt, M., Koch, M., Garbe-Schönberg, D., and Huber, R. (2002). Prokaryotic phylogenetic diversity and corresponding geochemical data of the brine–seawater interface of the Shaban Deep, Red Sea. Environ. Microbiol. 4, 758-763. https://doi.org/10.1046/j.1462-2920.2002.00351.x

Edgcomb, V.P., Pachiadaki, M.G., Mara, P., Kormas, K.A., Leadbetter, E.R. and Bernhard, J.M. (2016) Gene expression profiling of microbial activities and interactions in sediments under haloclines of E. Mediterranean deep hypersaline anoxic basins. ISME J. 10, 2643-2657. https://doi.org/10.1038/ismej.2016.58

Emery, J.P., Burr, D.M., Cruikshank, D.P., Brown, R.H., and Dalton, J.B. (2005). Near-infrared (0.8-4.0 µm) spectroscopy of Mimas, Enceladus, Tethys, and Rhea. Astron. Astrophys. 435, 353-362. https://doi.org/10.1051/0004-6361:20042482

Ehrhardt, A., Hübscher, C., and Gajewski, D. (2005). Conrad Deep, Northern Red Sea: Development of an early stage ocean deep within the axial depression. Tectonophysics, 411(1-4), 19-40. https://doi.org/10.1016/j.tecto.2005.08.011

Elsenousy, A., Vance, S., Bills, B.G., and Goodman, J. (2015). Modeling Heat and Salt Transfer at Europa's Ice-Ocean Interface. In Lunar and Planetary Science Conference vol. 46 (The Woodlands,TX, USA: Lunar and Planetary Institute), p. 1676.

Faber, E., Botz, R., Poggenburg, J., Schmidt, M., Stoffers, P., and Hartmann, M. (1998). Methane in Red Sea brines. Org. Geochem. 29, 363-379. https://doi.org/10.1016/S0146-6380(98)00155-7

Ferrer, M., Golyshina, O.V., Chemikova, T.N., Khachane, A.N., dos Santos V.A.P.M., Yakimov, M.M., Timmis, K.N., and Golyshin, P.N. (2005). Microbial enzymes mined from the Urania deep-sea hypersaline anoxic basin. Chem. Biol. 12, 895-904. https://doi.org/10.1016/j.chembiol.2005.05.020

Fiala, G., Woese, C.R., Langworthy, T.A., and Stetter, K.O. (1990). *Flexistipes sinusarabici*, a novel genus and species of eubacteria occurring in the Atlantis II Deep brines of the Red Sea. Arch. Microbiol. 154, 120-126. https://doi.org/10.1007/BF00423320

Fox-Powell, M.G., Hallsworth, J.E., Cousins, C.R., and Cockell, C.S. (2016). Ionic strength is a barrier to the habitability of Mars. Astrobiology 16, 427-442. https://doi.org/10.1089/ast.2015.1432

Funke, O. and Horneck, G. (2018). The Search for Signatures of Life and Habitability on Planets and Moons of Our Solar System. In Biological, Physical and Technical Basics of Cell Engineering, G. Artmann, A. Artmann, A. Zhubanova, and I. Digel, eds. (Singapore: Springer), pp. 457-481. https://doi.org/10.1007/978-981-10-7904-7_20

Garcia-Lopez, E., and Cid, C. (2017). Glaciers and Ice Sheets as Analog Environments of Potentially Habitable Icy Worlds. Front. Microbiol. 8, 1407. https://doi.org/10.3389/fmicb.2017.01407

Garrity, G.M., Holt, J.M., Huber, H., Stetter, K.O., Greene, A.C., Patel, B.K., Caccavo Jr., F., Allison, M.J., MacGregor, B.J., and Stahl, D.A. (2001). Phylum BIX. Deferribacteres phy. nov. In Bergey's Manual of Systematic Bacteriology vol. 1., D.R. Boone, R.W. Castenholz, and G.M., eds. (New York, NY, USA: Springer), pp. 465-47. https://doi.org/10.1007/978-0-387-21609-6_26

Geissler, P.E., Greenberg, R., Hoppa, G., Mcewen, A., Tufts, R., Phillips, C., Clark, B., Ockert-Bell, M., Helfenstein, P., Burns, J., Veverka, J., Sullivan, R., Greeley, R., Pappalardo, R.T., Head, J.W., Belton, M.J.S., and Denk, T. (1998). Evolution of lineaments on Europa: Clues from Galileo multispectral imaging observations. Icarus. 135, 107-126. https://doi.org/10.1006/icar.1998.5980

Gleeson, D.F., Pappalardo, R.T., Anderson, M.S., Grasby, S.E., Mielke, R.E., Wright, K.E., and Templeton, A.S. (2012). Biosignature Detection at an Arctic Analog to Europa. Astrobiology. 12, 135-150. https://doi.org/10.1089/ast.2010.0579

Glein, C.R., Baross, J.A., and Waite Jr, J.H. (2015). The pH of Enceladus' ocean. Geochim. Cosmochim. Acta. 162, 202-219. https://doi.org/10.1016/j.gca.2015.04.017

Glein, C.R., and Shock, E.L. (2010). Sodium chloride as a geophysical probe of a subsurface ocean on Enceladus. Geophys. Res. Lett. 37, L09204. https://doi.org/10.1029/2010GL042446

Gramain, A., Díaz, G.C., Demergasso, C., Lowenstein, T.K., and McGenity, T.J. (2011) Archaeal diversity along a subterranean salt core from the Salar Grande (Chile). Environ. Microbiol. 13, 2105-2121. https://doi.org/10.1111/j.1462-2920.2011.02435.x

Grant, W.D., Gemmell, R.T., and McGenity, T.J. (1998). *Halobacteria*: the evidence for longevity. Extremophiles. 2(3), 279-287.

Guan, Y., Hikmawan, T., Antunes, A., Ngugi, D., and Stingl, U. (2015). Diversity of methanogens and sulfate-reducing bacteria in the interfaces of five deep-sea anoxic brines of the Red Sea. Res. Microbiol. 166, 688-699. https://doi.org/10.1016/j.resmic.2015.07.002

Hallsworth, J.E., Yakimov, M.M., Golyshin, P.N., Gillion, J.L.M., D'Auria, G., de Lima Alves, F., La Cono, V., Genovese, M., McKew, B.A., Hayes, S.L., Harris, G., Giuliano, L., Timmis, K.N., and McGenity, T.J. (2007) Limits of life in $MgCl_2$-containing environments: chaotropicity defines the window. Environ. Microbiol. 9, 801-813. https://doi.org/10.1111/j.1462-2920.2006.01212.x

Han, L.J., and Showman, A.P. (2010). Coupled convection and tidal dissipation in Europa's ice shell. Icarus. 207, 834-844. https://doi.org/10.1016/j.icarus.2009.12.028

Hand, K.P., and Carlson, R.W. (2015). Europa's surface color suggests an ocean rich with sodium chloride. Geophys. Res. Lett. 42, 3174-3178. https://doi.org/10.1002/2015GL063559

Hand, K.P., Chyba, C.F., Priscu, J.C., Carlson, R.W., and Nealson, K.H. (2009). Astrobiology and the potential for life on Europa. In Europa, R.T. Pappalardo, W.B. McKinnon, and K.K. Khurana, eds. (Tucson, AZ, USA: University of Arizona Press), pp. 589-630.

Harrison, J.P., Gheeraert, N., Tsigelnitskiy, D., and Cockell, C.S. (2013). The limits for life under multiple extremes. Trends Microbiol. 21, 204-212. https://doi.org/10.1016/j.tim.2013.01.006

Heitzer, R.D. and Ottow, J.C.G. (1976). New denitrifying bacteria isolated from Red Sea sediments. Mar. Biol. 37, 1-10. https://doi.org/10.1007/BF00386773

Hendrix, A.R., Hurford, T.A., Barge, L.M., Bland, M.T., Bowman, J.S., Brinckerhoff, W., Buratti, B.J., Cable, M.L., Castillo-Rogez, J., Collins, G.C., and Diniega, S. (2019). The NASA Roadmap to Ocean Worlds. *Astrobiology* 19(1), 1-27. https://doi.org/10.1089/ast.2018.1955

Herschy, B., Whicher, A., Camprubi, E., Watson, C., Dartnell, L., Ward, J., Evans, J.R.G. and Lane, N. (2014). An origin-of-life reactor to simulate alkaline hydrothermal vents. J. Mol. Evol. 79, 213-227. https://doi.org/10.1007/s00239-014-9658-4

Hovland, M., Rueslåtten, H.G., and Johnsen, H.K. (2018). Large salt accumulations as a consequence of hydrothermal processes associated with 'Wilson cycles': A review Part 1: Towards a new understanding. Mar. Pet. Geol. 92, 128-148. https://doi.org/10.1016/j.marpetgeo.2017.12.029

Howett, C.J.A., Spencer, J.R., Pearl, J., and Segura, M. (2010). Thermal inertia and bolometric Bond albedo values for Mimas, Enceladus, Tethys, Dione, Rhea and Iapetus as derived from Cassini/CIRS measurements. Icarus. 206, 573-593. https://doi.org/10.1016/j.icarus.2009.07.016

Hsu, H.W., Postberg, F., Sekine, Y., Shibuya, T., Kempf, S., Horanyi, M., Juhasz, A., Altobelli, N., Suzuki, K., Masaki, Y., Kuwatani, T., Tachibana, S., Sirono, S., Moragas-Klostermeyer, G., and Srama, R. (2015). Ongoing hydrothermal activities within Enceladus. Nature. 519, 207. https://doi.org/10.1038/nature14262

less, L., Stevenson, D.J., Parisi, M., Hemingway, D., Jacobson, R.A., Lunine, J.I., Nimmo, F., Armstrong, J.W., Asmar, S.W., Ducci, M., and Tortora, P. (2014). The Gravity Field and Interior Structure of Enceladus. Science. 344, 78-80. https://doi.org/10.1126/science.1250551

Jaakkola, S.T., Ravantti, J.J., Oksanen, H.M., and Bamford, D.H. (2016). Buried alive: Microbes from ancient halite. Trends Microbiol. 24, 148-159. https://doi.org/10.1016/j.tim.2015.12.002

Jia, X.Z., Kivelson, M.G., Khurana, K.K., and Kurth, W.S. (2018). Evidence of a plume on Europa from Galileo magnetic and plasma wave signatures. Nat. Astron. 2, 459-464. https://doi.org/10.1038/s41550-018-0450-z

Jongsma, D., Fortuin, A.R., Huson, W., Troelstra, S.R., Klaver, G.T., Peters, J.M., van Harten, D., de Lange, G.J., and ten Haven, L. (1983). Discovery of an anoxic basin within the Strabo trench, eastern Mediterranean. Nature. 305, 795-797. https://doi.org/10.1038/305795a0

Kaartvedt, S., Antunes, A., Røstad, A., Klevjer, T.A., and Vestheim, H. (2016). Zooplankton at deep Red Sea brine pools. J. Plankton Res. 38, 679-684. https://doi.org/10.1093/plankt/fbw013

Kargel, J.S., Kaye, J.Z., Head, J.W., Marion, G.M., Sassen, R., Crowley, J.K., Prieto Ballesteros, O., Grant, S.A., and Hogenboom, D.L. (2000). Europa's crust and ocean: Origin, composition, and the prospects for life. Icarus. 148, 226-265. https://doi.org/10.1006/icar.2000.6471

Kaye, J. Z. and Baross, J. A. (2004). Synchronous effects of temperature, hydrostatic pressure, and salinity on growth, phospholipid profiles, and protein patterns of four *Halomonas* species isolated from deep-sea hydrothermal-vent and sea surface environments. Appl. Environ. Microbiol. 70(10), 6220-6229. https://doi.org/10.1128/AEM.70.10.6220-6229.2004

Kemp, B.L., Tabish, E.M., Wolford, A.J., Jones, D.L., Butler, J.K., and Baxter, B.K. (2018) The Biogeography of Great Salt Lake Halophilic Archaea: Testing the Hypothesis of Avian Mechanical Carriers. Diversity. 10, 124. https://doi.org/10.3390/d10040124

Khurana, K.K., Kivelson, M.G., Stevenson, D.J., Schubert, G., Russell, C.T., Walker, R.J., and Polanskey, C. (1998). Induced magnetic fields as evidence for subsurface oceans in Europa and Callisto. Nature. 395, 777-780. https://doi.org/10.1038/27394

Kivelson, M.G., Khurana, K.K., and Volwerk, M. (2002). The permanent and inductive magnetic moments of Ganymede. Icarus. 157, 507-522. https://doi.org/10.1006/icar.2002.6834

Kivelson, M.G., Khurana, K.K., Russell, C.T., Volwerk, M., Walker, R.J., and Zimmer, C. (2000). Galileo magnetometer measurements: A stronger case for a subsurface ocean at Europa. Science. 289, 1340-1343. https://doi.org/10.1126/science.289.5483.1340

La Cono, V., Bortoluzzi, G., Messina, E., La Spada, G., Smedile, F., Giuliano, L., Borghini, M., Stumpp, C., Schmitt-Kopplin, P., Harir, M., O'Neill, W.K., Hallsworth, J.E., and Yakimov M.M. (2019). The discovery of Lake Hephaestus, the youngest athalassohaline deep-sea formation on Earth. Sci. Rep. 9, 1679. https://doi.org/10.1038/s41598-018-38444-z

La Cono, V., Arcadi, E., and La Spada, G. (2015). A three-component microbial consortium from deep-sea salt-saturated anoxic Lake Thetis links anaerobic glycine betaine degradation with methanogenesis. Microorganisms. 3, 500-517. https://doi.org/10.3390/microorganisms3030500

La Cono, V., Smedile, F., Bortoluzzi, G., Arcadi, E., Maimone, G., Messina, E., Borghini, M., Oliveri, E., Mazzola, S., L'haridon, S., Toffin, L., Genovese, L., Ferrer, M., Giuliano, L., Golyshin, P.N., and Yakimov, M.M. (2011). Unveiling microbial life in new deep-sea hypersaline Lake Thetis. Part I: prokaryotes and environmental settings. Environ. Microbiol. 13, 2250-2268. https://doi.org/10.1111/j.1462-2920.2011.02478.x

Lima de Alves, F., Stevenson, A., Baxter, E., Gillion, J.L.M., Hejazi, F., Hayes, S., Morrison, I.E.G., Prior, B.A., McGenity, T.J., Rangel, D.E.N., Magan, N., Timmis, K.N., and Hallsworth, J.E. (2015). Concomitant osmotic and chaotropicity-induced stresses in *Aspergillus wentii*: compatible solutes determine the biotic window. Curr. Genet. 61, 457-477. https://doi.org/10.1007/s00294-015-0496-8

Lloyd, K.G., Lapham, L., and Teske, A. (2006). An anaerobic methane-oxidizing community of ANME-1b Archaea in hypersaline Gulf of Mexico sediments. Appl. Environ. Microbiol. 72, 7218-7230. https://doi.org/10.1128/AEM.00886-06

Lunine, J.I. (2017). Ocean worlds exploration. *Acta Astronaut.* 131, 123-130. https://doi.org/10.1016/j.actaastro.2016.11.017

Malinverno, E., Gasparoni, F., Gerber, H.W., and Corselli, C. (2006). The exploration of Eastern Mediterranean deep hypersaline anoxic basins with MODUS: a significant example of technology spin-off from the GEOSTAR program. Ann. Geophys. 49(2-3), 729-37. https://doi.org/10.4401/ag-3129

Marion, G.M., Fritsen, C.H., Eicken, H., Payne, M.C. (2003). The search for life on Europa: limiting environmental factors, potential habitats, and earth analogues. Astrobiology. 3, 785-811. https://doi.org/10.1089/153110703322736105

Martins, Z., Cottin, H., Kotler, J.M., Carrasco, N., Cockell, C.S., Noetzel, R.D., Demets, R., De Vera, J.P., D'hendecourt, L., Ehrenfreund, P., Elsaesser, A., Foing, B., Onofri, S., Quinn, R., Rabbow, E., Rettberg, P., Ricco, A.J., Slenzka, K., Stalport, F., Ten Kate, I.L., Van Loon, J., and

Westall, F. (2017). Earth as a Tool for Astrobiology-A European Perspective. Space Sci. Rev. 209, 43-81. https://doi.org/10.1007/s11214-017-0369-1

McGenity, T.J., Gemmell, R.T., Grant, W.D., and Stan-Lotter, H. (2000). Origins of halophilic microorganisms in ancient salt deposits. Environ Microbiol 2, 243-250. https://doi.org/10.1046/j.1462-2920.2000.00105.x

McGenity, T.J., Hallsworth, J.E., and Timmis, K.N. (2008). Connectivity between 'ancient' and 'modern' hypersaline environments, and the salinity limits of life. CIESM 2008, The Messinian Salinity Crisis from mega-deposits to microbiology – A consensus report. No. 33. In CIESM Workshop Monographs, F. Briand, ed. (Monaco: CIESM), pp. 115-120.

McGenity, T.J., and Sorokin, D. (2018). Methanogens and methanogenesis in hypersaline environments. In Handbook of Hydrocarbon and Lipid Microbiology: Biogenesis of Hydrocarbons 2nd Edition, A. Stams and D. Sousa, ed. (Berlin, Germany: Springer), pp. 665-680. https://doi.org/10.1007/978-3-319-53114-4_12-1

MEDRIFF Consortium. (1995). Three brine lakes discovered in the seafloor of the Eastern Mediterranean. EOS Trans. AGU. 76 (32), 313-318. https://doi.org/10.1029/95EO00189

Merlino, G., Barozzi, A., Michoud, G., Ngugi, D.K., and Daffonchio, D. (2018). Microbial ecology of deep-sea hypersaline anoxic basins. FEMS Microbiol. Ecol. 94, fiy085. https://doi.org/10.1093/femsec/fiy085

Messina, E., Sorokin, D.Y., Kublanov, I.V., Toshchakov, S., Lopatina, A., Arcadi, E., Smedile, F., La Spada, G., La Cono, V., and Yakimov, M.M. (2016). Complete genome sequence of *'Halanaeroarchaeum sulfurireducens'* M27-SA2, a sulfur-reducing and acetate-oxidizing haloarchaeon from the deep-sea hypersaline anoxic lake Medee. Stand. Genomic Sci. 11(1), 35. https://doi.org/10.1186/s40793-016-0155-9

Michaelis, W.A., Jenisch, A., and Richnow, H.H. (1990). Hydrothermal petroleum generation in Red Sea sediments from the Kebrit and Shaban Deeps. Appl. Geochem. 5, 103-114. https://doi.org/10.1016/0883-2927(90)90041-3

Miller, A.R. (1964). Highest salinity in the world ocean? Nature. 203, 590-591. https://doi.org/10.1038/203590a0

More-Mutch, P., Huber, R., and Antunes, A. (2020). Gel-stabilized gradient plates: a new cultivation approach for the laboratory recreation of naturally-occurring interfaces. Front. Microbiol. (submitted).

Mormile, M.R., Biesen, M.A., Gutierrez, M.C., Ventosa, A., Pavlovich, J.B., Onstott, T.C., and Fredrickson, J.K. (2003). Isolation of *Halobacterium salinarum* retrieved directly from halite brine inclusions. Environ. Microbiol. 5, 1094-1102.

Mwirichia, R., Alam, I., Rashid, M., Vinu, M., Ba-Alawi, W., Kamau, A.A., Ngugi, D.K., Göker, M., Klenk, H.P., Bajic, V., and Stingl, U. (2016). Metabolic traits of an uncultured archaeal lineage-MSBL1-from brine pools of the Red Sea. Sci. Rep. 6, 19181. https://doi.org/10.1038/srep19181

Ngugi, D.K., and Stingl, U. (2018). High-quality draft single-cell genome sequence belonging to the archaeal candidate division SA1, isolated from Nereus Deep in the Red Sea. Genome Announc. 6(19), e00383-18. https://doi.org/10.1128/genomeA.00383-18.

Ngugi, D.K., Blom, J., Alam, I., Rashid, M., Ba-Alawi, W., Zhang, G., Hikmawan, T., Guan, Y., Antunes, A., Siam, R., El Dorry, H., Bajic, V., and Stingl, U. (2015). Comparative genomics reveals adaptations of a halotolerant thaumarchaeon in the interfaces of brine pools in the Red Sea. ISME J. 9, 396-411. https://doi.org/10.1038/ismej.2014.137

Ngugi, D.K., Blom, J., Stepanauskas, R., and Stingl, U. (2016). Diversification and niche adaptations of Nitrospina-like bacteria in the polyextreme interfaces of Red Sea brines. ISME J. 10, 1383. https://doi.org/10.1038/ismej.2015.214

Nigro, L.M., Hyde, A. S., MacGregor, B.J., and Teske, A. (2016). Phylogeography, salinity adaptations and metabolic potential of the candidate division KB1 bacteria based on a partial single cell genome. Front. Microbiol. 7, 1266. https://doi.org/10.3389/fmicb.2016.01266

Nimmo, F., Spencer, J.R., Pappalardo, R.T., and Mullen, M.E. (2007). Shear heating as the origin of the plumes and heat flux on Enceladus. Nature. 447, 289-291. https://doi.org/10.1038/nature05783

Oliver, P. G., Vestheim, H., Antunes, A., and Kaartvedt, S. (2015). Systematics, functional morphology and distribution of a bivalve (*Apachecorbula muriatica* gen. et sp. nov.) from the rim of the 'Valdivia Deep' brine pool in the Red Sea. J. Mar. Biol. Assoc. U. K. 95, 523-535. https://doi.org/10.1017/S0025315414001234

Oren, A. (2011). Thermodynamic limits to microbial life at high salt concentrations. Environ. Microbiol. 13, 1908-1923. https://doi.org/10.1111/j.1462-2920.2010.02365.x

Pachiadaki, M.G., Taylor, C., Oikonomou, A., Yakimov, M.M., Stoeck, T., and Edgcomb, V. (2016). In situ grazing experiments apply new technology to gain insights into deep-sea microbial food webs. Deep Sea Res Part 2: Top Stud Oceanogr. 129, 223-231. https://doi.org/10.1016/j.dsr2.2014.10.019

Pachiadaki, M.G., Yakimov, M.M., LaCono, V., Leadbetter, E., and Edgcomb, V. (2014). Unveiling microbial activities along the halocline of Thetis, a deep-sea hypersaline anoxic basin. ISME J 8, 2478-2489. https://doi.org/10.1038/ismej.2014.100

Patthoff, D.A., and Kattenhorn, S.A. (2011). A fracture history on Enceladus provides evidence for a global ocean. Geophys. Res. Lett. 38, L18201. http://dx.doi.org/10.1029/2011GL048387

Pautot, G., Guennoc, P., Coutelle, A., and Lyberis, N. (1984). Discovery of a large brine deep in the northern Red Sea. Nature. 310, 133-136. https://doi.org/10.1038/310133a0

Payler, S.J., Biddle, J.F., Coates, A.J., Cousins, C.R., Cross, R.E., Cullen, D.C., Downs, M.T., Direito, S.O., Edwards, T., Gray, A.L., Genis, J., Gunn, M., Hansford, G.M., Harkness, P., Holt, J., Josset, J.L., Li, X., Lees, D.S., Lim, D.S.S., Mchugh, M., Mcluckie, D., Meehan, E., Paling, S.M. Souchon, A., Yeoman, L., and Cockell, C.S. (2017). Planetary science and exploration in the deep subsurface: results from the MINAR

Program, Boulby Mine, UK. Int. J. Astrobiol. 16(2), 114-129. https://doi.org/10.1017/S1473550416000045

Postberg, F., Kempf, S., Schmidt, J., Brilliantov, N., Beinsen, A., Abel, B., Buck, U., and Srama, R. (2009). Sodium salts in E-ring ice grains from an ocean below the surface of Enceladus. Nature. 459, 1098-1101. https://doi.org/10.1038/nature08046

Postberg, F., Schmidt, J., Hillier, J., Kempf, S., and Srama, R. (2011). A salt-water reservoir as the source of a compositionally stratified plume on Enceladus. Nature. 474, 620-622. https://doi.org/10.1038/nature10175

Prieto-Ballesteros, O., Rodriguez, N., Kargel, J.S., Kessler, C.G., Amils, R., and Remolar, D.F. (2003). Tirez lake as a terrestrial analog of Europa. Astrobiology 3, 863-877. https://doi.org/10.1089/153110703322736141

Quick, L.C., and Marsh, B.D. (2015). Constraining the thickness of Europa's water-ice shell: Insights from tidal dissipation and conductive cooling. Icarus 253, 16-24. https://doi.org/10.1016/j.icarus.2015.02.016

Robinson, J.M. and Mikucki, J.A. (2018). Occupied and empty regions of the space of extremophile parameters. In Habitability of the Universe Before Earth, R. Gordon, ed. (London, UK: Imperial College Press), pp. 199-230. https://doi.org/10.1016/B978-0-12-811940-2.00009-5

Rodondi, G., Andreis, C., Pellegrini, S., Brusa, T., Del Puppo, E., Ferrari, A., and Cita, M.B. (1996). Presenza di batteri in bacini anossici del Mediterraneo orientale: indagini preliminari. Rend. Fis. Acc.Lincei 7, 63-78. https://doi.org/10.1007/BF03001700

Ross, M.N. and Schubert, G. (1987). Tidal heating in an internal ocean model of Europa. Nature. 325, 133-134. https://doi.org/10.1038/325133a0

Roth, L., Saur, J., Retherford, K.D., Strobel, D.F., Feldman, P.D., Mcgrath, M.A., and Nimmo, F. (2014). Transient Water Vapor at Europa's South Pole. Science. 343, 171-174. https://doi.org/10.1126/science.1247051

Roveri, M., Flecker, R., Krijgsman, W., Lofi, J., Lugli, S., Manzi, V., Sierro, F.J., Bertini, A., Camerlenghi, A., De Lange, G., Govers, R., Hilgen, F.J., Hübscher, C., Meijer, P.T., and Stoica, M. (2014). The Messinian Salinity Crisis: past and future of a great challenge for marine sciences. Mar. Geol. 352, 25-58. https://doi.org/10.1016/j.margeo.2014.02.002

Røy, H., Kallmeyer, J., Adhikari, R.R., Pockalny, R., Jørgensen, B.B., and D'Hondt, S. (2012). Aerobic microbial respiration in 86-million-year-old deep-sea red clay. Science. 336, 922-925. https://doi.org/10.1126/science.1219424

Russell, M.J., Murray, A.E., and Hand, K.P. (2017). The possible emergence of life and differentiation of a shallow biosphere on irradiated icy worlds: the example of Europa. Astrobiology. 17, 1265-1273. https://doi.org/10.1089/ast.2016.1600

Rutishauser, A., Blankenship, D.D., Sharp, M., Skidmore, M.L., Greenbaum, J.S., Grima, C., Schroeder, D.M., Dowdeswell, J.A., and Young, D.A. (2018). Discovery of a hypersaline subglacial lake complex beneath Devon Ice Cap, Canadian Arctic. Sci. Adv. 4, eaar4353. https://doi.org/10.1126/sciadv.aar4353

Sagar, S., Esau, L., Hikmawan, T., Antunes, A., Holtermann, K., Stingl, U., Bajic, V.B., and Kaur, M. (2013a). Cytotoxic and apoptotic evaluations of marine bacteria isolated from brine-seawater interface of the Red Sea. BMC Complement. Altern. Med. 13, 29. https://doi.org/10.1186/1472-6882-13-29

Sagar, S., Esau, L., Holtermann, K., Hikmawan, T., Zhang, G., Stingl, U., Bajic, V.B., and Kaur, M. (2013b). Induction of apoptosis in cancer cell lines by the Red Sea brine pool bacterial extracts. BMC Complement. Altern. Med. 13, 344. https://doi.org/10.1186/1472-6882-13-344

Sass, A.M., Sass, H., Coolen, M.J., Cypionka, H., and Overmann, J. (2001). Microbial communities in the chemocline of a hypersaline deep-sea basin (Urania basin, Mediterranean Sea). Appl. Environ. Microbiol. 67(12), 5392-5402. https://doi.org/10.1128/AEM.67.12.5392-5402.2001

Sass, A.M., McKew, B.A., Sass, H., Fichtel, J., Timmis, K.N., and McGenity, T.J. (2008). Diversity of *Bacillus*-like organisms isolated from deep-sea hypersaline anoxic sediments. Saline Systems. 4, 8. https://doi.org/10.1186/1746-1448-4-8

Schmidt, M., Al-Farawati, R., and Botz, R. (2015). Geochemical classification of brine-filled Red Sea deeps. In The Red Sea, N.M.A. Rasul and I.C.F. Stewart, eds. (Berlin, Germany: Springer), pp. 219-233.

Schmidt, M., Botz, R., Faber, E., Schmitt, M., Poggenburg, J., Garbe-Schönberg, D., and Stoffers, P. (2003). High-resolution methane profiles across anoxic brine-seawater boundaries in the Atlantis-II, discovery, and Kebrit deeps (Red Sea). Chem. Geol. 200, 359-375. https://doi.org/10.1016/S0009-2541(03)00206-7

Schubert, B., Lowenstein, T., Timofeeff, M., and Parker, M. (2010). Halophilic Archaea cultured from ancient halite, Death Valley, California. Environ. Microbiol. 12, 440-454. https://doi.org/10.1111/j.1462-2920.2009.02086.x

Sekine, Y., Shibuya, T., Postberg, F., Hsu, H.W., Suzuki, K., Masaki, Y., Kuwatani, T., Mori, M., Hong, P.K., Yoshizaki, M., Tachibana, S., and Sirono, S. (2015). High-temperature water-rock interactions and hydrothermal environments in the chondrite-like core of Enceladus. Nat. Commun. 6, 8604. https://doi.org/10.1038/ncomms9604

Seyfried Jr, W.E. (1987). Experimental and theoretical constraints on hydrothermal alteration processes at mid-ocean ridges. Ann. Rev. Earth Planet. Sci. 15(1), 317-335. https://doi.org/10.1146/annurev.ea.15.050187.001533

Shokes, R.F., Trabant, P.K., Presley, B., and Reid, D.F. (1977). Anoxic, hypersaline basin in the northern Gulf of Mexico. Nature 196, 1443-1446. https://doi.org/10.1126/science.196.4297.1443

Showman, A.P., and Malhotra, R. (1999). The Galilean satellites. Science 286, 77-84. https://doi.org/10.1126/science.286.5437.77

Siam, R., Mustafa, G.A., Sharaf, H., Moustafa, A., Ramadan, A.R., Antunes, A., Bajic, V.B., Stingl, U., Marsis, N.G., Coolen, M.J., Sogin, M.,

Ferreira, A.J.S., and El Dorry, H. (2012). Unique prokaryotic consortia in geochemically distinct sediments from Red Sea Atlantis II and Discovery Deep brine pools. PloS one. 7, e42872. https://doi.org/10.1371/journal.pone.0042872

Sorokin, D.Y., Tourova, T.P., Lysenko, A.M., and Muyzer, G. (2006). Diversity of culturable halophilic sulfur-oxidizing bacteria in hypersaline habitats. Microbiology. 152, 3013-3023.

Sorokin, D.Y., Kublanov, I.V., Gavrilov, S.N., Rojo, D., Roman, P., Golyshin, P.N., Slepak, V.Z., Smedile, F., Ferrer, M., Messina, E., La Cono, V., and Yakimov, M.M. (2016). Elemental sulfur and acetate can support life of a novel strictly anaerobic haloarchaeon. ISME J. 10(1), 240-252. https://doi.org/10.1038/ismej.2015.79

Sorokin, D.Y., Makarova, K., Abbas, B., Ferrer, M., Golyshin, P.N., Galinski, E.A., Ciordia, S., Mena, M.C., Merkel, A.Y., Wolf, Y.I., van Loosdrecht, M.C.M., and Koonin, E.V. (2017). Discovery of extremely halophilic methyl-reducing euryarchaea provides insight into the evolutionary origin of methanogenesis. Nature Microbiol. 2, 17081. https://doi.org/10.1038/nmicrobiol.2017.81

Sparks, W.B., Hand, K.P., Mcgrath, M.A., Bergeron, E., Cracraft, M., and Deustua, S.E. (2016). Probing for evidence of plumes on Europa with HST/STIS. Astrophysical J. 829, 121. http://dx.doi.org/10.3847/0004-637X/829/2/121

Spencer, J.R., and Nimmo, F. (2013). Enceladus: An Active Ice World in the Saturn System. Annu. Rev. Earth Planet. Sci. 41, 693-717. https://doi.org/10.1146/annurev-earth-050212-124025

Spencer, J.R., Tamppari, L.K., Martin, T.Z., and Travis, L.D. (1999). Temperatures on Europa from Galileo photopolarimeter-radiometer: Nighttime thermal anomalies. Science. 284, 1514-1516. https://doi.org/10.1126/science.284.5419.1514

Spencer, J.R., Pearl, J.C., Segura, M., Flasar, F.M., Mamoutkine, A., Romani, P., Buratti, B.J., Hendrix, A.R., Spilker, L.J., and Lopes, R.M.C. (2006). Cassini encounters Enceladus: Background and the discovery of a south polar hot spot. Science. 311, 1401-1405. https://doi.org/10.1126/science.1121661

Speth, D. R., Lagkouvardos, I., Wang, Y., Qian, P. Y., Dutilh, B. E., and Jetten, M. S. (2017). Draft genome of *Scalindua rubra*, obtained from the interface above the discovery deep brine in the Red Sea, sheds light on potential salt adaptation strategies in anammox bacteria. Microb. Ecol. 74, 1-5. https://doi.org/10.1007/s00248-017-0929-7

Squyres, S.W., Reynolds, R.T., Cassen, P.M., and Peale, S.J. (1983a). The evolution of Enceladus. Icarus. 53, 319-331. https://doi.org/10.1016/0019-1035(83)90152-5

Squyres, S.W., Reynolds, R.T., Cassen, P.M., and Peale, S.J. (1983b). Liquid water and active resurfacing on Europa. Nature. 301, 225-226. https://doi.org/10.1038/301225a0

Steinle, L., Knittel, K., Felber, N., Casalino, C., Lange, G., Tessarolo, C., Stadnitskaia, A., Damsté, J.S.S., Zopfi, J., Lehmann, M.F., Treude, T., and Niemann, H. (2018). Life on the edge - active microbial communities in the Kryos $MgCl_2$-brine basin at very low water activity. ISME J. 12, 1414-1426. https://doi.org/10.1038/s41396-018-0107-z

Stevenson, A., Cray, J.A., Williams, J.P., Santos, R., Sahay, R., Neuenkirchen, N., McClure, C.D., Grant, I.R., Houghton, J.D., Quinn, J.P., Timson, D.J., Patil, S.V., Singhal, R.S., Antón, J., Dijksterhuis, J., Hocking, A.D., Lievens, B., Rangel, D.E.N., Voytek, M.A., Gunde-Cimerman, N., Oren, A., Timmis, K.N., McGenity, T.J., and Hallsworth, J.E. (2015). Is there a common water-activity limit for the three Domains of life? ISME J. 9, 1333-1351. https://doi.org/10.1038/ismej.2014.219

Stevenson, A., Hamill, P.G., O'Kane, C.J., Kminek, G., Rummel, J.D., Voytek, M.A., Dijksterhuis, J., and Hallsworth, J.E. (2016). *Aspergillus penicillioides* differentiation and cell division at 0.585 water activity. Environ. Microbiol. 19, 687-697. https://doi.org/10.1111/1462-2920.13597

Stoffers, P., Moammar, M., Abu-Ouf, M., Ackermand, D., Alassif, O., Al-Hazim, Y., Boldt, S., Botz, R., Eder, W., El-Garafi, A., El-Mamoney, M., Fleitmann, D., Garbe-Schönberg, D., Geiselhart, S., Goedecke, D., Hartmann, M., Klauke, S., Moussa, K., Mühlhan, N., Mühlstrasser, T., Poggenburg, J., Rehder, W., Schmidt, M., Schmitt, M., Schoeps, D., Scholten, J., Shbalaby, M., Wismann, A., and Yohannes, E. (1998). Cruise Report SONNE 121, Red Sea. Hydrography, Hydrothermalism and Paleoceanography in the Red Sea, No. 88. (Kiel, Germany: Geologisch-Paläontologisches Institut und Museum, Christian-Albrechts-Universität der Universität Kiel). https://doi.org/10.2312/reports-gpi.1998.88

Tamburini, C., Boutrif, M., Garel, M., Colwell, R.R., and Deming, J.W. (2013). Prokaryotic responses to hydrostatic pressure in the ocean – a review. Environ. Micro. 15, 1262-1274. https://doi.org/10.1111/1462-2920.12084

Taubner, R.S., Pappenreiter, P., Zwicker, J., Smrzka, D., Pruckner, C., Kolar, P., Bernacchi, S., Seifert, A.H., Krajete, A., Bach, W., and Peckmann, J. (2018). Biological methane production under putative Enceladus-like conditions. *Nat. Commun.* 9, 748. https://doi.org/10.1038/s41467-018-02876-y

Tsimplis, M.N., and Baker, T.F. (2000). Sea level drop in the Mediterranean Sea: an indicator of deep water salinity and temperature changes? Geophys. Res. Lett. 27, 1731–1734

Torreblanca, M., Rodriguez-Valera, F., Juez, G., Ventosa, A., Kamekura, M., and Kates, M. (1986). Classification of non-alkaliphilic halobacteria based on numerical taxonomy and polar lipid composition, and description of *Haloarcula* gen. nov. and *Haloferax* gen. nov. System. Appl. Microbiol. 8, 89-99. https://doi.org/10.1016/S0723-2020(86)80155-2

Trüper, H. G. (1969). Bacterial sulfate reduction in the Red Sea hot brines In Hot brines and recent heavy metal deposits in the Red Sea, E.T. Degens and D.A. Ross, eds. (New York, N.Y., USA: Springer-Verlag), pp 263-271.

Van der Wielen, P.W.J.J., Bolhuis, H., Borin, S., Daffonchio, D., Corselli, C., Giuliano, L., de Lange, G.J., Varnavas, S.P., Thompson, J., Tamburini, C., Marty, D., McGenity, T.J., Timmis, K.N., and BioDeep Scientific Party. (2005). The enigma of prokaryotic life in deep hypersaline anoxic basins. Science. 307, 121-123. https://doi.org/10.1126/science.1103569

Vance, S. and Goodman J. (2009). Oceanography of an ice-covered moon.In Europa, R.T. Pappalardo, W.B. McKinnon, and K.K. Khurana, eds. (Tucson, AZ, USA: University of Arizona Press), pp. 459-482.

Vandaele, A.C., Neefs, E., Drummond, R., Thomas, I.R., Daerden, F., Lopez-Moreno, J.J., Rodriguez, J., Patel, M.R., Bellucci, G., Allen, M., Altieri, F., Bolséea, D., Clancy, t., Delanoyea, S., Depiessea, C., Cloutisg, E., Fedorovah, A., Formisano, V., Funke, B., Fussen, D., Geminale, A., Gérard, J.-C., Giuranna, M., Ignatiev, N., Kaminski, J., Karatekin, O., Lefèvre, F., López-Puertas, M., López-Valverde, M., Mahieux, A., McConnell, J., Mumma, M., Neary, L., Renotte, E., Ristic, B., Robert, S., Smith, M., Trokhimovsky, S., Vander Auwera, J., Villanueva, G., Whiteway, J., Wilqueta, V., Wolff, M., and the NOMAD Team. (2015). Science objectives and performances of NOMAD, a spectrometer suite for the ExoMars TGO mission. Planet. Space Sci. 119, 233-249. https://doi.org/10.1016/j.pss.2015.10.003

Vengosh, A., de Lange, G.J., and Starinsky, A. (1998). Boron isotope and geochemical evidence for the origin of Urania and Bannock brines at the eastern Mediterranean: effect of water-rock interactions. Geochim. Cosmochim. Acta. 62, 3221-3228. https://doi.org/10.1016/S0016-7037(98)00236-1

Vestheim, H., and Kaartvedt, S. (2016). A deep sea community at the Kebrit brine pool in the Red Sea. Mar. Biodivers. 46, 59-65. https://doi.org/10.1007/s12526-015-0321-0

Vetriani, C., Crespo-Medina, M., and Antunes, A. (2014). The family *Salinisphaeraceae*. In The Prokaryotes: Gammaproteobacteria 4th Edition, E. Rosenberg, E.F. DeLong, S. Lory, E. Stackebrandt, and F. Thompson, eds. (Berlin, Germany: Springer), pp. 591-596. https://doi.org/10.1007/978-3-642-38922-1_296

Waite, J.H., Glein, C.R., Perryman, R.S., Teolis, B.D., Magee, B.A., Miller, G., Grimes, J., Perry, M.E., Miller, K.E., Bouquet, A., Lunine, J.I., Brockwell, T., and Bolton, S.J. (2017). Cassini finds molecular hydrogen in the Enceladus plume: Evidence for hydrothermal processes. Science. 356, 155-159. https://doi.org/10.1126/science.aai8703

Watson, S.W. and Waterbury, J.B. (1969). The sterile hot brines of the Red Sea. In Hot Brines and Recent Heavy Metal Deposits in the Red Sea. E.T. Degens D.A. and Ross, eds. (New York, NY, USA: Springer-Verlag), pp. 272-281. https://doi.org/10.1007/978-3-662-28603-6_27

Weber, W.W. and Gurskii, Y.N. (1982). Maltene formation in present sediments of the Kebrit brine depressions of the Red Sea. Geologiya Nefti Gaza. 1, 29-33.

Werner, J., Ferrer, M., Michel, G., Mann, A.J., Huang, S., Juarez, S., Ciordia, S., Albar, J.P., Alcaide, M., La Cono, V., Yakimov, M.M., Antunes,

A., Taborda, M., da Costa, M.S., Hai, T., Glöckner, F.O., Golyshina, O.V., Golyshin, P.N., Teeling, H., and the MAMBA Consortium. (2014). *Halorhabdus tiamatea*: Proteogenomics and glycosidase activity measurements identify the first cultivated euryarchaeon from a deep-sea anoxic brine lake as potential polysaccharide degrader. Environ. Microbiol. 16, 2525-2537. https://doi.org/10.1111/1462-2920.12393

Yao, F., and Hoteit, I. (2018). Rapid red sea deep water renewals caused by volcanic eruptions and the North Atlantic Oscillation. Science Adv. 4, 5637. https://doi.org/10.1126/sciadv.aar5637

Yakimov, M.M., Giuliano, L., Cappello, S., Denaro, R., Golyshin, P.N. (2007a). Microbial community of a hydrothermal mud vent underneath the deep-sea anoxic brine lake Urania (Eastern Mediterranean). Origins Life Evol. B 37, 177-188. https://doi.org/10.1007/s11084-006-9021-x

Yakimov, M.M., La Cono, V., Denaro, R., D'Auria, G., Decembrini, F., Timmis, K.N., Golyshin, P.N., and Giuliano, L. (2007b). Primary producing prokaryotic communities of brine, interface and seawater above the halocline of deep anoxic lake L'Atalante, Eastern Mediterranean Sea. ISME J. 1, 743-755. https://doi.org/10.1038/ismej.2007.83

Yakimov, M.M., La Cono, V., Slepak, V.Z., La Spada, G., Arcadi, E., Messina, E., Borghini, M., Monticelli, L.S., Rojo, D., Barbas, C., Golyshina, O.V., Ferrer, M., Golyshin, P.N., and Giuliano, L. (2013). Microbial life in the Lake Medee, the largest deep-sea salt-saturated formation. Sci. Rep. 3, 3554. https://doi.org/10.1038/srep03554

Yakimov, M.M., La Cono, V., Spada, G.L., Bortoluzzi, G., Messina, E., Smedile, F., Arcadi, E., Borghini, M., Ferrer, M., Schmitt-Kopplin, P., Hertkorn, N., Cray, J.A., Hallsworth, J.E., Golyshin, P.N., and Giuliano, L. (2015). Microbial community of the deep-sea brine Lake Kryos seawater-brine interface is active below the chaotropicity limit of life as revealed by recovery of mRNA. Environ. Microbiol. 17, 364-382. https://doi.org/10.1111/1462-2920.12587

Yancey, P.H. (2004). Compatible and counteracting solutes: protecting cells from the Dead Sea to the deep sea. Sci. Prog. 87(1), 1-24. https://doi.org/10.3184%2F003685004783238599

Yayanos, A.A. (1986). Evolution and ecological implications of the properties of deep-sea barophilic bacteria. Proc. Natl. Acad. Sci. USA 83, 9542-9546. https://doi.org/10.1073/pnas.83.24.9542

Zajc, J., Džeroski, S., Kocev, D., Oren, A., Sonjak, S., Tkavc, R., and Gunde-Cimerman, N. (2014). Chaophilic or chaotolerant fungi: a new category of extremophiles? Front. Microbiol. 5, 708. https://doi.org/10.3389/fmicb.2014.00708

Zhang, G., Haroon, M.F., Zhang, R., Hikmawan, T., and Stingl, U. (2016a). Draft genome sequence of *Pseudoalteromonas* sp. strain XI10 isolated from the brine-seawater interface of Erba Deep in the Red Sea. Genome Announc. 4, e00109-16. https://doi.org/10.1128/genomea.00109-16

Zhang, G., Haroon, M.F., Zhang, R., Hikmawan, T., and Stingl, U. (2016b). Draft genome sequences of two *Thiomicrospira* strains isolated from the brine-seawater interface of Kebrit Deep in the Red Sea. Genome Announc. 4, e00110-16. https://doi.org/10.1128/genomeA.00110-16

Zhang, G., Haroon, M.F., Zhang, R., Dong, X., Liu, D., Xiong, Q., Xun, W., Dong, X., and Stingl, U. (2017a). *Ponticoccus marisrubri* sp. nov., a moderately halophilic marine bacterium of the family *Rhodobacteraceae*. Int. J. Syst. Evol. Microbiol. 67, 4358-4364. https://dx.doi.org/10.1099/ijsem.0.002280

Zhang, G., Haroon, M.F., Zhang, R., Dong, X., Wang, D., Liu, Y., Xun, W., Dong, X., and Stingl, U. (2017b). *Ruegeria profundi* sp. nov. and *Ruegeria marisrubri* sp. nov., isolated from the brine-seawater interface at Erba Deep in the Red Sea. Int. J. Syst. Evol. Microbiol. 67, 4624-4631. https://dx.doi.org/10.1099/ijsem.0.002344

Zhang, G., Gu, J., Zhang, R., Rashid, M., Haroon, M.F., Xun, W., Ruan, Z., Dong, X. and Stingl, U. (2017c). *Haloprofundus marisrubri* gen. nov., sp. nov., an extremely halophilic archaeon isolated from a brine-seawater interface. Int. J. Syst. Evol. Microbiol. 67, 9-16. https://dx.doi.org/10.1099/ijsem.0.001559

Zhang, G., Dong, X., Sun, Y., Antunes, A., Hikmawan, T., Haroon, M.F., Wang, J., and Stingl, U. (2020). *Haloferax profundi* sp. nov. and *Haloferax marisrubri* sp. nov., isolated from the Discovery Deep brine-seawater interface in the Red Sea. Int. J. Syst. Evol. Microbiol. (submitted).

Zolotov, M.Y. (2007). An oceanic composition on early and today's Enceladus. Geophys. Res. Lett. 34, L23203. https://doi.org/10.1029/2007GL031234

Zolotov, M.Y. and Kargel, J.S. (2009). On the chemical composition of Europa's icy shell, ocean, and underlying rocks. In Europa, R.T. Pappalardo, W.B. McKinnon, and K.K. Khurana, eds. (Tucson, AZ, USA: University of Arizona Press), pp. 431-458.

Chapter 7

Exploring Microbial Activity in Low-pressure Environments

Petra Schwendner[1]* and Andrew C. Schuerger[1]

[1]University of Florida, 505 Odyssey Way, Space Life Sciences Lab, Exploration Park, Merritt Island, FL 32953, USA

*petra.schwendner@ufl.edu

DOI: https://doi.org/10.21775/9781912530304.07

Abstract

The importance of hypopiezophilic and hypopiezotolerant microorganisms (i.e., life that grows at low atmospheric pressures; see section 2) in the field of astrobiology cannot be overstated. The ability to reproduce and thrive at Martian atmospheric pressure (0.2 to 1.2 kPa) is of high importance to both modeling the forward contamination of its planetary surface and predicting the habitability of Mars. On Earth, microbial growth at low pressure also has implications for the dissemination of microorganisms within clouds or the bulk atmosphere. Yet our ability to understand the effect of low pressure on microbial metabolism, growth, cellular structure and integrity, and adaptation is still limited. We present current knowledge on hypopiezophilic and hypopiezotolerant microorganisms, methods for isolation and cultivation, justify why there should be more focus for future research, and discuss their importance for astrobiology.

1. Introduction

Earth's global average atmospheric pressure at sea level is 101.3 kPa (0.1 MPa) and can reach as high as 120 MPa at the bottom of the Mariana Trench 11 km below sea level (Picard and Daniel, 2013). On Earth, low-pressure environments below 101.3 kPa, are only present at high alpine sites in mountainous regions; the highest peak, Mt. Everest in Nepal, has a peak-height pressure of 33.0 kPa. With increasing altitude, the pressure drops to 1 kPa in the middle stratosphere and <1 Pa above 80 km. Therefore, life on Earth typically encounters pressures that range from 0.1 to 120 MPa (Oger and Jebbar, 2010; Yayanos, 1995). The pressure in interplanetary space is 1.32×10^{-8} kPa.

On Mars, the atmospheric composition and pressure differ dramatically compared to Earth. The average atmospheric pressure on the surface is approx. 0.7 kPa, which is equivalent to the atmospheric pressure at approx. 27 km altitude on Earth. The pressure on Mars varies between 0.1 kPa at the summit of Olympus Mons and 1.2 kPa in Hellas Basin (e.g., Barth et al., 1992; Rummel et al., 2014; Taylor et al., 2010). The gas-phase pressure (e.g., within interstitial spaces) increases very slowly in the lithosphere on Mars reaching 2.5 kPa at 13.8 km. In contrast, the lithographic overburden pressure can reach 2.5 kPa at only 19.5 cm below the surface in a confined niche (e.g., ice or salt inclusions) covered by rock or regolith (Schuerger et al., 2013).

Furthermore, the atmosphere on Mars consists mainly of CO_2 (approx. 96%) with low partial pressures of nitrogen (2%), argon (1.7%), and O_2 (0.13%; Mahaffy et al., 2013). In contrast, Earth's atmosphere is composed of nitrogen (78%), oxygen (21%), argon (1%), and trace amounts of CO_2 and other gases.

Mars, as a candidate for finding life elsewhere in the Solar System, has been of interest for space fairing nations for several decades, and remains a central goal in astrobiology. With data about the planet's geology, atmosphere, etc. returned from different Mars missions since the 1960s, and the in-depth knowledge about life in extreme terrestrial environments, Martian habitability has become a key focus (Cockell et al., 2016). To consider any environment or extraterrestrial body as habitable, a plethora of different requirements need to be met in order to allow life to survive and eventually thrive. Potential energy sources, ambient geochemical composition, the availability of a life-sustaining solvent, protection from biocidal factors, and the availability of carbon sources all need to be taken into account. In fact, at least 17 biocidal factors have been identified that potential life on Mars would encounter (Beaty et al., 2006; Rummel et al., 2014; Schuerger et al., 2012). Examples of global environmental hazards on Mars include a CO_2-dominated anoxic atmosphere, UV solar irradiation, hypobaria (0.1 – 1.2 kPa), low global temperatures (-61 °C), and extremely low water activity (a_w) of the surface regoliths. Other more episodic or randomly distributed factors include high salinity, low pH in certain soils, unknown or poorly described redox potentials (Eh) in hydrated brines, oxidizing compounds such as perchlorates prevalent in the regolith, the presence of heavy metals, UV-induced volatile oxidants (O^{2-}, O^-, H_2O_2, NO_x, O_3), solar particle events, and galactic cosmic rays.

Consequently, it has to be evaluated whether the combination of these factors provide environmental conditions suitable to build microbial biomass. Various environmental stressors have been extensively studied on different forms of life since the 1930s. These include simulation experiments (reviewed by Olsson-Francis and Cockell, 2010; Rothschild and Mancinelli, 2001), flight missions in Earth orbit using rockets and balloons (DasSarma et al., 2017; Pulschen et al., 2018), outside Earth's orbit during NASA's Apollo program (reviewed by Horneck et al., 2010) and exposure experiments on the EXPOSE platform mounted outside the International Space Station (Rabbow et al., 2009; 2012; 2017). The stressors tested include UV radiation, gamma-rays, galactic cosmic rays, vacuum, high and low temperatures, and freeze-thaw cycling and combinations thereof. In addition, exposure to oxidizing chemicals, vacuum and subsequent survivability has been explored in a myriad of experiments (e.g., Horneck 1981; Horneck et al., 1994; Paulino-Lima et al., 2010; 2011; Sancho et al., 2007). The model organisms were mostly bacteria (e.g., *Bacillus subtilis*, *Deinoccoccus radiodurans*), but also archaea, viruses, fungi, lichens, and tardigrades have been tested (see reviews by Mileikowsky et al., 2000; Horneck et al., 2010; Nicholson et al., 2000). While numerous publications exist on the effects of high pressures on microbial biology (see section 8), surprisingly, little information is published on the effects of hypobaria on microbial metabolism, growth, cell integrity, and adaptation.

Herein, we review the current status of knowledge on microorganisms capable of metabolism and cellular replication at pressures below the average atmospheric pressure on Earth of 101.3 kPa down to 0.7 kPa. The literature cited serves as a brief introduction to concepts of hypobaric microbiology. Furthermore, we propose a redefinition of the word hypobarophiles (coined by Schuerger et al., 2013). Next we give a short overview about microbial survival and growth at low pressures, and describe the results of the molecular studies done on the topic. We also briefly present the effects of high pressure on microorganisms focusing on the major adaptations of piezophiles to cope at extreme oceanic depths. Lastly, we discuss the implications of low-pressure microbiology for astrobiology (e.g., Des Marais et al., 2008). This review hopes to convince readers that studies on microbial activity in low-pressure environments are of high importance to gain insights into basic biological mechanisms, the factors involved in adaption to low-pressure environments, the likelihood of microbial growth in the upper atmospheric boundaries of Earth's biosphere (0.7 kPa is equivalent to 34 km in the middle stratosphere of Earth), and the potential to assess habitability of extraterrestrial planetary bodies like Mars. In addition, the information may allow us to improve the development of planetary protection guidelines for robotic and manned space missions to Mars.

2. Definitions: Piezophilic, hypopiezotolerant, and hypopiezophilic microorganisms

Low-pressure microbiology is a nascent field in extreme environments because there are no low pressure ecological niches on Earth, except for the possibility of mid-troposphere cloud microbiology. But even here, it is debatable whether the mid-troposphere (or stratosphere) can function as a unique ecological niche, with long-term microbial survival, growth, and adaptations, or merely as a conduit between ecosystems (DasSarma and DasSarma, 2018; Diehl, 2013; Griffin et al., 2018; Smith, 2013). Perhaps the best chance of finding a microbial community truly adapted to low-pressure niches will be the discovery of extraterrestrial life in the shallow subsurface of Mars (Schuerger et al., 2013) at a pressure range of 0.1 to 1.2 kPa (Rummel et al., 2014; Taylor et al., 2010). Here we would like to reevaluate the term *hypobarophile* for bacteria growing under low-pressure conditions between 0.7 and 1.2 kPa, and propose new terms that are consistent with microbial species able to reproduce under a wide spectrum of pressures on Earth from sea level (0.1 MPa) to the deep hadal regions in oceanic trenches (<120 MPa; see Yayanos, 1995).

The term *barophile* was coined by ZoBell and Johnson (1949), and has been used historically to refer to microorganisms growing above 10 MPa (Eisenmenger and Reyes-De-Corcuera, 2009; Picard and Daniel, 2013). However, Yayanos (1995) proposed the term *piezophile* (Greek: *piezo* = to press, and *philo* = love) instead to be consistent with the use of the prefix *piezo-* in physics and chemistry. Thus, piezophiles refer to high-pressure microbial species that optimally grow between 10 and < 60 MPa found in the deep lithosphere or oceanic benthic regions which are defined as the piezosphere (Jannasch and Taylor, 1984; Fang et al., 2010; Oger and Jebbar, 2010). The *piezosphere* which starts 1 km below sea level, excludes the upper 1 km because it is considered too well mixed (Fang et al., 2010; Oger and Jebbar, 2010). Yayanos (1995) further proposed adding the term *hyper-* to piezophile to refer to microbial species that optimally grow at pressures between 60 and 120 MPa.

To be consistent with the use of the term piezophile for a microorganism adapted to high-pressure environments, we will use the term *hypopiezophile* to refer to a microorganism that grows optimally under hypobaric conditions < 2.5 kPa (0.025 MPa). To date, no true hypopiezophile has been described. However, currently 62 bacterial isolates have been identified that can tolerate low-pressure conditions and grow at pressures down to 0.7 kPa (section 5, Nicholson et al., 2013a; Schuerger et al., 2013; Schuerger and Nicholson, 2016) and are therefore considered hypopiezotolerants. We propose to withdraw the term hypobarophile as being inaccurate because the bacteria described do not optimally grow at low pressures, but instead tolerate pressures down to 0.7 kPa in which growth rates are significantly retarded compared to normal growth at 101.3 kPa. Thus, the term *hypopiezotolerant* is more appropriate and consistent with the terms proposed by Yayanos (1995).

3. Experimental methods - how to isolate hypopiezotolerant microorganisms

3.1. Desiccators

To grow microorganisms at low pressures the construction of hypobaric systems holding pressure as low as 0.7 kPa at 0 °C are required (Figure 1). A simple system (Schuerger et al., 2013) was developed to simulate three conditions (pressure, P; temperature, T; and atmosphere, A) found on the surface of Mars, henceforth called *low-PTA* conditions, that are defined as: 0.7 kPa (close to the average surface pressure on Mars; Rummel et al., 2014), 0 °C (to maintain stable liquid water near its triple point; Haberle et al., 2001), and a CO_2-enriched anoxic atmosphere (96% CO_2 on Mars; Mahaffy et al., 2013). The systems can also accommodate Earth-normal pO_2/pN_2 atmospheres of 21% and 78%, respectively, by allowing room air to diffuse into the desiccators instead of flushing the chambers with CO_2 or including anaerobic pouches in the system.

Three approaches have been used to isolate and identify hypopiezotolerant bacteria from culture collections and environmental samples (see Schuerger et al., 2013; Schuerger and Nicholson; 2016, respectively). First,

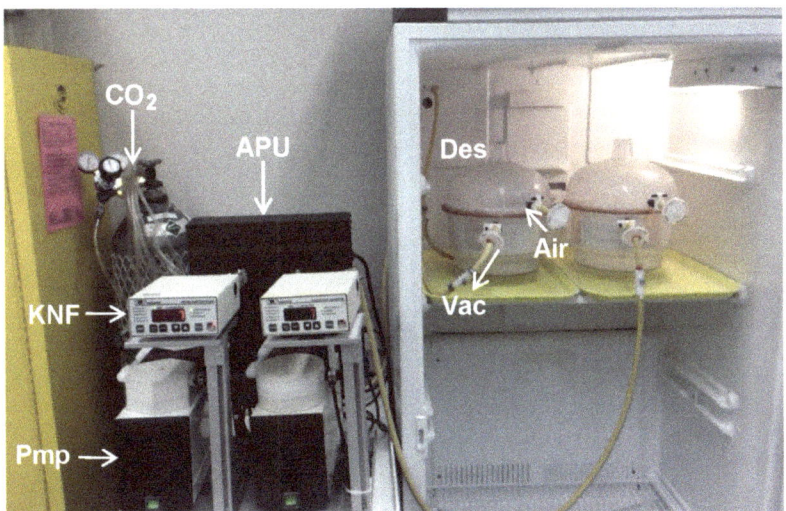

Figure 1. Two vacuum control systems (KNF) and pumps (Pmp, model PU-842, KNF Neuberger, Trenton, NJ, USA) were connected to individual 4-L desiccators (Des, model 08-642-7, Fisher Scientific, Pittsburgh, PA, USA) via vacuum (Vac) lines (described by Schuerger et al., 2013). The KNF controllers should be connected to a battery back-up (APU) unit to prevent losing the KNF controller program if any power glitches occur. The vent lines (Air) are used to repressurize the desiccators. Ultra-high purity carbon dioxide (CO_2) gas is used to flush room air out of the desiccators (2-3 minutes) prior to sealing the desiccators and connecting the vacuum lines.

the purified isolates are either streaked (Figure 2) or spotted on the media of choice (Nicholson et al., 2013a), placed in the hypobaric desiccators and either flushed with CO_2 or air, prior to closing the vent lines. When anoxic conditions are required, four anaerobic pouches are placed in each desiccator to continuously maintain low pO_2 < 0.1% (Van Horn et al., 1997). Pumping directly down to 0.7 kPa, would lead to cracks, bubbles, or distortions in the agar, and boiling of liquid media. Thus, a slow pump-down protocol was developed that sequentially reduces the internal atmospheric pressures down to 10, 5, 2.5, and finally 0.7 kPa in 15 minute increments. This procedure allows for adequate time to outgas internally trapped gases in agars or liquids, and permits a slow cool down to 0 °C of the media. When stabilized at 0 °C and 0.7 kPa, agar surfaces and liquid media can be stable for 28-35 days. When using semi-solid media, it is important to use double-thick layers of agar (≈ 30 mL) in deep petri dishes.

Up to 100 strains (Nicholson et al., 2013a; Schuerger et al., 2013) could be evaluated per petri dish at low-PTA conditions using the streak or spot technique. Each 4-L desiccator can hold up to 12-15 double-deep petri dishes which are stored at 4 °C until all plates are ready for insertion into the low-pressure desiccators. This process prevents strains from initiating growth at room temperature while other plates are prepared. It is important to include both positive hypopiezotolerant controls (e.g., *Carnobacterium* sp. in position #10 in Figure 2A and 2B; Nicholson et al., 2013a) and

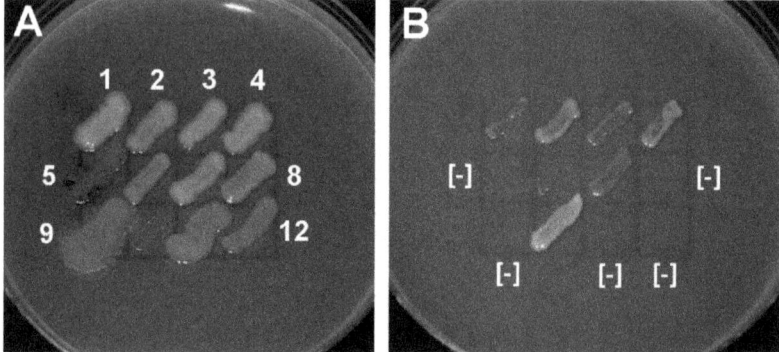

Figure 2. Example of the streak technique to check for growth of hypopiezotolerants at low-PTA conditions. Displayed is the growth of eight *Serratia* spp. and appropriate controls under Earth-normal conditions at 101.3 kPa and 30 °C (**A**) and low-PTA conditions used to simulate the Martian surface at 0.7 kPa (**B**). Positions 1 to 8 include (from left to right, and top to bottom): *S. ficaria* DSM4569, *S. fonticola* DSM4576, *S. grimesii* ATCC14460, *S. liquefaciens* ATCC27592, *S. marcescens* ATCC13880, *S. plymuthica* DSM4540, *S. quinivorans* DSM4597, and *S. rubidaea* ATCC27593. The bottom row was composed of negative [-] and positive controls that included (from #9 to #12): *Bacillus subtilis* 168 (negative), *Carnobacterium* sp. WN1359 (positive), *Escherichia coli* ATCC35218 (negative), and *Sporosarcina aquamarina* SAFN-008 (negative) (adapted from Schuerger and Nicholson, 2016).

negative controls (e.g., *Bacillus subtilis* 168 in position #9; Schuerger et al., 2013) in the assays.

In the second method, a soil-dilution protocol was developed to screen environmental samples such as arctic and alpine soils (Figure 3; Schuerger and Nicholson, 2016). The soil-dilution protocol combines 1.0 g of thawed soil and 25 mL of a pre-autoclaved 0.1% agar solution in a 125 mL Erlenmeyer flask. The soil suspension is vigorously agitated using a magnetic stirrer for 10-20 minutes to form a very dilute but semisolid agar matrix that will keep all soil particles in suspension for several minutes after stopping the agitation. Two hundred µL of the agar/soil suspension is then pipetted and spread onto the agar surface using sterile pre-cut 1000 µL pipette tips with the bottom 2-3 mm of the tips cut off prior to sterilization to avoid clogging of the tips by soil particles. The plates are then transferred into the desiccators and incubated as described above.

In the third method, the hypobaric desiccator set-up is used for growing hypopiezotolerant microbes in liquid medium down to 0.7 kPa. This protocol requires that the depth of the liquid medium be kept to a minimum

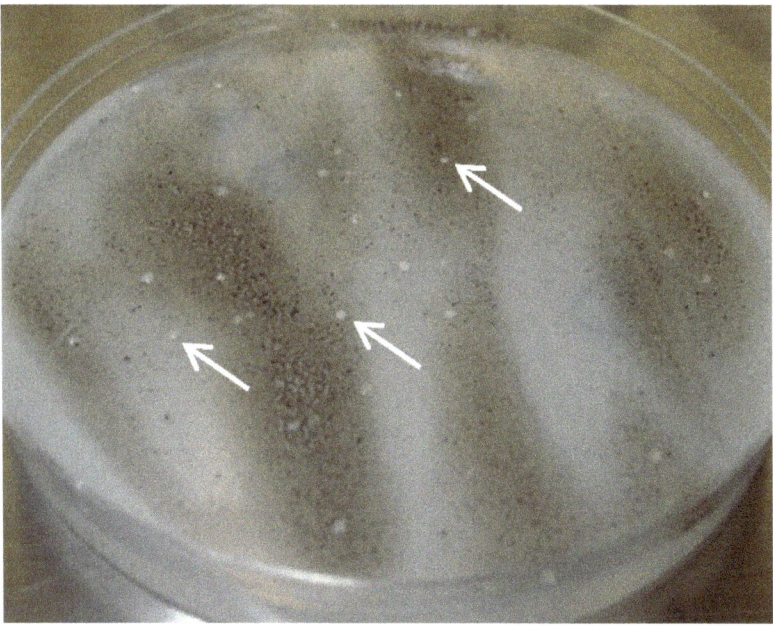

Figure 3. Isolation of hypopiezotolerant bacteria from environmental soil samples incubated under low-PTA conditions for 4 weeks. Hypopiezotolerant bacterial colonies (arrows) were recovered from an arctic soil from Colour Lake, Axel Heiberg, Canada (Schuerger and Nicholson, 2016).

(≤ 7 cm) to maintain pressure at the bottom of the culture tube consistent with the atmospheric pressure in the desiccator (e.g., Fajardo-Cavazos et al., 2018), and that the tubes are not sealed in order to permit the equalization of internal and external pressures in the desiccator. Another liquid medium protocol using 96-well plates was recently developed to examine the metabolic fingerprint of *Serratia liquefaciens* utilizing 95 carbon sources in Biolog GN2 microarray plates (Schwendner and Schuerger, 2018).

3.2. Hydrologic and thermodynamic issues related to growth of microbes at low pressures

The classic paper by Haberle et al. (2001) was the first to describe the triple point of water in the context of the Mars surface environment. It identified that the pressure (0.61 to 1.24 kPa) and temperature (0.1 to 10°C) ranges for stable liquid water on Mars severely constrain the habitability of the terrain to thin films of water plausibly formed during short-term melting of ice. Based on these findings we like to emphasize the limits of agar or liquid medium habitability in the hypobaric protocols described above.

As the pressure is lowered from 2.5 kPa to 0.7 kPa, temperature must also be concomitantly lowered from 30 to 0 °C (compare Schuerger and Nicholson, 2006 to Schuerger et al., 2013, respectively). Attempts to incubate double-thick agar plates at 30 °C and pressures below 2.5 kPa resulted in the agar surfaces splitting, pitting, and desiccating over only 48-72 hours (Schuerger and Nicholson, 2006). Within the stable liquid water "zone" of pressure and temperature on Mars (see above), the agar was stable for at least 35 days. In essence, incubations at low pressures near the surface "range" on Mars (0.1 to 1.2 kPa) are controlled by the thermodynamics of stable liquid water close to the triple point. At pressures lower than 0.5 kPa, liquid water cannot be maintained unless liquid brines are used with concurrent depression of the freezing point of water (Heinz et al., 2018), which in turn changes osmotic pressure and water activity that may also inhibit microbial activity for some microbes. Although extreme halophiles can tolerate extremes of osmotic pressure and low water activity (DasSarma and DasSarma, 2015; Fox-Powell et al., 2016).

Haberle et al. (2001) also cautioned that even though the pressure and temperature ranges in an ecological setting on Mars might fall within the range for stable liquid water, evaporation of liquid water will still occur. The best way to envision this process is to watch how the low-pressure control systems work near 0.7 kPa. After the samples are placed in the desiccators (or in Mars simulations chamber, see below), it generally takes 60-90 minutes to stabilize the pressure and temperature to 0.7 kPa and 0 °C, respectively. But over time, water from the samples evaporates and increases the pressure in the chamber/desiccator systems, which causes the hypobaric control systems to kick-in to lower the pressure. During this process, the water vapor is removed and consequently, the medium slowly desiccates. Thus, these hypobaric assays are a thermodynamic struggle in which adequate microbial growth must occur before the water reserves of the agar or liquid media are exhausted.

3.3. Other Mars simulation chambers

Many designs have been published for hypobaric Mars simulation chambers, but few have been used to attempt to grow bacteria, archaea, or fungi under low-PTA conditions. Most of the more complex and instrumented Mars chambers have been used to study the survival of microorganisms, biosignature molecules, or geochemical processes under diverse conditions found on Mars (e.g., dos Santos et al., 2016; Gomez et al., 2010; Schuerger et al., 2008; Stan-Lotter et al., 2003).

Table 1 lists 14 Mars simulation chambers, constructed of stainless steel tanks placed in either a vertical or horizontal orientation with numerous ports, electrical connector feeds, cooling systems, and UV illumination sources, that have been evaluated for the control of low pressure, gas composition, UV irradiation, and atmospheric composition. Most chambers can be adapted for microbial metabolism, growth, and adaptation experiments, but in general, the more complex Mars chambers are primarily relevant when UV exposures are required in the simulations. However, the same thermodynamic issues described above for agar and liquid media in the 4-L desiccators will also hold for maintaining hydrated growth media in the more complex chambers.

For studying active metabolism and growth with hydrated media, the systems need to be compatible with water vapor, otherwise the listed chambers only work effectively for microbial survival, desiccation, and UV irradiation experiments. Only one chamber was specifically designed to handle liquid medium under simulated Martian conditions (i.e., the Planetary Environmental Liquid Simulator (PELS) chamber; Martin and Cockell, 2015). The PELS chamber can accommodate six independently controlled liquid samples in isolated vessels at low pressures down to 0.1 kPa between −50 and +70 °C. To collect aliquots from *in situ* experiments without venting the PELS chamber, separate sampling lock-out chambers for each reaction vessel were installed. Most of the chambers described in Table 1 appear to have extra ports that could accommodate internal sampling lock-out systems.

The Mars chambers in Table 1 were selected in part because they are likely to be available for new research into the survival, metabolism, growth, and adaptation of hypopiezotolerant microorganisms relevant to near-term robotic and human missions. Thus, the details provided here are offered as a brief primer for new Mars astrobiology research.

Table 1. List of Mars simulation chambers described in the literature since 2000. These chambers were selected here due to their plausible adaptability to conduct microbial survival and growth experiments at pressures < 10 kPa, and the inclusion of drawings, photographs, and specifications of the chambers in the cited papers.

Chamber name	Lowest pressure reported (kPa)	Temperature range (°C)	UV bands	Location	References
Andromeda Chamber	0.07	−25 to 24	none	Univ. of Arkansas, Fayetteville, AR, USA	Sears et al., 2002; 2005
LISA and mini-LISA Mars Chambers	5×10^{-3}	−123 to 57	UVC, UVB, UVA	Astronomical Observatory of Padua, Italy	Galletta et al., 2011
Mars Chamber Simulator	0.6	−80 to 20	UVC, UVB, UVA	Open University, Milton Keynes, UK	dos Santos et al., 2016
Mars Environmental Simulation Chamber	0.01	−140 to 25	UVB, UVA	Univ. of Aarhus, Aarhus, Denmark	Jensen et al., 2008
Mars Simulation Chamber (MSC)	0.01	−85 to 70	UVC, UVB, UVA	Univ. of Florida, Kennedy Space Center, FL	Schuerger et al., 2008
Mars Simulation Chamber	1.5	−123 to 25	UVC, UVB, UVA	Univ. of Maryland, MD, USA	Ertem et al., 2017
Mars Simulation Facility	0.7	−0 to 20	UVC, UVB, UVA	German Aerospace Center (DLR), Berlin, Germany	De la Torre Noetzel et al., 2018
Mars Simulation Vacuum Chamber	5×10^{-4}	−196 to 24	UVC, UVB, UVA	Leiden Univ., Leiden, The Netherlands	ten Kate et al., 2003
MaSimKa Chamber	1×10^{-6}	−20 to 85	UVC, UVB, UVA	German Aerospace Center (DLR), Cologne, Germany	Rabbow et al., 2016
Pegasus Planetary Simulation Chamber	5	24 (ambient only)	none	Univ. of Arkansas, Fayetteville, AR, USA	Kral et al., 2011
PELS Chamber	0.1	−50 to 70	UVB, UVA	Univ. of Edinburgh, Edinburgh, UK	Martin and Cockell, 2015
Planetary Atmosphere Simulation Chamber	0.7	−120 to 24	UVC	Centro de Astrobiologia Centre, Madrid, Spain	Gomez et al., 2010
Shot Mars Chamber	5	−80 to 26	UVB, UVA	Techshot, Inc., Greenville, IN, USA	Thomas et al., 2008
SRI Mars Chamber	1×10^{-3}	−60 to 24	none	Austrian Academy of Science, Graz, Austria	Stan-Lotter et al., 2003

4. Microbial survival experiments in low pressure or space vacuum

The majority of experiments under low-pressure conditions have reported on the survivability of microorganisms rather than actual growth. Test organisms for survivability studies initially included mainly spores from bacteria and fungi, but later microbial vegetative cells were also investigated. Portner et al. (1961) was among the first to demonstrate the survival of microorganisms after exposure to ultra-high vacuum (down to 2.6×10^{-11} kPa). Additional Earth-based studies on bacterial spores of *Bacillus stearothermophilus*, *B. megaterium*, *B. subtilis*, *Clostridium sporogenes*, and *Aspergillus niger* as well as vegetative cells of *Staphylococcus epidermidis* and *Micrococcus sp.* exposed to ultra-high vacuum (10^{-3}-10^{-10} kPa) showed evidence of survival, but also revealed a temperature-effect (Brueschke et al., 1961; Hagen et al., 1971; Morelli et al., 1962; Silverman et al., 1964).

After Earth-based tests showed that microorganisms can survive ultra-high vacuum, space studies were conducted (see reviews by Horneck et al.,

2010; Nicholson et al., 2000). For example, following six years in space and shielded from solar UV irradiation, vacuum-only effects on monolayers of *Bacillus subtilis* showed that only 1-2% of spores survived; but multilayered aggregates of spores coated in either glucose or buffering salts revealed increased survival to 80% (Horneck et al., 1994). Similar effects on microbial survival for monolayers versus multilayers have been reported for other bacterial species exposed to low pressures (Mancinelli and Klovstad, 2000; Osman et al., 2008; Schuerger et al., 2005). In a series of experiments investigating the effects of several simulated Martian conditions on bacterial survival, Schuerger et al. (2003) observed ~30% inactivation of *B. subtilis* spores due to low pressure alone (0.7 kPa). Moeller et al. (2012) exposed *B. subtilis* and eight mutants to a simulated Martian atmosphere at 0.7 kPa and reported up to a 1.5 log reduction for certain mutants while the majority of tested *Bacillus* strains exhibited only minor reductions in spore survival. In all of these examples, spore survival rates under low pressures were enhanced if the cells were present as multi-layered aggregates.

A few studies have tested the survival of archaeal species to low-pressure environments, in combination with other simulated Martian conditions (e.g., Kral et al., 2011; Mickol and Kral, 2017; Mancinelli, 2015; Morozova et al., 2007). For example, *Methanothermobacter wolfeii*, *Methanosarcina barkeri*, and *Methanobacterium formicicum* survived up to 120 days of desiccation and *Methanococcus maripaludis* survived 60 days at 0.6 kPa.

Survival studies are of high importance and great value with regard to planetary protection and the prediction of habitability (see section 9). However, they do not give an indication whether the microorganisms have the ability to grow in low-pressure environments.

5. Current status of hypopiezotolerants

5.1. Microorganisms able to grow in low-pressure environments (101.3 < 2.5 kPa)
Low-pressure environments are defined as environments with pressures below the Earth-normal sea level pressure of 101.3 kPa. Initially, a low-pressure threshold of 2.5 kPa for microbial growth was proposed by Schuerger and Nicholson (2006) based on early results with the growth of seven *Bacillus* spp. Later experiments established the growth of hypopiezotolerant bacteria down to 0.7 kPa in which *Bacillus* spp. were in general unable to grow below 2.5 kPa (section 5.2; Nicholson et al., 2013a; Schuerger et al., 2013; Schuerger and Nicholson, 2016).

The current observed low-pressure threshold for microbial growth is 0.7 kPa, though this may not be the actual limit. Table 2 summarizes microbial species that have been observed to grow under hypobaric conditions (i.e., < 10 kPa) in various growth media and hypobaric chambers. The data were selected based on our interpretation that the presented data showed unequivocal evidence of microbial growth at pressures < 10 kPa. However, there were a few notable exceptions to the data in Table 2 that are worth mentioning.

First, Hawrylewicz et al. (1968) reported growth of *Staphylococcus aureus* in both O_2 and CO_2 atmospheres between 1 and 4 kPa, but failed to give details on the temperatures used during incubation. Furthermore, sealed tubes were used that were initially pressurized to 1, 2.5, or 4 kPa, but no mechanism was mentioned to either verify or adjust the pressure during the course of the experiments. This shortfall makes it difficult to assess whether growth occurred at the pressures indicated or at higher pressures resulting from evaporated water from the growth media.

Second, Pokorny et al. (2005) demonstrated growth of *Escherichia coli* and *Bacillus subtilis* at 33 kPa, much higher than the upper limit of hypopiezotolerant bacteria being considered here. These results were eventually superseded by several other studies in Table 2 growing both species down to 2.5 kPa.

Third, Pavlov et al. (2010) reported growth of a *Vibrio* sp. at 0.001 to 0.01 kPa. However, this study is problematic for three reasons: (1) starting cell densities (10^5 to 10^6 cells per assay) were always higher than the measured cell densities at the end of the Mars simulations ($\leq 2.63 \times 10^4$ cells per assay). (2) There was no data that could be used to verify that the pressures being measured were in fact present in sample. Often pressures measured by a gauge external to a chamber can be several kPa lower than the actual pressures present inside the hypobaric chambers due to local effects of water evaporation or ice sublimation. And (3), the purported pressure range used was significantly below the triple point of water on Mars (section 3.2), and thus, water could only persist as either ice or vapor in the assays. Based on these reasons we conclude that Pavlov et al. (2010) actually observed cell survival and not growth at the low pressures reported.

Fourth, several papers describe problems with maintaining low pressures due to evaporation of water from hydrated media at temperatures greater than the Mars "water stability window" described in section 3.2 (Haberle et al., 2001). For example, Bauermeister et al. (2014) used the MaSimKa chamber for experiments on the growth of *Acidithiobacillus ferrooxidans* at low pressures. The microbial incubations were planned for 20 °C and 0.7 kPa, but only a pressure of 1.5 kPa could be maintained at that temperature due to the evaporation of water from the growth matrix. Similar problems with desiccation of media at low pressures and elevated temperatures have been described by Thomas et al. (2008) and Schuerger and Nicholson (2006).

And lastly, we want to point out that in the study reported by Kral et al. (2011), growth was indirectly measured for *Methanothermobacter wolfeii*, *Methanosarcina barkeri* and *Methanobacterium formicicum* by monitoring methane evolution. Methane production rates which were linked to growth without cell counts being performed, were reported down to 5.0 kPa, but not tested at lower pressures. Methane production in archaea does not always correlate to actual growth and cell proliferation.

Additionally, a number of studies have explored the effects of both Low-Earth Orbit space and Martian conditions on the survival of a diversity of lichens (e.g., Brandt et al., 2015; de la Torre Noetzel et al., 2018; Meeßen et al., 2015; and citations within). However, no studies to date have measured growth of lichens directly under hypobaric conditions similar to the Martian surface. A few studies have tried to correlate photosynthetic activity to growth through monitoring the chlorophyll fluorescence of PS I and PS II systems, and have revealed varying results (increased fluorescence in de Vera et al., 2010; stable fluorescence in Sanchez et al., 2014; or decreased fluorescence over time in Sanchez et al., 2014) when exposed to 0.7 to 1.0 kPa and UVC fluence rates found on Mars. In general, lichen viability, photosynthesis, and cellular structures all appear to decrease over time when exposed to simulated conditions found on the surface of Mars (see Sanchez et al., 2014; Meeßen et al., 2014, de Vera et al., 2014). It remains to be shown if lichens can actually acquire hydrated nutrients, carry out metabolism, and increase cell numbers under simulated Martian conditions at 0.7 kPa.

As indicated above, other important factors to consider are the impacts of temperature on growth rates at low pressures, atmospheric composition in the assays, and whether spores or vegetative cells are investigated. Schuerger and Nicholson (2006) demonstrated that although vegetative cells of *B. pumilus* (SAFR-03, FO-36B), *B. subtilis* (HA-101, 42HS-1), *B. nealsonii*, and *B. licheniformis* were able to grow slowly at 2.5 kPa at 30 °C in O_2/N_2 or CO_2 atmosphere, at 20 °C growth was inhibited indicating a temperature effect. *B. megaterium* was not able to grow at 2.5 kPa. In addition, their endospores were, in general, only able to germinate and subsequently grow at atmospheric pressures higher ≥ 5.0 kPa in Earth-normal O_2/N_2 atmosphere at 30 °C. Interestingly, endospores of one strain, *B. subtilis* HA 101, were able to germinate and grow at 3.5 kPa, 30 °C, and Earth-normal pO_2/pN_2 conditions, but not in CO_2-enriched anoxic atmospheres at the same pressures and temperatures. In contrast, endospores of *B. nealsonii* and *B licheniformis* were able to germinate and grow at 3.5 kPa, 30 °C, and CO_2 atmospheres, but not in pO_2/pN_2 atmospheres (Schuerger and Nicholson, 2006). Thus, atmospheric composition during the low-pressure growth assays had a markedly different effect on the germination of endospores for seven *Bacillus* spp.

Table 2. Examples of low-pressure microbial growth at which clear signs of growth were reported. Examples given here were selected based on the testing of microbial growth at or below 10 kPa. Lowest pressures and temperatures are indicated.

Species	Pressure (kPa)	Temperature (°C)	Gas Mix
Chorella ellipsoidea[a]	10	−80 night +29 day	CO_2
Chroococcidiopsis sp[a]	10	−80 night +29 day	CO_2
Plectonema boryanum[a,b]	10	−80 night +29 day	CO_2
Anabaena sp.[a,b]	10-5	−80 night +29 day	CO_2
Synechocystis sp.[b,c,d,e]	10-5	32	5% CO_2
Bacillus subtilis[f,g,h]	5	27-37	O_2
Methanothermobacter wolfeii[i]	5	35	50:50 $H_2:CO_2$
Methanosarcina barkeri[i]	5	35	50:50 $H_2:CO_2$
Methanobacterium formicicum[i]	5	35	50:50 $H_2:CO_2$
Pseudomonas aeruginosa[j]	5	30	O_2 or CO_2
Synechococcus sp.[b]	5		5% CO_2
Bacillus (6 spp.)[k]	2.5	30	O_2 or CO_2
Enterococcus faecalis[j]	2.5	30	O_2 or CO_2
Escherichia coli[l]	2.5	20	CO_2
Staphylococcus aureus[j]	2.5	30	O_2 or CO_2
Bacillus sp.[m]	0.7	0	CO_2
Carnobacterium (10 spp.)[m,n]	0.7	0	CO_2
Clostridium sp.[m]	0.7	0	CO_2
Cryobacterium sp.[m]	0.7	0	CO_2
Exiguobacterium sibiricum[m]	0.7	0	CO_2
Paenibacillus (3 spp.)[m]	0.7	0	CO_2
Rhodococcus qingshengii[m]	0.7	0	CO_2
Serratia (6 spp.)[m]	0.7	0	CO_2
Serratia liquefaciens[j,m]	0.7	0	CO_2
Streptomyces (2 spp.)[m]	0.7	0	CO_2
Trichococcus (3 spp.)[m]	0.7	0	CO_2

[a]Thomas et al., 2008; [b]Thomas et al., 2005; [c]Kanervo et al., 2005; [d]Pokorny et al., 2005; [e]Sakon and Burnap, 2006; [f]Waters et al., 2014; [g]Nicholson et al., 2010; [h]Fajardo-Cavazos et al., 2012; [i]Kral et al., 2011; [j]Schuerger et al., 2013; [k]Schuerger and Nicholson, 2006; [l]Berry et al., 2010; [m]Schuerger and Nicholson, 2016; [n]Nicholson et al., 2013a

5.2. Bacteria able to grow at low-PTA conditions

A set of experiments investigated microbial growth under low-PTA conditions, i.e., 0.7 kPa, 0 °C, and a CO_2 atmosphere (Figure 4, Table 2; Schuerger et al., 2013; Nicholson et al., 2013a; Schuerger and Nicholson, 2016). In order to reduce the pressure to 0.7 kPa and maintain a stable hydrated growth medium, the temperature had to be concomitantly lowered to 0 °C (triple point of water; see section 3) and therefore the cells were exposed to two or more stresses simultaneously when determining the low-pressure limit for growth. Another factor that has a negative effect on growth in combination with low pressure and temperature is the atmospheric environment (e.g. oxygenated versus CO_2-enriched environments) in which the experiments are conducted. Results indicated that the majority of the tested microbial species remained in an inactive or dormant state at 0.7 kPa, but were not killed by the low-PTA conditions. When returned to optimal growth conditions at 30 °C, 101.3 kPa and Earth-normal atmosphere, microbial growth was resumed and colony development observed. The underlying mechanism for the inhibition at 0.7 kPa is currently not known. Thus, low-PTA conditions should be considered as non-lethal for the majority of microorganisms tested.

The first hypopiezotolerant microorganisms being described to grow at 0.7 kPa were isolates of the genera *Carnobacterium* (Nicholson et al., 2013a) and *Serratia* (Figure 2; Schuerger et al., 2013). When exposing nine type-strains of *Carnobacterium* (*C. alterfunditium*, *C. divergens*, *C. funditum*, *C. gallinarum*, *C. inhibens*, *C. maltaromaticum*, *C. mobile*, *C. pleistocenium*,

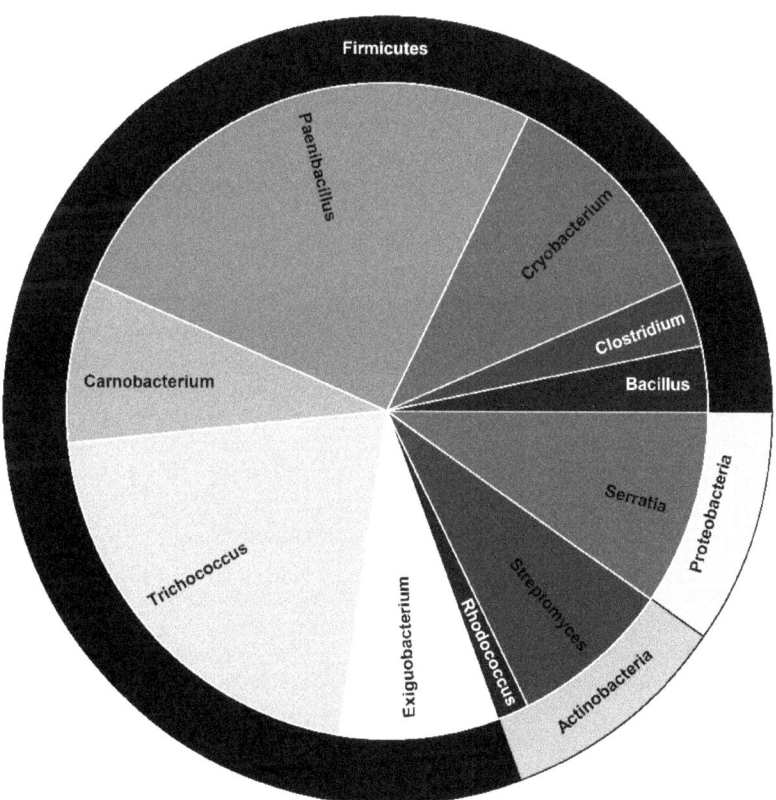

Figure 4. Microbial diversity of hypopiezotolerant bacteria on phylum and genus level (adapted from Schuerger and Nicholson, 2016).

and *C. viridans*) and eight type-strains of *Serratia* (*S. ficaria, S. fonticola, S. grimesii, S. liquefaciens, S. marcescens, S. plymuthica, S. quinivorans,* and *S. rubidaea*) to low-PTA conditions, all tested type strains of *Carnobacterium*, but only six of eight *Serratia* species were able to grow at 0.7 kPa (e.g., Figure 2 for *Serratia* spp.; Schuerger and Nicholson, 2016). In contrast, seven *Bacillus* spp. were not able to grow at pressures < 3.5 kPa in one study (Schuerger and Nicholson, 2006), while two undescribed *Bacillus* isolates were able to grow at 0.7 kPa in a second study (Schuerger and Nicholson, 2016). These results suggest that the capability of growing at 0.7 kPa may be species-specific and not ubiquitously manifested within a genus.

Schuerger and Nicholson (2016) described 62 bacterial isolates that grow at 0.7 kPa (Figure 4). The bacteria belonged to three different phyla and grouped within 10 bacterial genera. Fifty-eight percent of these isolates were identified on species level including *Paenibacillus antarcticus, P. macquariensis, Rhodococcus qingshengii, Streptomyces aureus, S. vinaceus, Exiguobacterium sibiricum, Trichococcus pasteurii, T. collinsii, Serratia liquefaciens, S. ficaria, S. fonticola, S. grimesii, S. plymuthica, S. rubidaea*. To date, no Archaea, fungi nor other eukaryotic organisms have been reported capable of growth in low-PTA conditions at 0.7 kPa. *Serratia liquefaciens* has become the most studied model organism for growth at low-PTA conditions, and had its whole genome sequenced (Nicholson et al., 2013b), paving the way for more complex molecular studies under low-PTA conditions (e.g., Fajardo-Cavazos et al., 2018).

Additional cultivation approaches to detect the total number of culturable hypopiezotolerants compared to the total viable microorganisms led to the enrichment of hypopiezotolerant bacteria, but not archaea or fungi, from a range of soils including permafrost and a nonglacial high arctic lake (Table 3). Samples from mesophilic environments were negative for indigenous hypopiezotolerant bacteria (Schuerger and Nicholson, 2016). The results indicated a general low percentage of culturable hypopiezotolerant microorganisms in these samples compared to the total culturable fraction incubated at 101.3 kPa, 25 °C and Earth-normal atmosphere.

6. Exploring the microbial "dark matter" of hypopiezophiles and -tolerants

The portion of microorganisms that cannot be cultivated in the laboratory are described as the microbial "dark matter". The problems associated with soil assay protocols (section 3.2) illustrate the severe limitations to predicting the amount/number of hypopiezophiles/-tolerants in samples using only cultivation basal methods. The primary limitation to the agar media assays to date is that it only works effectively for culturable microorganisms with visually observable colonies that grow under low-PTA conditions (e.g., Figure 3). It is plausible that many other culturable strains did, in fact, grow under low-PTA conditions but were lost in the soil particles present on the agar because their colony sizes were below the limits of

Table 3. Portion of hypopiezotolerant bacteria in various soil samples including permafrost and a non-glacial high arctic lake incubated at low-PTA conditions (0 °C, CO_2, 0.7 kPa) in the hypobaric chamber compared to the total viable microorganism count at 25 °C, 21% pO_2, 101.3 kPa. Cultivation was done on agar plates [adapted from Nicholson et al., 2013a; Schuerger and Nicholson, 2016]. Only bacterial colonies were observed.

Sample	Hypopiezotolerants	Total viable cells at 25°C, O_2, 101.3 kPa
Siberian permafrost	6[a]	9.3×10^3
Mt. Baker, Washington	1.9×10^2	1.2×10^8
Devon Island, Canada	2.5×10^2	3.4×10^4
Siberian Permafrost, Russia	2.8×10^4	1.5×10^8
Colour Lake, Axel Heiberg Island	5.1×10^4	1.0×10^7

[a]Given as number of colony forming units (cfu) per gram of soil

visual detection using 5x jeweler glasses (Schuerger and Nicholson, 2016). In addition, estimates revealed that 85-99% of bacteria and archaea, respectively, cannot yet be grown in the laboratory and the numbers can vary highly depending on the environment sampled and media used (Lok, 2015). Furthermore, the hypobaric protocols are not yet being fine-tuned to finding non-culturable hypopiezotolerant microorganisms. Thus, there is a crucial need for new assay protocols that can explore the so-called microbial dark matter of non-culturable microorganisms in environmental samples, and to identify slow-growing fastidious hypopiezotolerant microorganisms.

7. Cultivation and molecular approaches to unravel influences of low-pressure environments on and within the microbial cell

Microorganisms have evolved abilities to sense, respond, and adapt to a variety of physical parameters in the environment. However, little is known about how bacteria sense low pressure, how they acclimatize to pressure alterations and whether they possess pressure-specific adaptations, marker genes, or metabolic pathways. The following studies investigated the effects of hypobaria on the genome, gene expression, protein synthesis, lipid composition and metabolism of bacteria grown under low-pressure conditions. Furthermore, currently it is unknown whether these potential adaptations are species-specific or ubiquitous.

7.1. Adaptation experiments and genomic changes

The findings that microorganisms, which did not grow at low pressure but resumed growth when returned to Earth-normal pressure (Schuerger et al., 2013; Schuerger and Nicholson, 2006) led to the hypothesis that target molecules might exist within cells that are reversibly inactivated at low pressure. To understand the response and adaptation of microorganisms to low pressure, adaptation experiments with *Bacillus subtilis* were conducted for over 1,000 generations at the near-inhibitory low pressure of 5.0 kPa. Populations of the evolving strains were sampled every 50 generations, and led to the isolation of a low-pressure evolved strain that had developed higher growth rates at 5.0 kPa compared to wild-type strains maintained at 101.3 kPa (Nicholson et al., 2010). Compared to its ancestral strain, the evolved strain rapidly acquired increased fitness for higher growth rates at 5.0 kPa starting at 200 generations. The adaptations were akin to steps in punctuated equilibrium (defined by Gould, 2002) for the evolution of a low-pressure adapted *B. subtilis* strain over time.

To identify genomic alterations, like changes or mutations that were induced by the low-pressure treatments, whole-genome sequencing of the adapted strain and respective mutants was performed (Waters et al., 2015). The genomic adaptations to low pressure were found to be a dynamic process and revealed complex kinetics, i.e., different patterns of mutations that appeared in either early or late stages of the experiment with some of the earlier mutations not being detected in the end. During the 1,000 generations, final amino acid-altering mutations of seven genes and a single 9-bp in-frame deletion in a RNA degradosome encoding gene were detected (Waters et al., 2015). However, data on genomic changes are still scarce and only available from one strain at 5.0 kPa. There is still a lack of data from other hypopiezotolerant microorganisms grown at low-PTA conditions between 0.7 and 1.0 kPa. Apart from changes in the whole genome, adaptation to low pressure can be expressed on multiple levels, as for example gene expression.

7.2. Gene expression studies
Another step towards the identification of molecular mechanisms in hypopiezotolerant microorganisms adapted to novel environmental stresses at low pressure is to investigate the gene expression under hypobaric conditions. Currently, there are two separate studies available on two different bacterial strains.

Fajardo-Cavazos et al. (2012) investigated the gene expression of a low-pressure-adapted *B. subtilis* strain (section 7.1) to explore the mechanisms that enabled the previously low-pressure inhibited strain to grow at 5.0 kPa. A cluster of three candidate genes (*des*, *desK*, and *desR*) was up-regulated. The *des* gene encoding a Des membrane fatty acid (FA) desaturase, the *desK* encoding a DesK sensor kinase and *desR* genes encoding a DesR response regulator are involved in the maintenance of membrane fluidity. Inactivation of the *des* gene achieved by a knockout mutation, resulted in decreased fitness of the evolved strain to 5.0 kPa.

The first transcriptomics study under low-PTA conditions (0.7 kPa, 0 °C, CO_2 atmosphere) was performed using the hypopiezotolerant bacterium, *Serratia liquefaciens* ATCC 27592 (Fajardo-Cavazos et al., 2018) to search for new insights into the molecular mechanisms responsible for microbial adaptation to alterations in their pressure environment. RNA-seq and subsequent transcriptome analyses revealed significant differentially expressed transcripts. Up-regulation in genes that encode transporters (ABC and PTS transporters), genes that are involved in translation (ribosomes and their biogenesis, biosynthesis of tRNAs and aminoacyl-tRNAs), DNA repair and recombination, and non-coding RNAs were observed. More specifically, several amino acid, purine, and sugar carbohydrate pathways were up-regulated at 0.7 kPa. Genes down-regulated included transporters (mostly ABC transporters), flagellar and motility proteins, transcription factors, and two-component systems. Despite the observed changes, the alterations did not reflect a major rearrangement of the transcriptome nor were pressure-specific genes identified. However, the data rather indicated that the transcriptome profile from *S. liquefaciens* at 0.7 kPa was a complex process altering gene expression and resulting in a stress response triggered by several environmental stressors acting simultaneously.

7.3. Membrane structure

The gene expression studies (section 7.2) suggest an effect of low pressure on the cell membrane. At high hydrostatic pressures (> 10 MPa), membranes become more rigid due to increased packing of membrane fatty acids (FAs) and proteins, leading to cellular adaption in which unsaturated FAs increase and saturated FAs decrease in an attempt to maintain membrane fluidity and functionality (Oger and Jebbar, 2010). In contrast, at low pressures a less ordered packing of membrane components is assumed when compared to Earth-normal atmospheric pressure, which likely increases the fluidity of the membrane. Consequently, it might affect the FA composition in an opposite manner to the effects reported under high pressure. In turn, the membrane processes including proton pumping, nutrient and ion transport as well as protein translocation are altered due to the increased disordering.

For example, analysis of membrane FA composition of *Bacillus subtilis* vegetative cells grown at the Earth-normal pressure of 101.3 kPa compared to 5.0 kPa revealed a decrease in the ratio of unsaturated to saturated FAs but an increase in the ratio of anteiso- to iso-FAs (Fajardo-Cavazos et al., 2012). Currently, it is not known whether these effects are widespread in the domains Bacteria, Archaea, and Eukarya, and valid for all hypopiezotolerants/philes, or whether they are species-specific. In addition, there are no data available on the membrane composition of other bacteria grown at low-PTA-conditions near 0.7 kPa. More research is required to unravel the effects of hypobaria on membrane structure.

7.4. Carbon source utilization

The changes of the metabolic fingerprint of *Serratia liquefaciens* ATCC 27592, a hypopiezotolerant model organism, was investigated under decreasing atmospheric pressures to 2.5 kPa and low-PTA conditions

(Schwendner and Schuerger, 2018). Currently, this is the only study that has investigated microbial metabolism under low pressures. To outline the metabolic changes, Biolog GN2 microarray plates were used to test the utilization of 95 different single carbon sources. Apart from temperature and atmospheric composition, decreasing the atmospheric pressure revealed a distinct effect on the metabolic fingerprint. More specifically, above 10 kPa *S. liquefaciens* utilized the various carbon sources in a similar manner as observed at an Earth-normal pressure of 101.3 kPa; whereas below 10 kPa, significant changes were observed indicating that the cells may have undergone physiological alterations. In particular, *S. liquefaciens* preferred to utilize a range of carbohydrates while the cells lost the ability to metabolize the majority of the provided carbon sources with a significant decrease in the oxidation of amino acids. These alterations were suggested to be the result of several potential stress-induced reactions such as potential physiological changes, the alteration of gene expression, or changes in the membrane which in turn affected the uptake of nutrients. Data on the metabolic responses to different carbon sources are of great value to identify nutritional constraints that support cellular replication in low-pressure habitats.

8. Looking at the other end of the spectrum - what can we learn from piezophiles?

We believe that we have not yet determined the limits to life in high- and low-pressure environments. An upper limit of growth at pressures between 130-150 MPa was described for *Pyrococcus yayanosii* (Zeng et al., 2009), but most other hyperpiezophiles are only able to grow up to 60-110 MPa (Oger and Jebbar, 2010; Picard and Daniel, 2013). Piezophiles and hyperpiezophiles have been found in both bacterial and archaeal domains. Strains have been isolated from the subseafloor oceanic crusts, deep subterranean environments such as the Mariana Trench with a maximum depth of 11 km (~ 110 MPa), and many other high-pressure marine locations on Earth (Fang et al., 2010; 2017; Inagaki et al., 2015; Margosch et al., 2006; Nogi et al., 1998; Russell et al., 2016). Studies on how microorganisms adapt to high pressures reveal alterations in gene expressions, protein synthesis, and membrane lipid composition (reviewed in Bartlett et al., 1995; Kato and Bartlett, 1997; Oger and Jebbar 2010).

With increasing pressure, double stranded DNA becomes more rigid which negatively affects the transition into single strands; a prerequisite for replication, transcription, and translation. Consequently, protein and nucleic acid synthesis is hindered (Oger and Jebbar, 2010; Simonato et al., 2006). In addition, increased packing occurs in lipid membranes leading to decreased fluidity; which in turn affects the permeability of cells for water and nutrient uptake, and protein-lipid interactions (Oger and Jebbar, 2010; Winter and Jeworrek, 2009). Similarly, proteins adapt their conformations according to volume restrictions caused by high pressures, which negatively affect multimer associations, stability, and catalytic sites, leading to a loss of enzymatic function and metabolic activity (Balny et al., 2002; Northtrop, 2002). Without adaptation mechanisms to cope with high-pressures and its induced alterations of cellular architectures, these effects would eventually cause cell death.

Three mechanisms have been proposed to counteract the damages experienced at high pressures: (1) upregulation of gene expression to compensate for the loss of biological activity leading to an increase in specific components to manage energy and osmotic stability as well as chaperons aiding protein folding (Campanero et al., 2005; Fernandes et al., 2004); (2) expression of pressure-inducible genes such as those involved in the ToxR/S two–component system (Bartlett, 1991; Kato and Qureshi, 1999); and (3), structural changes of biomolecules by increasing the proportion of unsaturated fatty acids in cell membranes at high pressures (Chilukuri and Bartlett, 1997; Fang and Bazylinksi, 2008). Microorganisms not adapted to but able to withstand high pressures can synthesize pressure inducible proteins (PIPs) which are "stress" proteins previously identified as heat shock, cold shock, or ribosomal proteins or proteins with unknown function (Bartlett et al., 1995).

We predict that the genomic, metabolic, and physiological adaptive trends for piezophiles growing at lower pressures (down to 0.1 MPa) would be similar to microorganisms which are optimally adapted to Earth-normal pressure growing at hypobaric conditions close to 0.7 kPa. For example, saturated fatty acid (FA) levels in general decrease and unsaturated FA levels increase in piezophiles as the hydrostatic pressure is increased (see reviews by Oger and Jebbar, 2010; Picard and Daniel, 2013). Conversely, we would expect to see saturated FAs to increase and unsaturated FAs to decrease at low pressures < 10 kPa in bacteria ecologically adapted to sea-level pressures of 0.1 MPa. Such a situation was reported for *B. subtilis* strains grown at low pressures down to 5.0 kPa (Fajardo-Cavazos et al., 2012). Thus, the study of piezophiles and (hyper-)piezophiles growing at pressures lower than their normal ecological niches may provide insights on the genomic, metabolic, and cellular adaptations of mesophilic bacteria, normally adapted to sea-level pressures of 0.1 MPa growing in hypobaric environments down to 0.7 kPa.

9. Implications of hypopiezotolerants to Mars astrobiology

Despite the current planetary protection regulations and the vigorous cleaning efforts taken place before launch, spacecraft launched into space carry finite amounts of viable microorganisms. In fact, on a Category IVa mission to Mars (i.e., soft-landed spacecraft without life-detection payloads) the bioburden on the vehicle is required to be < 3×10^5 spores per spacecraft; and on Category IVb and IVc missions (i.e., life-detection payloads present), the bioburden reductions are even more stringent requiring the total surface bioburden of the spacecraft to be < 30 spores (Frick et al., 2014). Both, vegetative cells and spores are present on spacecraft with spores beingthe most likely to survive interplanetary transfer to Mars, and therefore might pose the greatest potential forward contamination risks (Kempf et al., 2005; Moissl-Eichinger et al., 2013; Nicholson et al., 2009). However, results of investigating the effects of low pressure on spores indicates that spores fail to germinate at low pressure (Schuerger and Nicholson, 2006). Once launched, spacecraft microorganisms face harsh environmental conditions including high UV irradiation, extreme desiccation, ionizing radiation, vacuum, and extreme thermal cycling; which leads to reductions of survivability between 50-70% for spores and up to two orders of magnitude for vegetative cells during a typical 6 to 8-month cruise-phase to Mars (Dose and Klein, 1996; Hagen et al., 1971; Horneck et al., 1994; Koike and Oshima, 1993).

Thus, a small number of viable spacecraft microorganisms are likely to survive the cruise phase to Mars (Horneck et al., 2010; Nicholson et al., 2000) and may be present on spacecraft surfaces after landing. The initial evidence indicated that a few bacteria (e.g., Figure 4), but so far no fungi or archaea, have been shown to grow at pressures down to 0.7 kPa (i.e., the average surface pressure on Mars). However, the astrobiology community has barely scratched the surface characterizing the effects of hypobaria on microbial survival, metabolism, growth, and adaptation under relevant Martian conditions. Nevertheless, low pressure is an emerging extreme environmental factor which needs to be addressed when discussing habitability of Mars and planetary protection issues related to new missions with life-detection payloads to Mars. It is plausible that microorganisms originating from Earth with the capability of growing at 0.7 kPa may have an unwanted or previously unpredicted impact on the search for life on Mars by creating a risk of false positives in the assays.

Some of the studies cited above (e.g., Nicholson et al., 2013a; Schuerger et al., 2013) suggest that some Earth microorganisms possess the physiological range to initiate microbial activity and grow on Mars if they are dispersed to hydrated niches that might support microbial activity (e.g., recurring slope lineae; Ojha et al., 2015). For example, growth of *S. liquefaciens* on Biolog GN2 organics at 0.7 kPa (Schwendner and Schuerger, 2018) suggests that if similar organics are found on Mars from accreted interplanetary dust particles and carbonaceous chondrites (e.g. Sephton and Botta, 2005) or *in situ* organics (Eigenbrode et al., 2018), then heterotrophic metabolic activity using *in situ* organics might occur on Mars if the microorganisms are dispersed to stable, UV-protected, and hydrated niches. Data on the ecological settings and nutrient requirements of microbial hypopiezotolerant strains in a simulated Martian environment are key factors to both ascertain the potential risk of forward contamination and to determine whether Mars is or might have been habitable.

Another aspect to consider is that in the past the atmospheric pressure was much higher than it is now (Brain et al., 2010). The observation that *B.*

subtilis was able to evolve and gain fitness over 1,000 generations when grown at 5.0 kPa (Nicholson et al., 2010) suggests that other microorganisms may have the capability to adapt to low-pressure environments on a short timescale; perhaps even to adapt to low-PTA conditions at 0.7 kPa. Thus, the study of hypopiezotolerant microorganisms growing at pressures similar to the surface of Mars (0.1 to 1.2 kPa) not only allows predictions for forward contamination in general and of Special Regions in particular (i.e., locations that might support the growth of Earth microorganisms, or harbor an extant Mars microbiota; Rummel et al., 2014), but also may help us locate terrains in which to search for an extant microbiota and provide insights into the habitability of Mars. In addition, hypopiezotolerant microorganisms can serve as positive controls for the development of life-detection experiments and payloads.

Consequently, experiments searching for hypopiezophilic and hypopiezotolerant microorganisms on Earth, and investigating their growth and adaptation to Martian conditions, may provide data that strengthens the hypothesis of an extant microbiota on Mars. However, more research is required to examine the ability of Earth microbes to grow and proliferate on Mars under low-PTA conditions that are relevant to future exploration missions.

Acknowledgements
This research was supported by a grant from NASA's Planetary Protection Office (#80NSSC17K0263) for both P.S. and A.C.S.

References
Barth, C.A., Stewart, A.I.F., Bougher, S.W., Hunten, D.M., Bauer, S.J., and Nagy, A.F. (1992). Aeronomy of the current Martian atmosphere. In Mars: H.H., Kieffer, B.M., Jakosky, C.W., Snyder, M.S., Matthews, ed, (Tuscon, USA: University of Arizona Press), pp. 1054-1089.

Balny, C., Masson, P., and Heremans, K. (2002). High pressure effects on biological macromolecules: from structural changes to alteration of cellular processes. Biochim. Biophys. Acta. 1595, 3-10.

Bartlett, D.H., Kato, C., and Horikoshi, K. (1995). High pressure influences on gene and protein expression. Res. Microbiol. 146, 697-706.

Bartlett, D.H. (1991). Pressure sensing in deep-sea bacteria. Res. Microbiol. 142, 923-925.

Beaty, D., Buxbaum, K., Meyer, M., Barlow, N., Boynton, W., Clark, B., Deming, J., Doran, P.T., Edgett, K., Hancock, S., Head, J., Hecht, M., Hipkin, V., Kieft, T., Mancinelli, R., McDonald, E., McKay, C., Mellon, M., Newsom, H., Ori, G., Paige, D., Schuerger, A.C., Sogin, M., Spry, J.A., Steele, A., Tanaka, K., and Voytek, M. (2006). Findings of the Mars Special Regions Science Analysis Group. Astrobiology. 6, 677-732

Berry, B.J. Jenkins, D.G., and Schuerger, A.C. (2010). Effect of simulated Mars conditions on the survival and growth of *Escherichia coli* and

Serratia liquefaciens. Appl. Environ. Microbiol. 76, 2377-2386. doi: 10.1128/AEM.02147-09.

Brain, D., Barabash, S., Boesswetter, A., Bougher, S., Brecht, S., Chanteur, G., Hurley, D., Dubinin, E., Fang, X., Fraenz, M., Halekas, J., Harnett, E., Holmstrom, M., Kallio, E., Lammer, H., Ledvina, S., Liemohn, M., Liu, K., Luhmann, J., Ma, Y., Modolo, R., Nagy, A., Motschmann, U., Nilsson, H., Shinagawa, H., Simon, S., and Terada, N. (2010). A comparison of global models for the solar wind interaction with Mars. Icarus. 206, 139-151. doi.org/10.1016/j.icarus.2009.06.030.

Bauermeister, A., Rettberg, P., and Flemming, H.-C. (2014). Growth of the acidophilic iron-sulfur bacterium *Acidithiobacillus ferrooxidans* under Mars-like geochemical conditions. Planet. Space Sci. 98, 205-215. doi.org/10.1016/j.pss.2013.09.009.

Brandt, A., de Vera, J.-P., Onofri, S., and Ott, S. (2015). Viability of the lichen *Xanthoria elegans* and its symbiont after 18 months of space exposure and simulated Mars conditions on the ISS. Int. J. Astrobiol. 14, 411-425. doi.org/10.1017/S1473550414000214.

Brueschke, E.E., Suess, R.H., and Willard, M. (1961). The viability of microorganisms in ultra-high vacuum. Planet. Space Sci. 8, 30-34. doi.org/10.1016/0032-0633(61)90115-5.

Campanero, S., Vezzi, A., Vitulo, N., Lauro, F.M., D'Angelo, M., Simonato, F., Cestaro, A., Malacrida, G., Bertoloni, G., Valle, G., and Bartlett, D.H. (2005). Laterally transferred elements and high pressure adaptation in *Photobacterium profundum* strains. BMC Genom. 6:122. doi.org/10.1186/1471-2164-6-122.

Chilukuri, L.N., and Bartlett, D.H. (1997). Isolation and characterization of the gene encoding single-stranded-DNA-binding protein (SSB) from four marine *Shewanella* strains that differ in their temperature and pressure optima for growth. Microbiology. 143, 1163-1174. doi: 10.1099/00221287-143-4-1163.

Cockell, C.S., Bush, T., Bryce, C., Direito, S., Fox-Powell, M., Harrison, J.P., Lammer, H., Landenmark, H., Martin-Torres, J., Nicholson, N., Noack, L., O'Malley-James, J., Payler, S.J., Rushby, A., Samuels, T., Schwendner, P., Wadsworth, J., and Zorzano, M.P. (2016). Habitability: A review. Astrobiology. 16, 89-117. doi: 10.1089/ast.2015.1295.

DasSarma, S., and DasSarma, P. (2015). Halophiles and their enzymes: Negativity put to good use. Curr. Opin. Microbiol. 25, 120-126. doi: 10.1016/j.mib.2015.05.009.

DasSarma, P., Laye, V.J., Harvey, J., Reid, C., Shultz, J., Yarborough, A., Lamb, A., Koske-Phillips, A., Herbst, A., Molina, F., Grah, O., Phillips, T., and DasSarma, S. (2017). Survival of halophilic Archaea in Earth's cold stratosphere. Int. J. Astrobiol. 16, 321-327. doi.org/10.1017/S1473550416000410.

DasSarma, P., and DasSarma, S. (2018). Survival of microbes in Earth's stratosphere. Curr. Opin. Microbiol. 43, 24-30. doi.org/10.1016/j.mib.2017.11.002

de la Torre Noetzel, R., Miller, A.Z., de la Rosa, J., Pacelli, C., Onofri, S., Sancho, L.G., Cubero, B., Lorek, A., Wolter, D., and de Vera, J.P. (2018). Cellular responses of the lichen *Circinaria gyrosa* in Mars-like conditions. Front. Microbiol. 9, 308. doi:10.3389/fmicb.2018.00308.

de Vera, J.-P., Mohlmann, D., Butina, F., Lorek, A., Wernecke, R., and Ott, S. (2010). Survival potential and photosynthetic activity of lichens under Mars-like conditions: A laboratory study. Astrobiology. 10, 215-227. doi: 10.1089/ast.2009.0362.

de Vera, J.-P., Schulze-Makuch, D., Khan, A., Lorek, A., Koncz, A., Mohlmann, D., and Spohn, T. (2014). Adaptation of an Antarctic lichen to Martian niche conditions can occur within 34 days. Planet. Space Sci. 98, 182-190. doi.org/10.1016/j.pss.2013.07.014.

Des Marais, D.J., Nuth, J.A., III, Allamandola, L.J., Boss, A.P., Farmer, J.D., Hoehler, T.M., Jakosky, B.M., Meadows, V.S., Pohorille, A., Runnegar, B., and Sporman, A.M. (2008). The NASA Astrobiology Roadmap. Astrobiology. 8, 715-730. doi: 10.1089/ast.2008.0819.

dos Santos, R., Patel, M.R., Cuadros, J., and Martins, Z. (2016). Influence of mineralogy on the preservation of amino acids under simulated Mars conditions. Icarus. 277, 342-353. doi.org/10.1016/j.icarus.2016.05.029.

Diehl, R.H. (2013). The airspace is habitat. Trends Ecol. Evol. 28, 377-379. doi.org/10.1016/j.tree.2013.02.015.

Dose, K., and Klein, A. (1996). Response of *Bacillus subtilis* spores to dehydration and UV irradiation at extremely low temperatures. Orig. Life Evol. Biosph. 26, 47-59.

Eigenbrode, J.L., Summons, R.E., Steele, A., Freissinet, C., Millan, M., Navarro-González, R., Sutter, B., McAdam, A.C., Franz, H.B., Glavin, D.P., Archer, P.D., Jr., Mahaffy, P.R., Conrad, P.G., Hurowitz, J.A., Grotzinger, J.P., Gupta, S., Ming, D.W., Sumner, D.Y., Szopa, C., Malespin, C.A., Buch, A., and Coll, P. (2018). Organic matter preserved in 3-billion-year-old mudstones at Gale Crater, Mars. Science. 360, 1096-1101. doi: 10.1126/science.aas9185.

Eisenmenger, M.J., and Reyes-De-Corcuera, J.I. (2009). High pressure enhancements of enzymes: A review. Enzyme Microb. Technol. 45, 331-347. doi.org/10.1016/j.enzmictec.2009.08.001.

Ertem, G., Ertem, M.C., McKay, C.P., and Hazen, R.M. (2017). Shielding biomolecules from effects of radiation by Mars analogue minerals and soils. Int. J. Astrobiol. 16, 280-285. doi.org/10.1017/S1473550416000331.

Fajardo-Cavazos, P., Waters, S.M., Schuerger, A.C., George, S., Marois, J.J., and Nicholson, W.L. (2012). Evolution of *Bacillus subtilis* to enhanced growth at low pressure: Up-regulated transcription of *des-desKR*, encoding the fatty acid desaturase system. Astrobiology. 12, 258-270. doi: 10.1089/ast.2011.0728.

Fajardo-Cavazos, P., Morrison M.D., Miller, K.M., Schuerger, A.C., and Nicholson, W.L. (2018). Transcriptomic responses of *Serratia liquefaciens* cells grown under simulated Martian conditions of low temperature, low pressure, and CO_2-enriched anoxic atmosphere. Sci. Rep. 8, 14938. doi: 10.1038/s441598-018-33140-4.

Fang, J., and Bazylinski, D.A. (2008). Deep sea microbiology. In High-Pressure Microbiology, C., Michiels, D., Bartlett, and A., Aertsen, ed. (Washington DC, USA: ASM Press), pp. 2387-264.

Fang, J., Zhang, L., and Bazylinski, D.A. (2010). Deep-sea piezosphere and piezophiles: geomicrobiology and biogeochemistry. Trends Microbiol. 18, 413-422. doi: 10.1016/j.tim.2010.06.006.

Fang, J., Kato, C., Runko, G.M., Nogi, Y., Hori, T., Li, J., Morono, Y., and Inagaki, F. (2017). Predominance of viable spore-forming piezophilic bacteria in high-pressure enrichment cultures from ~1.5 to 2.4 km-deep coal-bearing sediments below the ocean floor. Front. Microbio. 8, 137. doi: 10.3389/fmicb.2017.00137.

Fernandes, P.M.B., Domitrovic, T., Kao, C.M., and Kurtenbach, E. (2004). Genomic expression pattern in *Saccharomyces cerevisiae* cells in response to high hydrostatic pressure. FEBS Lett. 556, 153-160.

Flynn, G.J. (1996). The delivery of organic matter from asteroids and comets to the early surface of Mars. Earth, Moon, and Planets. 72, 469-474.

Frick, A., Mogul, R., Stabekis, P.D., Conley, C.A., and Ehrenfreund, P. (2014). Overview of current capabilities and research and technology developments for planetary protection. Adv. Space Res. 54, 221-240. doi.org/10.1016/j.asr.2014.02.016.

Fox-Powell, M.G., Hallsworth, J.E., Cousins, C.R., and Cockell, C.S. (2016). Ionic strength is a barrier to the habitability of Mars. Astrobiology. 16, 427-442. http://doi.org/10.1089/ast.2015.1432.

Galletta, G., Bertoloni, G., and D'Allessandro, M. (2011). New approaches to the exploration: planet Mars and bacterial life. Proceedings IAU Symposium 269, 1-4. arxiv.org/abs/1102.4438.

Gillmore, J.D., and Gordon, F.B. (1975). Effect of exposure to hyperoxic, hypobaric, and hyperbaric environments on concentration of selected and aerobic and anaerobic fecal flora of mice. Appl. Microbiol. 3, 358-367.

Griffin, D.W., Gonzalez-Martin, C., Hoose, C., and Smith, D.J. (2018). Global-scale atmospheric dispersion of microorganisms. In Microbiology of Aerosols. A.-M. Delort, P. Amato, ed, (John Wiley & Sons, Inc.) Chapter 2.3. doi.org/10.1002/9781119132318.ch2c.

Goméz, F., Mateo-Martí, E., Prieto-Ballesteros, O., Martin-Gago, J., and Amils, R. (2010). Protection of chemolithoautotrophic bacteria exposed to simulated Mars environmental conditions. Icarus. 209, 482-487. doi.org/ 10.1016/j.icarus.2010.05.027.

Gould, S.J. (2002). The Structure of Evolutionary Theory. Harvard University Press, Cambridge, MA, USA.

Haberle, R.M., McKay, C.P., Schaeffer, J., Cabrol, N.A., Grin, E.A., Zent, A.P., and Quinn, R. (2001). On the possibility of liquid water on present-day Mars. J. Geophys. Res. 106, 23317-23326. doi: 10.1029/2000JE001360.

Hagen, C.A., Godfrey, J.F., and Green, R.H. (1971). The effect of temperature on the survival of microorganisms in a deep space vacuum. Space Life Sci. 3, 108-117.

Hawrylewicz, E.J., Hagen, C.A., Tolkacz, V., Anderson, B.T., and Ewing, M. (1968). Probability of growth (p_G) of viable microorganisms in Martian environments. Life Sci. Space Res. 6, 147-156.

Heinz, J., Schirmack, J., Airo, A., Kounaves, S.P., and Schulze-Makuch, D. (2018). Enhanced microbial survivability in subzero brines. Astrobiology. 18, 1171-1180. doi: 10.1089/ast.2017.1805.

Horneck, G. (1981). Survival of microorganisms in space; a review. Adv. Space Res. 1, 39-48.

Horneck, G., Bücker, H., and Reitz, G. (1994). Long-term survival of bacterial spores in space. Adv. Space Res. 14, 41-45.

Horneck, G., Klaus D.M., and Mancinelli, R.L. (2010). Space microbiology. Microbiol. Mol. Biol. Rev. 74, 121-156. doi: 10.1128/MMBR.00016-09.

Inagaki, F., Hinrichs, K.-U., Kubo, Y., Bowles, M.W., Heuer, V.B., Hong, W.-L., Hoshino, T., Ijiri, A., Imachi, H., Ito, M., Kaneko, M., Lever, M.A., Sin, Y.-S., Methé, B.A., Morita, S., Morono, Y., Tanikawa, W., Bihan, M., Bowden, S.A., Elvert, M., Glombitza, C., Gross, D., Harrington, G.J., Hori, T., Li, K., Limmer, D., Liu, C.-H., Murayama, M., Ohkouchi, N., Ono, S., Park, Y.-S., Phillips, S.C., Prieto-Mollar, X., Purkey, M., Riedinger, N., Sanada, Y., Sauvage, J., Snyder, G., Susilawati, R., Takano, Y., Tasumi, E., Terada, T., Tomaru, H., Trembath-Reichert, E., Wang, D.T., and Yamada, Y. (2015). Exploring deep microbial life in coal-bearing sediment down to ~2.5 km below the ocean floor. Science. 349, 420-424. doi: 10.1126/science.aaa6882.

Jannasch, H.W., and Taylor, C.D. (1984). Deep-sea microbiology. Ann. Rev. Microbiol. 38, 487-514. doi.org/10.1146/annurev.mi.38.100184.002415.

Jensen, L.L., Merrison, J., Hansen, A.A., Mikkelsen, K.A., Kristoffersen, T., Nornberg, P., Lomstein, B.A., and Finster, K. (2008). A facility for long-term Mars simulation experiments: The Mars Environmental Simulation Chamber (MESCH). Astrobiology. 8, 537-548. doi: 10.1089/ast.2006.0092.

Kato, C., and Bartlett, D.H. (1997). The molecular biology of barophilic bacteria. Extremophiles. 1, 111-116.

Kato, C., and Qureshi, M.H. (1999). Pressure response in deep-sea piezophilic bacteria. J. Mol. Microbiol. Biotechnol. 1, 87-92.

Kanervo, E., Lehto, K., Stahle, K., Lehto, H., and Maenpaa, P. (2005). Characterization of growth and photosynthesis of *Synechocystis* sp. PCC6803 cultures under reduced atmospheric pressures and enhanced CO_2 levels. Int. J. Astrobiol. 4, 97-100. doi.org/10.1017/S1473550405002466.

Kempf, M.J., Chen, F., Kern, R., and Venkateswaran, K. (2005). Recurrent isolation of hydrogen peroxide-resistant spores of *Bacillus pumilus* from a spacecraft assembly facility. Astrobiology. 5, 391-405. doi: 10.1089/ast.2005.5.391.

Koike, J., and Oshima, T. (1993). Planetary quarantine in the Solar System: Survival rates of some terrestrial microorganisms under simulated space conditions by proton irradiation. Acta Astronaut. 29, 629-632. doi.org/10.1016/0094-5765(93)90080-G.

Kral, T.A., Altheide T.S., Lueders, A.E., and Schuerger, A.C. (2011). Low pressure and desiccation effects on methanogens: Implications for life on Mars. Planet. Space Sci. 59, 264-270. doi.org/10.1016/j.pss.2010.07.012.

Lok, C. (2015). Mining the microbial dark matter. Nature. 522, 270-273. doi: 10.1038/522270a.

Mahaffy, P.R., Webster, C.R., Atreya, S.K., Franz, H.B., Wong, M., Conrad, P.G., Harpold, D., Jones, J.J., Leshin, L.A., Manning, H., Owen, T., Pepin, R.O., Squyres, S.W., Trainer, M., and MSL Science Team (2013). Abundance and isotopic composition of gases in the Martian atmosphere from the Curiosity rover. Science. 341, 263-266. doi: 10.1126/science.1237966.

Mancinelli, R.L., and Klovstad, M. (2000). Martian soil and UV radiation: Microbial viability assessment on spacecraft surfaces. Planet. Space Sci. 48, 1093-1097. doi.org/10.1016/S0032-0633(00)00083-0.

Mancinelli, R.L (2015). The affect of the space environment on the survival of *Halorubrum chaoviator* and *Synechococcus* (Nägeli): data from the Space Experiment OSMO on EXPOSE-R. Int. J. Astrobiol. 14, 123-128. doi.org/10.1017/S147355041400055X

Margosch, D., Ehrmann, M.A., Buckow, R., Heinz, V., Vogel, R.F., and Ganzle, M.G. (2006). High-pressure-mediated survival of *Clostridium botulinum* and *Bacillus amyloliquefaciens* endospores at high temperature. Appl. Environ.l Microbiol. 72, 3476-3481. doi: 10.1128/AEM.72.5.3476-3481.2006

Martin, D., and Cockell, C.S. (2015). PELS (Planetary Environmental Liquid Simulator): A new approach of simulation facility to study extraterrestrial aqueous environments. Astrobiology. 15, 111-118. doi: 10.1089/ast.2014.1240.

Meeßen, J., Backhaus, T., Sadowsky, A., Mrkalj, M., Sánchez, F.J., de la Torre, R., and Ott, S. (2014). Effects of UVC-254nm on the photosynthetic activity of photobionts from the astrobiologically relevant lichens *Buella figida* and *Circinaria gyrosa*. Int. J. Astrobiol. 13, 340-352. doi: 10.3389/fmicb.2018.00308.

Meeßen, J., Wuthenow, P., Schille, P., de Vera, J.-P., and Ott, S. (2015). Resistance of the lichen *Buella frigida* to simulated space conditions during the preflight tests for BIOMEX — Viability assay and morphological stability. Astrobiology. 15, 601-615. doi: 10.1089/ast.2015.1281.

Mickol, R.L., and Kral, T. (2017). Low pressure tolerance by methanogens in an aqueous environment: Implications for subsurface life on Mars. Orig. Life Evol. Biosph. 47, 511-532. doi:10.1007/s11084-016-9519-9.

Mileikoswky, C., Cucinotta, F.A., Wilson, J.W., Gladman, B., Horneck, G., Lindegren, L., Melosh, J., Rickman, H., Valtonen, M., and Yheng, J.Q.

(2000). Natural transfer of viable microbes in space. Icarus. 145, 391-427.

Moeller, R., Schuerger, A.C., Reitz, G., and Nicholson, W.L. (2012). Protective role of spore structural components in determining *Bacillus subtilis* spore resistance to simulated Mars surface conditions. Appl. Environ. Microbiol. 78, 8849-8853. doi: 10.1128/AEM.02527-12.

Moissl-Eichinger, C., Pukall, R., Probst, A.J., Stieglmeier, M., Schwendner, P., Mora, M., Barczyk, S., Bohmeier, M., and Rettberg, P. (2013). Lessons learned from the microbial analysis of the Herschel spacecraft during assembly, integration, and test operations. Astrobiology. doi: 10.1089/ast.2013.1024.

Morelli, F.A., Fehlner, F.P., and Stembridge, C.H. (1962). Effect of ultra-high vacuum on *Bacillus subtilis* var *niger*. Nature. 196, 106-107.

Morozova, D., Mohlmann, D., and Wagner, D. (2007). Survival of methanogenic archaea from Siberian permafrost under simulated martian thermal conditions. Orig. Life Evol. Biosph. 37, 189-200. doi: 10.1007/s11084-006-9024-7.

Nicholson, W.L., Munakata, N., Horneck, G., Melosh, H.J., and Setlow, P. (2000). Resistance of *Bacillus* endospores to extreme terrestrial and extraterrestrial environments. Microbiol. Mol. Biol. Rev. 64, 548-572. doi: 10.1128/MMBR.64.3.548-572.2000.

Nicholson, W.L., Schuerger, A.C., and Race, M.S. (2009). Migrating microbes and planetary protection. Trends Microbiol. 17, 389-392. doi: 10.1016/j.tim.2009.07.001.

Nicholson, W.L., Fajardo-Cavazos, P., Fedenko, J., Ortiz-Lugo, J.L., Rivas-Castillo, A., Waters, S.M., and Schuerger, A.C. (2010). Exploring the low-pressure growth limit: Evolution of *Bacillus subtilis* in the laboratory to enhanced growth at 5 kilopascals. Appl. Environ. Microbiol. 76, 7559-7565. doi: 10.1128/AEM.01126-10.

Nicholson, W.L., Krivushin, K., Gilichinsky, D., and Schuerger, A.C. (2013a). Growth of *Carnobacterium spp.* isolated from Siberian permafrost under simulated Mars conditions of pressure, temperature and atmosphere. PNAS. 110, 666-671. doi.org/10.1073/pnas.1209793110.

Nicholson, W.L., Leonard, M.T., Fajardo-Cavazos, P., Panayotova, N., Farmerie, W.G., Triplett, E.W., and Schuerger, A.C. (2013b). Complete genome sequence of *Serratia liquefaciens* strain ATCC 27592. Genome Announce. 1(4):e00548-13. doi:10.11288/genomeA.00548-13.

Nogi, Y., Kato, C., and Horikoshi, K. (1998). Taxonomic studies of deep-sea barophilic *Shewanella* strains and description of *Shewanella violacea* sp. nov. Arch. Microbiol. 170, 331-338. doi: 10.1007/s002030050650.

Northrop, D.B. (2002). Effects of high pressure on enzymatic activity. Biochim. Biophys. Acta. 1595, 71-79.

Ojha, L., Wilhelm, M.B., Murchie, S.L., McEwen, A.S., Wray, J.J., Hanley, J., Masse, M., and Chojnacki, M. (2015). Spectral evidence for hydrated salts in recurring slope lineae on Mars. Nat. Geosci. 8, 829-832. doi: 10.1038/NGEO2546.

Olsson-Francis, K., and Cockell, C.S. (2010). Experimental methods for studying microbial survival in extraterrestrial environments. J. Microbiol. Methods. 80, 1-13. doi: 10.1016/j.mimet.2009.10.004.

Oger, P.M., and Jebbar, M. (2010). The many ways of coping with pressure. Res. Microbiol. 161, 799-809. doi: 10.1016/j.resmic.2010.09.017.

Osman, S., Peeters, Z., La Duc, M.T., Mancinelli, R.L., Ehrenfreund, P., and Venkateswaran, K. (2008). Effect of shadowing on survival of bacteria under conditions simulating the martian atmosphere and UV radiation. Appl. Environ. Microbiol. 74, 959-970. doi: 10.1128/AEM.01973-07.

Paulino-Lima, I.G., Pilling, S., Janot-Pacheco, E., de Brito, A.N., Barbosa, J.A.R.G., Leitão, A.C., and Lage, C.D.A.S. (2010). Laboratory simulation of interplanetary ultraviolet radiation (broad spectrum) and its effects on *Deinococcus radiodurans*. Planet. Space. Sci. 58, 1180-1187. doi.org/10.1016/j.pss.2010.04.010.

Paulino-Lima, I.G., Janot-Pacheco, E., Galante, D., Cockell, C., Olsson-Francis, K., Brucato, J.R., Baratta, G.A., Strazzulla, G., Merrigan, T., McCullough, R., Mason, N., and Lage, C. (2011). Survival of *Deinococcus radiodurans* against laboratory-simulated solar wind charged particles. Astrobiology. 11, 875-882. doi: 10.1089/ast.2011.0649.

Pavlov, A.K., Shelegedin, V.N., Vdovina, M.A., and Pavlov, A.A. (2010). Growth of microorganisms in Martian-like shallow subsurface conditions: Laboratory modeling. Int. J. Astrobiol. 9, 51-58. doi.org/10.1017/S1473550409990371.

Picard, A., and Daniel, I. (2013). Pressure as an environmental parameter for microbial life - a review. Biophys. Chem. 183, 30-41. doi: 10.1016/j.bpc.2013.06.019.

Pokorny, N.J., Boulter-Bitzer, J.I., Hart, M.M., Storey, L., Lee, H., and Trevors, J.T. (2005). Hypobaric bacteriology, growth cytoplasmic membrane polarization and total cellular fatty acids in *Escherichia coli* and *Bacillus subtilis*. Int. J. Astrobiol. 4, 187-193. doi.org/10.1017/S1473550405002727.

Portner, D.M., Spiner, D.R., Hoffman, R.K., and Phillips, C.R. (1961). Effect of ultrahigh vacuum on the viability of micro-organisms. Science. 134, 2,047. doi: 10.1126/science.134.3495.2047.

Pulschen, A.A., Guarany de Araujo, G., Souza Ramos de Carvalho, A.C., Cerini, M.F., de Mendonça Fonseca, L., Galante, D., and Rodrigues, F. (2018). Survival of extremophilic yeasts in the stratospheric environment during balloon flights and in laboratory simulations. Appl. Environ. Microbiol. 84, e01942-18. doi: 10.1128/AEM.01942-18.

Rabbow, E., Horneck, G., Rettberg, P., Schott, J.-U., Panitz, C., L'Afflitto, A., von Heise-Rotenburg, R., Willnecker, R., Baglioni, P., Hatton, J., Dettmann, J., Demets, R., and Reitz, G. (2009). EXPOSE, an astrobiological exposure facility on the International Space Station – from proposal to flight. Orig. Life Evol. Biosph. 39, 581. doi.org/10.1007/s11084-009-9173-6.

Rabbow, E., Rettberg, P., Barczyk, S., Bohmeier, M., Parpart, A., Panitz, C., Horneck, G., von Heise-Rotenburg, R., Hoppenbrouwers, T., Willnecker,

R., Baglioni, P., Demets, R., Dettmann, J., and Reitz, G. (2012). EXPOSE-E: An ESA astrobiology mission 1.5 years in space. Astrobiology. 12, 374-386. doi.org/10.1089/ast.2011.0760.

Rabbow, E., Parpart, A., and Reitz, G. (2016). The planetary and space simulation facilities at DLR Cologne. Microgravity Sci. Technol. 28, 215-229.

Rabbow, E., Rettberg, P., Parpart, A., Panitz, C., Schulte, W., Molter, F., Jaramillo, E., Demets, R., Weiß, P., and Willnecker, R. (2017). EXPOSE-R2: The astrobiological ESA Mission on board of the International Space Station. Front. Microbiol. doi.org/10.3389/fmicb.2017.01533.

Rothschild, L J., and Mancinelli, R L. (2001). Life in extreme environments. Nature. 209, 1092-1101.

Rummel, J.D., Beaty, D.W., Jones, M.A., Bakermans, C., Barlow, N.G., Boston, P.J., Chevrier, V.F., Clark, B.C., de Vera, J.-P., Gough, R.V., Hallsworth, J.E., Head, J.W., Hipkin, V.J., Kieft, T.L., McEwen, A.S., Mellon, M.T., Mikucki, J.A., Nicholson, W.L., Omelon, C.R., Peterson, R., Roden, E.E., Lollar, B.S., Tanaka, K.L., Viola, D., and Wray, J.J. (2014). A new analysis of Mars "Special Regions": Findings of the Second MEPAG Special Regions Science Analysis Group (SR-SAG2). Astrobiology. 14, 887-968. doi: 10.1089/ast.2014.1227.

Russell, J.A., León-Zayas, R., Wrighton, K., and Biddle, J.F. (2016). Deep subsurface life from North Pond: enrichment, isolation, characterization and genomes of heterotrophic bacteria. Front. Microbiol. 7, 678. doi: 10.3389/fmicb.2016.00678.

Sakon, J.J., and Burnap, R.L. (2006). An analysis of potential photosynthetic life on Mars. Int. J. Astrobiol. 5, 171-180. doi.org/10.1017/S1473550406003144.

Sánchez, F.J., Meessen, J., Ruiz, M.C., Sancho, L.G., Ott, S., Vilchez, C., Horneck, G., Sadowsky, A., and de la Torre, R. (2014). UV-C tolerance of symbiotic *Trebouxia* sp. in the space-tested lichen species *Rhizocarpon geographicum* and *Circinaria gyrosa*: role of the hydration state and cortex/screening substances. Int. J. Astrobiol. 13, 1-18. doi.org/10.1017/S147355041300027X.

Sancho, L.G., de la Torre, R., Horneck, G., Ascaso, C., de los Rios, A., Pintado, A., Wierzchos, J., and Schuster, M. (2007). Lichens survive in space: results from the 2005 LICHENS experiment. Astrobiology. 7, 443-454. doi: 10.1089/ast.2006.0046.

Schuerger, A.C., Mancinelli, R.L., Kern, R.G., Rothschild, L.J., and McKay, C.P. (2003). Survival of endospores of *Bacillus subtilis* on spacecraft surfaces under simulated martian environments: Implications for the forward contamination of Mars. Icarus. 165, 253-276.

Schuerger, A.C., Richards, J.T., Hintze, P.E., and Kern, R.G. (2005). Surface characteristics of spacecraft components affect the aggregation of microorganisms and may lead to different survival rates of bacteria on Mars landers. Astrobiology. 5, 545-559. doi: doi.org/10.1089/ast.2005.5.545.

Schuerger, A.C. and Nicholson, W.L. (2006). Interactive effects of hypobaria, low temperature, and CO_2 atmospheres inhibit the growth of mesophilic *Bacillus spp.* under simulated martian conditions. Icarus. 185, 143-152. doi.org/10.1016/j.icarus.2006.06.014.

Schuerger, A.C., Fajardo-Cavazos, P., Clausen, C.A., Moores, J.E., Smith, P.H., and Nicholson, W.L. (2008). Slow degradation of ATP in simulated martian environments suggests long residence times for the biosignature molecule on spacecraft surfaces on Mars. Icarus. 194, 86-100. doi.org/10.1016/j.icarus.2007.10.010.

Schuerger, A.C., Golden, D.C., and Ming, D.W. (2012). Biotoxicity of Mars soils: 1. Dry deposition of analog soils on microbial colonies and survival under Martian conditions. Planet. Space Sci. 72, 91-101. doi.org/10.1016/j.pss.2012.07.026.

Schuerger, A.C., Ulrich, R., Berry, B.J., and Nicholson, W.L. (2013). Growth of *Serratia liquefaciens* under 7 mbar, 0 °C and CO_2-enriched anoxic atmospheres. Astrobiology. 13, 115-131. doi: 10.1089/ast.2011.0811.

Schuerger, A.C., and Nicholson, W.L. (2016). Twenty species of hypobarophilic bacteria recovered from diverse soils exhibit growth under simulated martian conditions at 0.7 kPa. Astrobiology. 16, 964-976. doi: 10.1089/ast.2015.1394.

Schwendner, P., and Schuerger, A.C. (2018). Metabolic fingerprints of *Serratia liquefaciens* under simulated Martian conditions using Biolog GN2 microarrays. Sci. Rep. 8, 15721. doi: 10.1038/s41598-018-3356-3.l

Sears, D.W.G., Benoit, P.H., McKeever, S.W.S., Banjerjee, D., Kral, T., Stites, W., Roe, L., Jansma, P., and Mattioli, G. (2002). Investigation of biological, chemical and physical processes on and in planetary surfaces by laboratory simulation. Planet. Space Sci. 50, 821-828. doi.org/10.1016/S0032-0633(02)00056-9.

Sears, D.W.G., and Chittenden, J.D. (2005). On laboratory simulations and the temperature dependence of the evaporation rate of brine on Mars. Geophys. Res. Lett. 32, L23204. doi: 10.1029/2005GL024154.

Sephton, M. A. and Botta, O. (2005). Recognizing life in the Solar System: Guidance from meteoritic organic matter. Int. J. Astrobiol. 4, 269-276. doi: 10.1017.S1473550405002806.

Silverman, G.J., Davis, N.S., and Keller, W.H. (1964). Exposure of microorganisms to simulated extraterrestrial space ecology. Life Sci. Space Res. 2, 372-383.

Simonato, F., Campanaro, S., Lauro, F.M., Vezzi, A., D'Angelo, M., Vitulo, N., Valle, G., and Bartlett, D.H. (2006). Piezophilic adaptation: a genomic point of view. J. Biotechnol. 126, 11-25. doi: 10.1016/j.jbiotec.2006.03.038.

Smith, D.J. (2013). Microbes in the upper atmosphere and unique opportunities for astrobiology research. Astrobiology. 13, 981-990. doi: 10.1089/ast.2013.1074.

Stan-Lotter, H., Radax, C., Gruber, C., Legat, A., Pfaffenhuemer, M., Wieland, H., Leuko, S., Weidler, G., Komle, N. I., and Kargl, G. (2003).

Astrobiology with haloarchaea from Permo-Triassic rock salt. Int. J. Astrobiol. 1, 271-284. doi.org/10.1017/S1473550403001307.

Taylor, P.A., Kahanpää, O., Weng, W., Akingunola, A., Cook, C., Daly, M., Dickinson, C., Harri, A.-M., Hill, D., Hipkin, V., Polkko, J., and Whiteway, J.A. (2010). On pressure measurements and seasonal variations during the Phoenix mission. J. Geophys. Res. 115, E00E15. doi: 10.1029/2009JE003422.

ten Kate, I.L., Ruiterkamp, R., Botta, O., Lehmann, B., Gomez-Hernandez, C., Boudin, N., Foing, B.H., and Ehrenfreund, P. (2003). Investigating complex organic compounds in a simulated Mars environment. Int. J. Astrobiol. 1, 387-399. doi.org/10.1017/S1473550403001277.

Thomas, D.J., Sullivan, S.L., Price, A.L., and Zimmerman, S.M. (2005). Common freshwater cyanobacteria grow in 100% CO_2. Astrobiology. 5, 66-74. doi.org/10.1089/ast.2005.5.66.

Thomas, D.J., Eubanks, L.M., Rector, C., Warrington, J., and Todd, P. (2008). Effects of atmospheric pressure on the survival of photosynthetic microorganisms during simulations of ecopoesis. Int. J. Astrobiol. 7, 243-249. doi.org/10.1017/S1473550408004151.

Van Horn, K.G., Warren, K., and Baccaglini, E.J. (1997). Evaluation of the AnaeroPack System for growth of anaerobic bacteria. J. Clin. Microbiol. 35, 2170-2173.

Waters, S.M., Robles-Martinez, J.A., and Nicholson, W.L. (2014). Exposure of *Bacilllus subtilis* to low pressure (5 kilopascals) induces several global regulons, including those involved in the SigB-mediated general stress response. Appl. Environ. Microbiol. 80, 4788-4794. doi: 10.1128/AEM.00885-14.

Waters, S.M., Zeigler, D.R., and Nicholson, W.L. (2015). Experimental evolution of enhanced growth by *Bacillus subtilis* at low atmospheric pressure: genomic changes revealed by whole-genome sequencing. Appl. Environ. Microbiol. 81, 7525-7532. doi: 10.1128/AEM.01690-15.

Winter, R., and Jeworrek, C. (2009). Effect of pressure on membranes. Soft Mat. 5, 3157-3173.

Yayanos, A.A. (1995). Microbiology to 10,500 meters in the deep sea. Ann. Rev. Microbiol. 49, 777-805. doi.org/10.1146/annurev.mi.49.100195.004021.

Zeng, X., Birrien, J.-L., Fouquet, Y., Cherkashov, G., Jebbar, M., Querellou, J., Oger, P., Cambon-Bonavita, M.-A., Xiao, X., and Prieur, D. (2009). *Pyrococcus* CH1, an obligate piezophilic hyperthermophile: extending the upper pressure-temperature limits for life. The ISME J. 3, 873-876. doi.org/10.1038/ismej.2009.21.

ZoBell, C.E., and Johnson, F.H. (1949). The influence of hydrostatic pressure on the growth and viability of terrestrial and marine bacteria. J. Bacteriol. 57, 179-189.

Chapter 8

Earth's Stratosphere and Microbial Life

Priya DasSarma[1], André Antunes[2], Marta Filipa Simões[2] and Shiladitya DasSarma[1]*

[1]Institute of Marine and Environmental Technology, Department of Microbiology and Immunology, University of Maryland School of Medicine, Baltimore, Maryland, USA
[2]State Key Laboratory of Lunar and Planetary Sciences, Macau University of Science and Technology (MUST), Avenida Wai Long, Taipa, Macau SAR, China

shiladityadassarma@gmail.com

DOI: https://doi.org/10.21775/9781912530304.08

Abstract

The Earth's atmosphere is an extremely large and sparse environment which is quite challenging for the survival of microorganisms. We have long wondered about the limits to life in the atmosphere, starting with Leeuwenhoek's observation of "animalcules" collected from the air. In the past century, significant progress has been made to capture and identify biological material from varying elevations, from a few meters above ground level, to the clouds near mountaintops, and the jet streams, the ozone layer, and even higher up in the stratosphere. Collection and detection techniques have been developed and advanced in order to assess the potential diversity of life from very high altitudes. Studies of microbial life in the stratosphere with its multiple stressors (cold, dry, irradiated, with low pressure and limited nutrients), have recently garnered considerable attention. Here, we review studies of Earth's atmosphere, with emphasis on the stratosphere, addressing implications for astrobiology, the dispersal of microbes around our planet, planetary protection, and climate change.

Introduction

Why is it important to study and understand the limits to life at the highest elevations in the atmosphere? Elevation is a limit to life that has not yet been fully explored. It is relevant for space exploration, the search for life outside our planet, as well as more down-to-Earth concerns such as climate and disease.

In this review, we focus on conditions in the upper atmosphere, especially the stratosphere (Figure 1), and primarily on what is known about how microorganisms can survive there, under extreme conditions. The stratosphere is the highest elevation of the atmosphere where life has been found and hence its importance in studies of microbiology in our atmosphere. Astrobiology research has utilized Earth's stratosphere and its conditions as an analogue to conditions found on Mars, addressing whether extremophilic and other hardy microorganisms that survive in the stratosphere may tolerate conditions on the surface of Earth's sister planet. Here, we explore both the stressors encountered and the associated microbial responses. However, since studies of the stratosphere are relatively few, we also refer to a set of atmospheric analyses conducted in the troposphere or closer to Earth's surface for comparison and contrast. We present a summary of the studies which have contributed to our current level of understanding as well as the implications for astrobiology, climate, and health.

A relatively limited number of microbes have been collected from or exposed to the stratosphere, and most of them were either dormant or otherwise metabolically inactive, such as spores and lyophilized cells (Moeller and Horneck, 2004). In a sole instance, metabolically active microbes in growth media were launched into the stratosphere and returned to Earth (DasSarma and DasSarma, 2018). These cells retained viability, suggesting that ice entrapment may enhance microbial survival. Sampling and experimentation missions in the stratosphere are technically challenging and further work needs to be done to catalog and characterize cells that successfully survive. Technical issues of these studies, such as potential contamination during sample collection, can be a significant concern. Nevertheless, it is imperative to understand microbial survival strategies from the perspective of life in multiple extreme conditions like the stratosphere.

From Aerobiology to Exobiology
Aerobiology is the scientific field that studies the passive transport of biological particles through the atmosphere and its effect on living systems and the environment. In general, it has focused on particles of biological origin (bioaerosols), ranging from 0.2 to 2.5 µm in diameter, which make up 25% of atmospheric particles (Griffin et al., 2018). With higher elevation and the complete disappearance of the atmosphere, aerobiology transitions to space biology or astrobiology (used by several authors as a synonym for exobiology). As early as 1960, Lederberg first used the term "exobiology" to describe the exploration of life in higher elevations, from the stratosphere into the realm of space, during his presentation at the 1st

International Space Science Symposium sponsored by the international Committee on Space Research (COSPAR) (Lederberg, 1960).

Research in aerobiology began with Earth-based questions, focusing on the spread of diseases. In 1546, Girolamo Fracastoro suggested that disease was transmitted not only by direct contact with a sick person, or contagion through contaminated objects, but also by transmission through the air at a distance or *ad distans* (Dubos, 1986). After the discovery of the microscope, Antoine van Leeuwenhoek determined, in 1702, that "animalcules" could be carried by the wind, together with dust floating in the air. John P. Ehrenberg examined air and dust specimens, collected by Charles Darwin on his trip on the H.M.S. Beagle in the 1830s, and found that it was composed of a multitude of "infusoria" (Darwin, 1846), a term commonly used at that time to describe microscopic life. In another study, Ehrenberg concluded the existence of an atmospheric "kingdom" of life, detected 6 km high in the Himalayas (Cunningham, 1873).

In one of the first experimental studies of microbial composition versus elevation, Louis Pasteur used swanneck flasks to demonstrate that the number of microbes found in the air diminished with increasing altitude and also varied by location, time and atmospheric conditions. He found that there were fewer microbes at higher altitudes, such as at the peak of Montan Verte at 2 km height (Pasteur, 1860). Soon after that, H.G. Dyar examined air in New York, in the 1890s, and reported the presence of microorganisms, predominantly *Micrococcus*, *Bacillus* and *Sarcina*, albeit at much more moderate elevations as these were collected closer to the surface (Dyar, 1894).

A number of studies starting in 1921 addressed the spread of fungal diseases of wheat and other agricultural grains in the United States (US). In 1934, a survey conducted dozens of flights over Boston and detected bacteria, molds, yeast and pollen at a height of 5-6 km (Proctor, 1934). Subsequently, the US Department of Agriculture commissioned further studies on the epidemiology of rusts and other plant diseases, leading F.C. Meier and Charles Lindbergh to collect bioaerosols from about 3 km elevation above sea level (ASL). For sample collection, they used an oiled microscope slide extended from the plane by a metal arm. This study collected samples from Maine to Copenhagen (via the Arctic) and tentatively identified microbes belonging to *Macrosporium*, *Cladosporium*, *Leptosphaeria*, *Mycosphaerella*, *Trichothecium*, *Helicosporium*, *Uromyces*, *Camasosporium*, and *Venturia*. They also found diminishing numbers as collections progressed over the sea ice cap of Greenland (Meier and Lindbergh, 1935).

Around the same time, a manned US high-altitude balloon, Explorer II, became the first air sampling mission to reach the stratosphere (up to 21 km ASL), and several viable microbes were isolated within the genera

Bacillus, Macrosporium, Aspergillus, Penicillium and *Rhizopus*, using autoclaved collection tubes (Rogers and Meier, 1936). In 1965, G.A. Soffen flew balloons even higher, up to 40 km ASL, and used an ethylene oxide-sterilized impactor for isolation, but only found *Penicillium* species (Soffen, 1965).

In the 1970s, A. A. Imshenetsky and colleagues collected samples of air from even higher elevations, from the stratosphere to the mesosphere (48-85 km ASL), using γ-radiation sterilized meteorological rockets and investigated the characteristics of the bacterial and fungal strains isolated. They also studied the effects of the various atmospheric stressors, which continues to be one of the most remarkable investigations of microbial isolates from the highest elevations ever reported (Imshenetsky et al., 1976; Imshenetsky et al., 1977; Imshenetsky et al., 1978; Imshenetsky et al., 1979).

These early studies set the stage for more recent endeavors, including collection from the stratosphere by both planes and balloons (DeLeon-Rodriguez et al., 2013; Griffin, 2004; Smith et al., 2010). Balloons continue to be valuable for stratospheric studies as they can remain in place for sampling, exposure and experimentation, and may be flown to higher altitudes carrying larger payloads than most planes. In addition, balloons are advantageous because they can be maneuvered to many different locations and do not require expensive landing strips when returning samples to Earth (Smith et al., 2010; Smith and Sowa, 2017).

The Atmosphere
Many recent atmospheric studies have resulted in significant implications for health, including providing a better understanding of epidemiology, climate patterns and change, and planetary protection. However, the challenges continue to be considerable, with the atmosphere constituting the largest fraction of the biosphere. The mass of the Earth's atmosphere is 5.1×10^{18} kg (or ~1/1,200,000th of the planet), with 50 % of atmospheric mass located above 5.6 km, 10 % above 16 km, and 0.1 % above 100 km (Lutgens and Tarbuck, 1995). The diversity and distribution of atmospheric life are considerably sparse when compared to terrestrial and aquatic environments. Answers to fundamental questions about the nature of life in the atmosphere and its survival are likely to lead to very significant results for both human health and the health of the planet.

Evolution of Earth's atmosphere
Earth's atmosphere has functioned for eons as a protective buffer for life on our planet. After the initial Hadean era, more than 3.4 billion years ago, before the evolution of life, the atmosphere primarily consisted of hydrogen and helium gases. As the atmosphere cooled, it became rich in nitrogen and carbon dioxide gases. During this very early period of the Earth's history, the sun was also dimmer by about 30% compared to the present

day. Around 2.3 billion years ago, the *Great Oxidation Event* resulted in the atmosphere becoming oxygen-rich and evolving into its modern composition over time. Solar radiation catalyzed the conversion of molecular oxygen into ozone in the upper atmosphere, which resulted in heating and temperature inversion, providing the planet with an UV-protective layer in the stratosphere, critical for subsequent development and successful spread of life (Henderson and Salem, 2016; DasSarma and DasSarma, 2018; DasSarma and Schwieterman, 2018).

In the Anthropocene, many environmental factors contribute to further changes in the atmosphere. Jet travel, rocketry, military activities, and industrial and household pollution are all contributing to these changes and the natural barriers that prevent the transport and mixing between the layers of the atmosphere, e.g. troposphere and stratosphere, are being disrupted. Nevertheless, there are still some important questions that remain to be answered: a) Are anthropogenic activities responsible for the eutrophication of clouds and the atmosphere?, b) To what extent are anthropogenic activities contributing to climate change?, c) To what degree is the higher atmosphere, especially the stratosphere, subject to anthropogenic factors?, and d) Is it a habitat/place where life is expanding and serving as conduit for the dispersal of bioaerosols?

Layers of the Atmosphere
The atmosphere is divided into five main layers (from closest to furthest from the Earth's surface) (Figure 1), with this review focusing only on the lowest two (i.e. the ones where life seems to thrive).

Troposphere
The lowest level of the atmosphere, the troposphere, clearly contrasts with the stratosphere in terms of the necessary components for life maintenance: nutrients, in the form of dust particles and chemical compounds, air with all its components, water and water vapor. Like Earth's atmosphere in general, it is composed of 78% nitrogen, 21% oxygen and 1% of trace gases (argon, carbon dioxide, and most of the atmospheric water vapor). Temperature in this layer decreases with increasing height by ~7 °C/km, ranging from 15 to -60 °C (Vargin et al., 2015). The troposphere is where most weather events occur and is a characteristically turbulent and well-mixed layer. Jet streams, associated with frontal weather (where two different air masses meet), occur at or below the tropopause.

Most studies carried out thus far on life in the atmosphere have been within the troposphere. These studies provide the foundation for stratospheric analyses, addressing many of the challenges found at the higher altitudes. Tropospheric research has shown not only the presence of microbes, but also their metabolic activity (e.g. Klein et al., 2016). Culturing and metagenomic studies have identified a range of microbes that may be residents or have been transported into the troposphere where they may

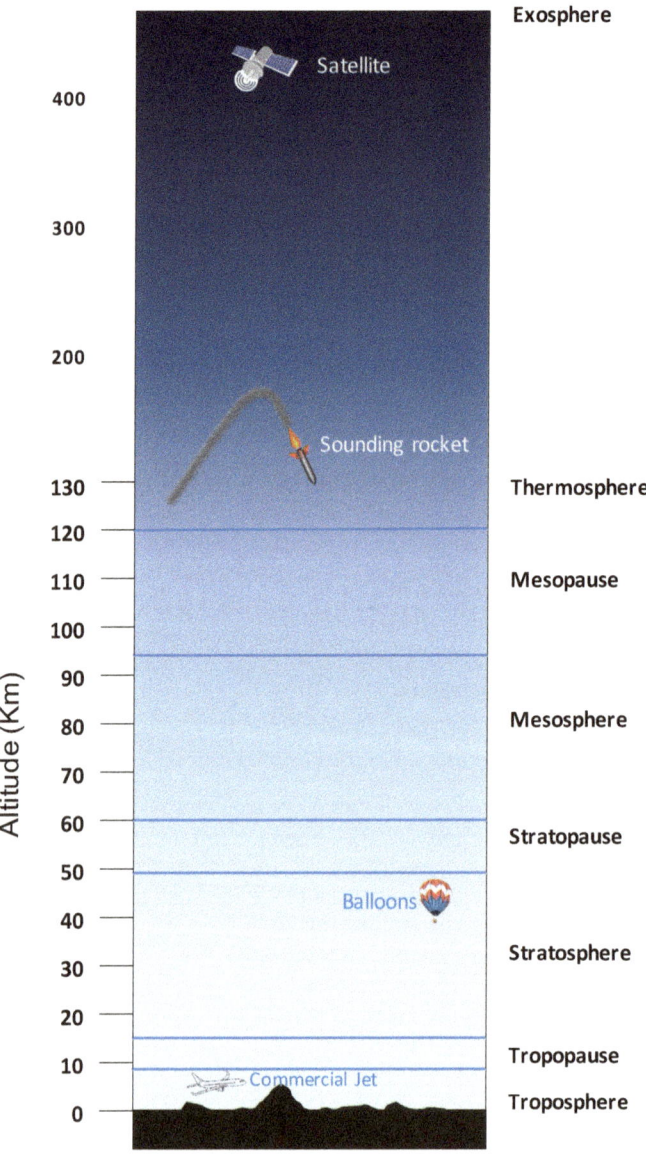

Figure 1. Overview of the different layers of Earth's atmosphere and their characteristics.

> **Exosphere**
> The top layer is the exosphere, located above the thermosphere and also considered to be part of space. It extends to at least 10,000 km, where it merges with the Solar wind (a constant stream of plasma and charged particles released from the corona, the Sun's outer layer). The exosphere is where most Earth-orbiting satellites are located.
>
> **Thermosphere**
> The thermosphere is located at heights above the mesopause with temperatures up to 1700 oC. It has a very low density of particles, is cloudless, and lacks water vapor altogether. This is where the International Space Station is located (at 350-420 km). The thermosphere ends with the thermopause, also called the exobase, ~600-1000 km, depending on solar activity.
>
> **Mesosphere**
> The mesosphere occupies heights from the top of the stratopause to 95-120 km, where the temperature decreases with height, at which point the mesopause, known as the coldest place on Earth with average temperature of -85oC is located. This is the layer where meteors burn up, sounding rockets and rocket-powered aircraft travel, and where the highest clouds, noctilucent, or night-glowing, made of ice crystals, and seen only during astronomical twilight exist.
>
> **Stratosphere**
> The stratosphere occupies heights from the top of the troposphere up to 50-60 km. Here, the temperature increases with height. This is the highest region where jet planes fly (~10-13 km). The stratosphere ends at the stratopause (Figure 1).
>
> **Troposphere**
> The troposphere occupies 15-18 km heights in the tropics and 10-12 km at the Polar Regions, with the temperature decreasing with increasing height. At the top of the troposphere, the tropopause separates the troposphere from the stratosphere.

be dormant or dead. The troposphere is considered a microbial ecological niche and ecosystem that requires study with efforts similar to that used in terrestrial and aquatic environments.

Clouds
About 15% of the first 6 km of the atmosphere consists of clouds, and several studies have focused on the importance of nucleation in both cloud and fog formation (Delort et al., 2010). Both homogeneous and heterogeneous particles are found near clouds and close to the tropopause. Ultrafine particles (3-15 nm in diameter) can serve as cloud condensation nuclei, which can grow to 100 nm within a few days (Kulmala et al., 2004).

Culturing, phylogenetics, and metagenomics have been used to extensively analyze cloud matter (Xu et al., 2017). Culture-based studies have identified bacteria (including α-, β- and γ-*Proteobacteria*, *Bacteroidetes*, *Firmicutes* and *Actinobacteria*), and fungi (Basidiomycetes and Ascomycetes) 1.5 km above the volcanic dome of Puy de Dôme. They found that less than 1% of bacteria and about 10% of fungi (e.g. *Dioszegia* and *Udeniomyces*) were reported to be culturable (Vaïtilingom et al., 2012). Another study reported that ~17% of known fungal species can be grown in culture (Amato et al., 2017).

Microorganisms are also able to act as cloud condensation nuclei and their presence is conditioned by several factors, including local presence of carbon sources and *in situ* pH. Data from the troposphere includes carboxylic acids and alcohols at concentrations of up to 1 mg/L, and a variety of hydrocarbons at concentrations ≤4 ng/L and pH ranging from 3-7, with acidity resulting from dissolved gases and compounds from aerosols in cloud water. Sulfate and nitrate nutrients in cloud water and rainwater can be relatively high and even reach levels typically found in oligotrophic lakes. The microorganisms previously detected in clouds include: *Micrococcus agilis*, *Mycoplana bullata*, and *Brevundimonas diminuta*, as well as plant pathogens such as *Erwinia carotovora* (Pérez-Díaz et al., 2017). Furthermore, species of the genus *Pseudomonas* detected in clouds have been shown to have nucleation properties, as they produce biosurfactants that facilitate the condensation of water on the surface of their cells. As a result, they are able to induce cloud formation, enhancing precipitation in the form of rain or snow (Amato, 2012).

Studies quantifying ATP concentrations and using differential staining have found that as much as ~1 million tons/year of organic carbon is metabolized by bacteria in the clouds (Vaïtilingom et al., 2013). Among these are the metabolically active oligotrophic, pigmented *Sphingomonas* spp. which have a high resistance to UV, cold and salinity, and can tolerate relatively high concentrations of oxidants. In addition, *Pseudomonas* spp. were found to use a variety of carbon compounds, though they were found to be less resistant to UV and oxidants. These microbes may use the atmosphere and clouds for both residence as well as for transport, until rain or snow brings them back to the Earth's surface (Delort et al., 2010; Xu et al., 2017). In another study, metagenomic analysis of cloud water samples from Mt. Tai (~1.5 km elevation), in China, also identified *Proteobacteria*, *Bacteroidetes*, *Firmicutes*, and *Actinobacteria* as predominant bacterial groups. Here, researchers also concluded that ozone and sulfur dioxide (SO_2) contributed to the variability of populations in these environments (Xu et al., 2017).

Dust
While microbes may be expected in the humid environment of clouds, studies into the transport of microbes by dust have led to some unexpected discoveries. Sources of aerosolized particulate matter can originate from e.g. vehicular pollution, construction, and industry as well as wind erosion (Griffin et al., 2018). Erosion sources include the ~10 million acres of farmland annually that are lost due to poor agricultural techniques and weather, dust storms from deserts, overgrazing, and deforestation. Most

wind-borne bacteria may be transported for relatively short distances, <1 km from their source, although some seem to be transported for extremely long distances, reportedly over 5,000 km (Kellogg and Griffin, 2006). Each year, more than 90x10^9 kg of dust are lifted from the Sahara Desert into the atmosphere, making up the Saharan air layer at about 6 km ASL, and crossing the Atlantic in 5–7 days (Di Liberto, 2018; Griffin et al., 2003). As an additional example, each year, ~8 million metric tons of lake-bed sediment from the dry Lake Owens bed in California are transported into the atmosphere and make up the primary source of atmospheric dust in the continental USA (Griffin et al., 2002).

It may be possible for microorganisms to take a ride in the tropospheric and stratospheric wind currents (Creamean et al., 2013; Barberán et al., 2015). Dust-associated bacterial and fungal spores have been reported to be transported across the Atlantic, from Africa to the Caribbean (Kellogg and Griffin, 2006). It has been calculated that the Sahara and Sahel regions of North Africa account for approximately 50-75% of the annual total atmospheric dust load (Griffin et al., 2018). In one study, halophilic endospores associated with Asian dust, or KOSA (the Japanese term for yellow sand), particles and other bioaerosols were mentioned as being transported across Asia from the Gobi Desert to Japan by dust storms (Echigo et al., 2005). Furthermore, it was found that these KOSA particles can act as enhancers of microbial growth (Maki et al., 2011).

Bacillus and *Microbacterium* were reported as the predominant bacteria isolated in several studies from African dust storms (Kellogg and Griffin, 2006). Metagenomic analyses of Caribbean and African dust storm events found that the microbes identified were the same. For example, the 18S sequences of *Cladosporium* isolates from the Caribbean samples were 99–100% identical to an isolate from African dust, indicating that these microorganisms may have been transported through the atmosphere (Kellogg and Griffin, 2006).

Results of such studies vary depending on numerous factors, including sampling methods, analysis and likely variability in the atmosphere at any given location and weather condition. For example, in two studies from Taiwan, dust from various time points was collected, including storms that transport dust from China and Mongolia to Taiwan, and spores were morphologically identified using microscopy-based analysis. In one study, ascospores and spores from *Cladosporium*, *Penicillium*, and *Aspergillus*, were found to be dominant in dust; though, only *Cladosporium* spore levels increased during dust storms. In addition, *Ganoderma*, *Arthrinium*, *Papularia*, *Cercospora*, *Periconia*, *Alternaria, and Botrytis* spores were also identified (Ho et al., 2005). In another study, the dominant spores of *Penicillium*, *Aspergillus*, *Nigrospora*, *Arthrinium*, *Curvularia*, *Stemphylium*, *Cercospora*, and *Pithomyces*, were found 15 m above ground level in calm conditions; and, only *Penicillium*, *Aspergillus*, *Nigrospora*, and some

Figure 2. View from the window of a commercial airliner at 12 km altitude. Below are tropospheric clouds, above, the clear zone of the stratosphere.

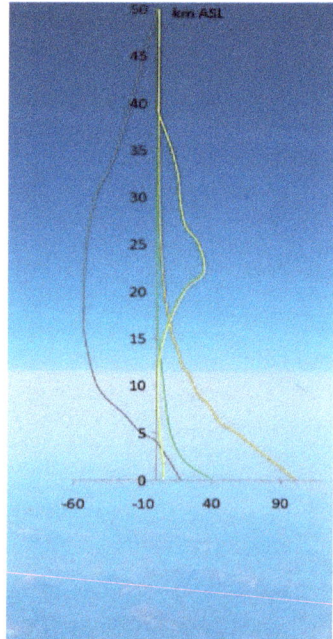

Figure 3. Selected characteristics of the troposphere and stratosphere. Average temperature (°C), gray; average water vapor (g/m^3), green; barometric pressure (kPa), orange; and ozone (mPa), yellow on x-axis, plotted against km above sea level (ASL) on y-axis.

unidentified spores were found to have increased concentrations during dust storm events (Wu et al., 2004).

The numbers of culturable airborne microorganisms were found to increase 2 to 3-fold during African dust-events. Direct microbial counts of air samples using epifluorescent microscopy, determined that bacteria- and virus-like particle counts were ~10-fold greater during these events than during clear conditions. Also, autofluorescence was exhibited by bacteria-like particles during an African dust-event, further supporting the presence of microorganisms (Griffin et al., 2001). Also, epifluorescent microscopy of nucleic acid stained filters of material from dust events showed that the bacterial and viral counts were the same, in contrast to results from soil and marine environments, where the viral counts were an order of magnitude higher. This suggests that viral particles are more susceptible to the high UV radiation and dry air associated with long-distance transport in dust events (Kellogg and Griffin, 2006).

At the top of the troposphere, the tropopause acts as a barrier to particulate matter, with peak heights over the equator and minimum heights over Polar Regions. Temperature inversion occurs in the tropopause, from decreasing with increasing height in the troposphere, to increasing with increasing height in the stratosphere (negative lapse rate) (Mohanakumar, 2008; Gettelman et al., 2011). The temperature in the tropopause is isothermal, and effectively stops the transfer of most aerosols from the troposphere into the stratosphere. It is also the region where the atmosphere becomes exceedingly dry (Gettelman et al., 2011).

Stratosphere
There is some variability in the height of different layers in the atmosphere, depending on factors such as e.g. latitude, season, atmospheric conditions and solar activity. The stratosphere (Figure 2), located above the troposphere and tropopause, can start as low as ~7 km near the poles and as high as 20 km at the equator and extends up to approximately 50-60 km ASL. Conditions here are among the most extreme on Earth and comparable to conditions found on the surface of Mars, making it valuable as a Mars analog (Figure 3). This Martian analogue status is supported by data from several studies. Indeed, the stratosphere has low nutrient availability, reduced atmospheric pressure (between 0.1 to 10 kPa), low temperatures (around -50 °C), presence of toxic chemical species, and intense solar radiation and is extremely dry (relative humidity ~23%) (Smith et al., 2011; Smith et al., 2013).

There is some drastic variation in conditions within the stratosphere. Near the top of this layer, the pressure is extremely hypobaric (~0.1% of that at sea level). Furthermore, its high concentrations of ozone and ozone layer restrict weather-producing turbulence and mixing. Temperature increases from -70°C at the tropopause, to 0°C at the top of the stratosphere due to

Table 1. Overview of microbes isolated from and/or tested in the stratosphere*.

Isolate name	Domain	Height (km ASL)	GC-composition*	Spore former?
Actinobacteria	Bacteria	18-29[I]	Variable	Some
Actinomyces sp.	Bacteria	19[T]	High	May form endospores
Aspergillus fumigatus, Aspergillus niger	Eukarya	11-21[I], 48-77[I], 19-22[T]	~ 50%	Yes
Bacillus endophyticus, Bacillus luciferensis, Bacillus pumilus SAFR-032, Bacillus simplex, Bacillus (Lysinibacillus) sphaericus, Bacillus subtilis	Bacteria	20-77[I], 20-30[T]	Low	Endospore forming
Brachysporium sp.	Eukarya	22[T]	ND	Spore-forming
Brevibacterium luteolum	Bacteria	20[I]	High	No
Circinella muscae	Eukarya	48-77[I]	ND	Spore-forming
Cladosporium sp.	Eukarya	22[T]	~ 50%	Spore-forming
Deinococcus aetherius, Deinococcus aerius TR0125	Bacteria	10-12[I,T]	High	No
Diplodia sp.	Eukarya	22[T]	High	ND
Engyodontium albus	Eukarya	41[I]	ND	Spore-forming
Escherichia coli	Bacteria	40[T]	~ 50%	No
Eurotiomycetes sp.	Eukarya	20[I]	ND	ND
Exophiala sp. 15LV1	Eukarya	25-30[T]	ND	ND
Fusarium sp.	Eukarya	19[T]	~ 50%	ND
Halobacterium species NRC-1	Archaea	36[T]	High	No
Halorubrum lacusprofundi	Archaea	36[T]	High	No
Helminthosporium sativum	Eukarya	22[T]	ND	ND
Hysterium sp.	Eukarya	22[T]	~ 50%	ND
Macrosporium sp.	Eukarya	11-21[I], 19[T]	ND	ND
Micrococcus albus	Bacteria	48-77[I]	High	No
Monilia sitophila	Eukarya	19[T]	ND	ND
Mycobacterium luteum	Bacteria	48-77[I]	ND	May form spores
Naganishia (Cryptococcus) friedmannii 16LV2	Eukarya	25-30[T]	ND	ND
Paenibacillus sp.	Bacteria	12-35[I]	~ 50%	Spore-forming
Papulaspora anomala	Eukarya	48-77[I]	ND	ND
Penicillium cyclopium, Penicillium chrysogenum (formerly notatum), Penicillium sp.	Eukarya	11-77[I], 19[T]	~50%	
Pestalozzia sp.	Eukarya	19[T]	ND	ND
Proteobacteria	Bacteria	18-29[I]	ND	ND
Proteus mirabilis	Bacteria	40[T]	Low	No
Pseudomonas aeruginosa	Bacteria	40[T]	High	No
Puccinia graminis	Eukarya	19[T]	Low	ND
Rhizopus sp.	Eukarya	11-21[I], 19[T], 22[T]	Low	ND
Salmonella enterica Serovar Typhimurium	Bacteria	40[T]	~50%	No
Staphylococcus pasteuri, Staphylococcus aureus MRSA, Staphylococcus aureus	Bacteria	41[I], 40[T]	Low	No

ozone absorbing UV rays and consequently releasing heat (Smith et al., 2011).

The stratosphere is almost devoid of clouds, with the exception of the tall cumulonimbus clouds, also called thunderheads, observed during storms, which can penetrate from the troposphere through the tropopause into the stratosphere. In the coldest polar regions, nacreous clouds may also be observed (Henderson and Salem, 2016). Cloud formation can sometimes occur at the tip of the boundary layer under extremely polluted conditions or during the winter in polar regions at temperatures below -78°C and form polar stratospheric clouds at 15-25 km (Ursem, 2016).

Chemical components of the stratosphere include: water vapor, methanol, nitrogen oxides and bromide, as well as a background aerosol layer consisting mainly of binary sulfuric acid-water aerosol droplets (Vargin et al., 2015; Smith and Sowa, 2017). Molecular hydrogen (H_2) is an atmospheric trace gas and acts as a source of water vapor in the stratosphere (Meredith et al., 2017). Most of the known aerosol nanoparticles in the stratosphere are found at 17-50 km ASL, and result from large volcanic eruptions and major meteorite impacts. They are known to have global effects for months or years and have been directly related to the Earth's climate alterations (Ursem, 2016).

For nearly 100 years, humans have been able to enter the stratosphere. In 2014, the Air Transport Action Group estimated the number of scheduled flights to be close to 100,000 per day. Adding to this number, other human sources of stratospheric breaches include orbital launches (which in 2017 consisted of 29 launches from the United States, 20 from Russia and 18 from China; Leary, 2018), and launches of weapons and other military material. All of these can contribute to the dispersal of hitchhiking microorganisms and organic elements into the upper troposphere, lower stratosphere, or higher, and is termed "artificial panspermia" (Lederberg, 1960).

In addition to potential microbial "hitchhikers" on these vehicles, there may also be microbial matter that made its way there via natural means. Natural sources include volcanic activity, storms, wind, fires, and clouds. Whether any of these microbes are metabolically active or even reproduce in the stratosphere is yet to be determined. To date, bacterial and fungal spores dominate the isolates found (Table 1). In addition, experiments that involve launches of microbes into the stratosphere have shown that a trip into the stratosphere is indeed non-fatal to metabolically inactive spores, lyophilized cultures of bacteria and fungi, and active cultures of halophilic archaea (Table 1) (DasSarma and DasSarma, 2018).

Extreme conditions in the Stratosphere
There are a multitude of extreme conditions that make it difficult for most microbes to survive stratospheric conditions, including: irradiation, desiccation, freeze-thaw cycles, low pressure, and lack of water and nutrients. Some of these stressors can also be observed on the Earth's surface, but there is a complex combination of factors in the stratosphere that more closely mimics conditions on Mars than alternative analogue surface sites. This makes the study of the stratosphere of great interest to astrobiologists, who use this natural environment as a laboratory to study both a) the effects of Martian-like conditions (as proxy), e.g.: by testing the viability of organisms, and b) native microbial communities and inhabitants found in these environments. Research into microbial viability in simulations has shown that sun-illuminated bioaerosols are quickly inactivated in the laboratory through application of these stressors (Griffin et al., 2018).

Among the deadliest stressors in the stratosphere are the cosmic rays that constantly bombard our planet. Most of these are either deflected by the Earth's magnetic field or interact with air molecules. Cosmic rays include galactic cosmic rays (from outside the solar system), anomalous cosmic rays (from interstellar space at the edge of the heliopause), solar energetic particles from solar flares and other energetic solar events, and other types of cosmic rays, which include X-rays, γ-rays, and the short wave ultra-violet (UV) portion of the electromagnetic spectrum (Christian, 2012). Cosmic rays are primary sources of ionizing radiation (IR) and carry enough energy to liberate electrons from atoms or molecules (UNSCEAR, 2008).

Damage from IR includes radiolysis of water that generates reactive oxygen species [ROS; e.g.: hydroxyl radicals (HO·), superoxide (O_2^-) and hydrogen peroxide (H_2O_2)], which cause damage to nucleic acids, generating oxidized DNA bases and sugar moieties, abasic (apurinic or/and apyrimidinic) sites, and single-stranded breaks (SSBs) (Hutchinson, 1985; Imlay, 2006). With increasing doses of IR, the linear density of DNA base damage and SSBs increases on both strands and gives rise to double stranded breaks (DSBs). ROS damages proteins by introducing carbonyl residues, amino acid radical chain reactions, cross-linking, and ultimately results in protein inactivation and denaturation (Stadtman and Levine, 2003). O_2^- does not easily cross membranes, and is not metabolized by the cell or react with DNA or most proteins; but it damages and inactivates enzymes with exposed 2Fe–2S or 4Fe–4S clusters causing release of Fe^{2+}, which in turn reacts with H_2O_2 and catalyzes Fenton reactions (oxidation of organic substrates) (Imlay, 2006). IR can also cause 40 times more SSBs than DSBs when macromolecules absorb X-ray and γ-ray photons, (von Sonntag, 1987; Daly et al., 1994).

Of all the stressors present in the stratosphere, short-wave UV radiation causes the greatest damage to cells. It causes an increase in ROS production, resulting in damage of biomolecules such as proteins, lipids, DNA and RNA. Stratospheric ozone reduces penetration of wavelengths <320 nm and completely excludes those <290 nm. Thus, UV-A (315-400 nm) which constitutes 95% of total energy of UV spectrum that reaches the Earth's surface does not damage organic material. The remaining 5% is UV-B (280-315 nm), which has the greatest biological impact on Earth. Some studies have predicted that aerosols are crucial factors involved in blocking UV-B radiation (Bais et al., 2018). UV-C (100-280 nm) is completely blocked by Earth's ozone layer in the stratosphere. However, above the ozone layer in the stratosphere, any bioaerosols present would experience high fluxes of UV-C radiation.

The destructive power of UV radiation was first determined to be an order of magnitude higher in air than in liquids (Wells and Fair, 1935). Whisler found that, effectiveness of UV to kill was determined not only by the source of radiation, but also by the type of organisms irradiated, with some air-borne microorganisms like *Micrococcus luteus, Staphylococcus aureus, and Bacillus subtilis* found to be considerably more tolerant of UV than *Escherichia coli*. *M. luteus* was found to be 100 times more resistant than *E. coli*, potentially due to clustering of their cells and shielding effects. *S. aureus* and sporulating cultures of *B. subtilis* were 3 and 8 times more resistant than *E. coli*, respectively. This lethal effect of UV radiation was also remarkably dependent on the relative humidity of the air, with cells being an order of magnitude more sensitive in dry air when compared to humid air (Whisler, 1940).

As conditions in the stratosphere also include low temperatures, low pressures and desiccating conditions, microorganisms in this environment would need to have cold- and freeze-adaptations, be able to tolerate hypobaric conditions and be xerotolerant (Rothchild and Mancinelli, 2001; Fletcher et al., 2014). Due to the sparseness of material in the stratosphere, it would be difficult for any living cells to find the building blocks of life needed to metabolize and reproduce, though attachment to particulate matter and the potential for a buildup of such materials, from natural or anthropogenic sources, may provide some resources.

Entry and return of material
Some of the natural uplift mechanisms, capable of driving vertical transport of aerosols into the atmosphere, include convective overshooting and strong up-drafting forces generated by processes associated with thunderstorms, typhoons, monsoons, hurricanes, and cyclones. Large-scale storms, especially hurricanes and cyclones, may reach into the lower stratosphere as they cross oceans. Blue jets, which are optical flashes above thunderclouds, propagate upwards from thunderclouds to ~70 km. Stratosphere-troposphere exchange occurs by deep convection in the tropics, tropopause folding, convective overshooting, as well as by meteors (Wilson et al., 1978; Pasko et al., 2002; Smith and Sowa, 2017; Berera,

2017, Griffin et al., 2018). Volcanic eruptions can send ash and other material 2-45 km and possibly higher (55 km) into the atmosphere, with a dust load of 4-25x10^6 tons in a single eruption (Griffin et al., 2018). Particle loads, from large eruptions, are known to force climate change through the resulting changes in planetary solar irradiance and contribute to bioaerosols being swept into the stratosphere (Rohatschek, 1984).

Aerosol particles' lifetime (such as those from volcanic eruptions) in the stratosphere, has been calculated to be 1 to 2 years and results in reduction of solar radiation to the Earth's surface and, in turn, reduction of surface temperature. This is unlike the troposphere, where particles are rapidly removed via precipitation (Vargin et al., 2015). Other calculations indicate that bacteria remain aloft for 2-10 days, and can travel over thousands of kilometers, while viruses are believed to be associated with particles in the nano- to micrometer range, with a predicted 2-188 days residency time (Cuthbertson and Pearce, 2017; Amato et al., 2017). Eventually, in months to years, depending on the aerosol size, stratospheric aerosols return to the surface by Brewer Dobson circulation which is believed to move air masses towards the poles (Smith and Sowa, 2017).

Micron scale particles can be elevated to altitudes of 80 km, due to irradiation of particles by sunlight through gravitophotophoretic effects and electrostatic levitation. In photophoresis, small particles, suspended in liquid or gas, start to migrate when illuminated by a sufficiently intense beam of light because of non-uniform temperature distribution. This levitation occurs only with negative photophoresis, and it has been predicted that pointing towards the sun would create a lifting component exceeding the gravitational force. Furthermore, it was also concluded that some particles in the stratosphere may rise against the force of gravity (Orr Jr and Keng, 1964; Rohatschek, 1996). Experimental evidence indicated that gravitophotophoresis in sunlight causes the ascent or the suspension in the middle atmosphere of carbonaceous, mineral, and metallic particles, mainly in the 1-10 μm size range, which otherwise would fall with considerable velocity (Rohatschek, 1984). Particles in the 1-100 μm size range were shown to levitate due to photophoretic forces in a laboratory simulation of solar and middle atmospheric air densities (Rohatschek, 1984). The studies of gravitophotophoresis suggested it may be the reason why microorganisms are found in the stratosphere and mesosphere (Imshenetsky et al., 1977). They also indicated that radiation absorbed by pigments produces photophoretic forces sufficient to overcome gravity (Rohatschek, 1984).

Bioaerosols
Materials of organic origin in the atmosphere are called bioaerosols, and can include microbes, spores and pollen. Some of this material is free-floating, while other can be found attached, trapped in, or on larger

particles or aggregates (Smith et al., 2012; Smith et al., 2013). Primary biological aerosols represent 5-10% of the total number of atmospheric particles >0.2 μm in diameter (Amato et al., 2017). It has been predicted that 40-1800 Gg of organic matter is aerosolized annually (Cuthbertson and Pearce, 2017). Bioaerosols may consist of raindrop-bubbles from soil which has been calculated to aerosolize ~0.01% of soil surface bacteria (Joung et al., 2017). In this phenomenon, raindrops impacting the soil, form tiny bubbles and rupture at the raindrop/air interface, and cause emission of tiny water jets that are subsequently broken into aerosols containing soil-associated bacteria that are released into the air column (Joung et al., 2017; Jang et al., 2018). Microbes and viruses can also be transferred from the ocean surface to the marine aerosol (Rastelli et al., 2017). Some bioaerosol microbes in the troposphere are believed to be metabolically active and capable of reproduction, with a calculated doubling time of 3.6-19.5 days (Cuthbertson and Pearce, 2017). The metabolic processes identified include nitrogen processing, sulfur oxidation and reduction, and photosynthesis (Cuthbertson and Pearce, 2017).

A survey of virus-like particles in the Sierra Nevada Mountains, Spain, above the atmospheric boundary layer (1.7 ± 0.5 km ASL), determined that the flux of viruses ranges from 0.26 to >7 × $10^9/m^2$ per day. These deposition rates were considerably greater than the rates for bacteria, which ranged from 0.3 to >8 × $10^7/m^2$ per day. The highest relative deposition rates for viruses were associated with atmospheric transport from marine rather than terrestrial sources. Virus deposition rates were positively correlated with organic aerosols <0.7 μm, whereas, bacteria were primarily associated with organic aerosols >0.7 μm, implying that viruses could have longer residence times in the atmosphere and, consequently, may be dispersed further (Reche et al., 2018). A similar survey of stratospheric material would be useful to expand our understanding of this phenomenon.

Studies of aerosolized material in the atmosphere have been undertaken in recent years using metagenomic techniques. In a study of aerosol material before and after heavy rains in Korea, researchers used pyrosequencing to identify *Actinobacteria*, *Firmicutes*, *Proteobacteria*, and *Bacteroidetes*, and lesser quantities of *Planctomycetes*, *Chloroflexi*, *Gemmatimonadetes*, and *Cyanobacteria* in post-rain samples (Jang et al., 2018). Surprisingly, no archaeal sequences were reported; and, levels of the marine bacterial sequences (α-*Proteobacteria*, *Actinobacteria*, and *Firmicutes*) decreased after rainfalls. Sequences corresponding to the non-spore forming *Actinobacteria* and the phytopathogens *Clavibacter michiganensis*, *Staphylococcus saprophyticus,* and the human pathogen, *Propionibacterium acnes*, were often observed in post-rain air samples, whereas the *Firmicutes* decreased in abundance. The *Firmicutes* were more abundant in coarse particles than in fine particles and the study considered that precipitation may selectively remove particle-associated bacterial groups, causing changes in airborne bacterial community composition after rainfall, despite increases in overall bioaerosol concentration.

Collection and analysis of material for microbiological studies

Over 10^{21} cells/year are estimated to be lofted into the atmosphere, with only a small fraction (<0.1%) believed to survive (Aguilera et al., 2018). How many of these actually exist in the stratosphere is still to be determined. Thus far, only a limited number of studies have been performed to address this question, and they include direct isolation from the stratosphere and simulations in the laboratory (DasSarma and DasSarma, 2018).

Analysis of collected material is conducted in two main ways: a) culturing and characterization, and b) DNA extraction and identification (Cuthbertson and Pearce, 2017; Griffin et al., 2018). Culturing results are dependent on both the collection method and media used, while DNA extraction will vary based on the methods used, which may selectively be easier for some microbes and more difficult for others. Polymerase chain reaction (PCR) and sequencing methods may also introduce biases based for example on the primer sequences used. Given the limited number of studies thus far, no preferred methods for stratospheric collection studies have arisen.

One of the earliest methods used, the plate-fall/drop-plate technique, involves leaving agar plates open to allow material to drop into it. This method is hard to implement when material is scarce, or gravity is less effective. It is probably not feasible for the stratosphere, since centrifugation would be a necessary step difficult to perform at this height, but has been used widely for work done in the past for lower levels of the atmosphere, as early as 1894, when the air of New York was sampled (Dyar, 1894). It limits the findings to only those microorganisms which are viable and can grow in the media used.

Impaction has also been used since early investigations (Meier and Lindbergh, 1935; Solomon et al., 1983). In this method, material is collected by collision of aerosols onto Petri dishes or other surfaces, which may be coated with sticky material (e.g. glycerol). For example, the Life's Atmospheric Microbial Boundary (LAMB) balloon payload collected ~100 cells per glycerol-coated rod, at ~38 km, in a passive sampling mission (Bryan et al., 2014; Griffin et al., 2018). The NASA Cosmic Dust Group were able to collect viable bacteria and fungi, on an impactor plate, in a housing located on the underside of a Lockheed Martin plane that was flown at 20 km ASL for 2.5 hours from New Mexico to California, USA (Griffin et al., 2004). The method allows for both culturing, as long as the media is suitable, and DNA analysis.

Filtration has been a more common isolation method performed on air samples and lower atmospheric materials. For example, in a study of tropospheric dust from dust storms, volumes of 139.5 and 269.7 liters were vacuumed during specific times (15 and 29 minutes, respectively) through pre-sterilized nylon filters with 0.2 µm pore size, and a portion of each filter was placed on agar plates for culturing. Using 0.02 µm pore size filters, air was also vacuumed for 10 to 15 minutes at a rate of 14.65 to 29.98 L/minute, for direct counting of average numbers of microbe-like particles using SYBR Gold staining (Griffin et al., 2001). Some problems with this method include possible damage of cellular material during the vacuum stage, and loss of cell viability from impaction with filter materials or media that does not support growth. This method is, at present, hard to perform at the stratospheric altitudes, though collected stratospheric material can be processed over filters. The method allows for both culturing as well as DNA analysis. Radosevich and colleagues (2002) collected air in Utah for which they developed an apparatus based on membrane filtration. From the particles collected, and through bacterial genomic DNA analysis, they were able to find a wide variety of species, many belonging to the phylum Proteobacteria. In other studies, sterile quartz and glass fiber filters were used to collect aerosols from the troposphere over the period of one year and identified archaeal DNA from *Thaumarchaeota* or *Euryarchaeota* with seasonal variation (Fröhlich-Nowoisky et al., 2014). Fungal spores had already been determined as being a large part of air particulate matter, and it had already been described a rate higher than initially expected, with a species richness of 368. These studies concluded that airborne diversity was closely linked to soil diversity, and that soil and soil dust might be the primary source of airborne microorganisms (Fröhlich-Nowoisky et al., 2009; Fröhlich-Nowoisky, 2014).

Cryosampling has also been used in the modern era with cooling of samples during the collection process. This process is expected to increase cell viability and facilitate both culturing and DNA extraction. In one study, sampling of air above Hyderabad, India, employed a cryosampler made of a 16-probe stainless steel manifold with remote-controlled motorized valves (Wainwright et al., 2003). The payload had a 2-meter-long intake tube which was tethered by a rope, both previously sterilized, behind the balloon gondola, in order to avoid the collection of any materials from the balloon surface. Throughout the ascent trip (up to 41 km), in order to create a cryopump effect, sequence probes were immersed in liquid neon allowing surrounding air to be collected when the valves were opened. Once all the collections were made, the cryosampler manifold was parachuted back to the ground. Collection of 38.4 and 18.5 liters of air at normal temperature and pressure, was accomplished at 30-39 and 40-41 km elevations, respectively. The aerosols were aseptically extracted by sequentially filtering the air through 0.45 and 0.22 µm micropore cellulose nitrate filters, and then injecting the probes with sterile phosphate buffer solution and agitating in order to release particles adhered to the walls. The subsequent liquid was then sequentially filtered through 0.7 µm glass, 0.45 µm and 0.2 µm cellulose acetate filters before analysis. Several different shaped bacteria were found to be present and one fungal species was isolated (Wainwright et al., 2003).

Another recent method, impingement, involves air collection by suction and then impinged/forced into liquid solutions such as water, buffer or oils. This method allows for the retrieval of the collection media for both microbial culturing (of selected microbes) as well as DNA extraction and subsequent analysis. However, this method is selective for only the cells that can survive in the collection medium. Many studies of the troposphere have used this method since the collection step allows for rapid sampling of large volumes and potential collection of both viable and nonviable material. Though more expensive than membrane filtration, high-flow rate impingers have the advantage of speed. When comparing different air sampling methods, impingers were considered as not appropriate or efficient for the cultivation of certain microorganisms (Cooper et al., 2019). Nevertheless, a study of atmospheric dust collection showed that impingers allow for microbial recovery rates 20x higher than membrane filtration (Griffin et al., 2011). In order to verify the absence of contaminations, impingers – like other air sampling equipment – can initially be analyzed in a sterile test chamber (Cooper et al., 2019).

The more complex cyclonic methods are designed to separate airborne dust into fractions. Air is drawn into a cylindrical chamber by rotation of the air flow, without the use of filters. As a result, aerosols of given sizes move toward the walls by centrifugal force, where they can be rapidly collected (Huard et al., 2010). The force of the collection process may result in some cellular damage and the composition of the walls may further impact microbial survival. This collection method is therefore more suited for DNA sequence analysis of samples than culturing. Furthermore, different designs will have different efficiency values (e.g.: Saunders et al., 2003; Chen et al., 2012).

Sterilization and aseptic techniques are obviously key to any microbiological work, as well as in the field of high-throughput sequencing (Spring et al., 2018). Many methods have been used to clean and sterilize collection vessels and materials, such as use of alcohol wipes, UV-, and ionizing-radiation. For example, in a cryosampling mission to the stratosphere, probes were flushed with acetone and heat sterilized to temperatures of 180 °C for several hours (Wainwright et al., 2003). More recently, enhanced techniques including the use of sodium hypochlorite for sterilization of refined sampling devises and other culturing and non-culturing methods have also been described (Bryan et al., 2014; 2019).

Some studies of the troposphere have involved using multiple approaches, e.g. both impingement and membrane-filter methods to collect material. For example, Jang et al. (2018) utilized such an approach before and after rainfalls in South Korea, albeit at the low altitude of just 16 m. Combination approaches may have the advantage of being more effective than using only a single method to gain broader insight. Nevertheless, collection and culturing of samples from high altitudes remain challenging and further studies are needed to establish standard protocols in high altitude microbiology.

Culturing has been used to identify microbes collected from the stratosphere. Some additional work has been done in comparing phenotypic traits of microbes collected from or exposed to the stratosphere (Table 1). For example, in the early 1930s, several microorganisms were sent up to the stratosphere, on the Explorer II balloon, in order to test the effect of drying, extreme cold, ozone, strong light rays, and low air pressure in the stratosphere on them. These microorganisms were isolated at various times, from the stratosphere (*Brachysporium* sp., *Hysterium* sp., *Rhizopus* sp. from 1.2 km; *Diplodia* sp., from 1.3 km; and, *Aspergillus niger*, and *Cladosporium* sp., from ~3 km), as well as from diseased sorghum plants (*Helminthosporium sativum*). When they returned to Earth, after a flight that lasted ca. 8 hours (Kennedy, 2018), most were able to grow to varying degrees, but one, *Hysterium* sp., did not (Meier, 1936). This early study illustrated differential effects of the stratosphere on different organisms.

As mentioned before, the equipment used for collection of aerosols may select for and against certain organisms as will culture media and cultivation conditions. Once samples have been collected, they can be inoculated into various media, which so far have mainly been restricted to common bacteriological growth conditions, rather than a wide range needed for culturing diverse microbes, especially those with specific or unusual requirements, such as anaerobic conditions or high salinity. For example, cryosampling of the stratosphere above Hyderabad (India), resulted in the isolation of four new species from the genus *Bacillus*: *B. aerius* sp. nov., *B. aerophilus* sp. nov., *B. stratosphericus* sp. nov., and *B. altitudinis* sp. nov. (Shivaji et al., 2006). In many studies, such as this one, the selection of microbes cultivated was dependent on the method of collection (cryotubes were flushed with buffer that was then spread on culture media) in this case and the medium used (e.g. Luria-Bertani agar or Nutrient agar), and effectively allowed for the recovery and cultivation only of specific microbes.

A common method for analysis of recovered samples is the use of light, phase, electron and epifluorescence microscopy. In addition to SYBR gold, another analysis used to determine cell viability of material collected, involves using 0.45 μm micro-pore filters, treated with either a fluorescent cationic carbocyanine or an anionic oxonol dye, both membrane potential-sensitive (or voltage sensitive) probes (Harris et al., 2002). While determining the viability of cells using dyes may sometimes lead to ambiguous results, generally cationic dyes will penetrate the cell membranes of viable, but not of dead cells, while anionic dyes will penetrate the membranes of only non-viable cells (Johnson et al., 2013). Viable cells can usually be visualized using epifluorescence microscopy. Harris and colleagues (2002), discovered clumps of cocci-shaped sub-micron-sized particles on filters recovered from an earlier stratospheric probe. They were identified as prokaryotic microorganisms, using both scanning electron and epifluorescence microscopy. The epifluorescence microscopy used a membrane potential-sensitive dye (carbocyanine), and fluorescence was suggestive of the presence of viable cells (Wainwright et al., 2003; DasSarma and DasSarma, 2018).

Sequencing analyses of the 16S rRNA gene have been largely used in microbial identification. These allow to simultaneously estimate gene abundances and diversity between samples (Smets et al., 2016). In isolates from the stratosphere, *Bacillus luciferensis*, collected at 10 km ASL, was 99% similar to a volcanic soil *B. luciferensis* isolate, suggesting that the source of the stratospheric isolate might have been from volcanic eruption (Griffin, 2004; Griffin et al., 2018). In another large-scale study of aerosols from 20 km ASL, in which samples were collected on an impactor Petri dish, four fungal isolates were observed, all of which belonged to the genus *Penicillium* (Soffen, 1965). In another study, a cryosampler elevated up to 41 km ASL was used to collect air samples, and viable, but initially non-cultivable, microbes were found. In subsequent work, two Gram-positive bacteria, *Bacillus simplex* and *Staphylococcus pasteuri*, and one fungus, *Engyodontium album*, were cultured and identified by: 16S rRNA sequencing – for bacteria, and morphological traits – for fungi (Wainwright et al., 2003). Griffin (2008), used a sterile impacter device with a layer of glycerol to collect air, from 20 km ASL, during a flight of 3.6 hours. Several bacteria were isolated after a long incubation period of 7 weeks. And, interestingly, most of the isolated *Micrococcaceae* and several of the Microbacteriaceae strains were similar to strains previously identified in volcanic soils.

Recently, atmospheric projects have included metagenomic analysis on bioaerosols, but they have been restricted to the troposphere. Metagenomic analysis may not strictly prove the existence of living organisms, since isolated DNA may be associated with dormant or nonviable cell types. However, it is consistent with their presence in the environment and results in a catalog of operational taxonomic units (OTUs) found in a particular sampling of the atmosphere. Moreover, metagenomic analysis can be used on material from nearly all methods of collection and provides considerable depth of data for analysis. For example, filtration may allow for DNA extraction directly from the filter without the need for isolation or culturing. However, those that require growth of microbes, such as the plate-drop method, only provide the opportunity for genomic analysis of cultivatable isolates, but not whole community analysis.

Effect of stratospheric stressors on microbes

Several studies have been performed to determine the effects of stratospheric stressors on isolates from the stratosphere as well as terrestrial ones. *Bacillus* spores could be shielded from damaging UV radiation either by neighboring spores, or by microniches in the platforms they were placed on (mainly metal coupons) or dust particles (Khodadad et al., 2017). Resistance to stratospheric conditions was also studied by exposing two *B. subtilis* strains, one isolated from 20 km ASL over the Pacific Ocean, and the other from a desert basal outcrop in Arizona (USA). For these two strains, no differences in survival were observed after exposure (Smith et al., 2011). Monolayers of *B. pumillus* spores (from a spacecraft clean room) were tested for the effects of UV exposure in the stratosphere. The exposure occurred sequentially from 2 to 8 hours, at ~31 km ASL, with half the spores exposed to sunlight and the other half shielded from it. Spores with increasing sunlight exposure were more affected, presenting an increase of inactivation directly related to the exposure time. After 8 hours of exposure with direct sunlight, viability decreased more than 99.9%, suggesting that UV was the determining factor. Multi-layers of spores resulted in greater survival, due to shielding, and this finding was later confirmed in the laboratory (Khodadad et al., 2017).

Other research has involved exposing desiccated or lyophilized strains, including several potential pathogens to the stratosphere on balloons. Though many survived, changes in protein expression and metabolic pathways were observed (Chudobova et al., 2015). In a study on the impact of sending microbes in liquid media into the stratosphere, two halophilic archaea, the mesophilic *Halobacterium* sp. NRC-1 and the biofilm-forming psychrotolerant Antarctic *Halorubrum lacusprofundi*, both survived launches into the stratosphere on weather balloons. Laboratory experiments showed that the mesophile, was more UV resistant than the psychrophile, while the psychrotolerant strain was more resistant to low temperatures (Anderson et al., 2016; DasSarma et al., 2017). Interestingly, in this study, the psychrotolerant strain, *H. lacusprofundi*, was found to have a 10-fold better survival when compared to the UV tolerant strain NRC-1 overall. Moreover, other investigations have addressed resistance of halophilic archaea to UV-radiation (including the use of both light and dark repair systems) (Crowley et al., 2006; Boubriak et al., 2008), and the high degree of resistance found has been associated to the presence of multiple copies of the genome within each cell, and the high intracellular concentration of halide ions that act as chemical chaperones. The intracellular salts are able to scavenge ROS, protecting the cells against radiation damage (Oren, 2014). Additional studies of genes involved in IR protection, including extensive molecular biological examinations via knockout and overexpression, have shown the importance of ssDNA binding for survival (DeVeaux et al., 2007; Karan et al., 2014).

A recent study of the effect of multiple stratospheric stressors used two balloon flights, up to ~25-30 km, to expose yeast desiccated onto polytetrafluoroethylene strips and *B. subtilis* spores as a control. One yeast strain, *Naganishia* (*Cryptococcus*) *friedmannii* 16LV2, found in volcanic soils, was shown to grow through freeze-thaw conditions down to -6.5°C, and belongs to a clade previously isolated from the troposphere. Another, *Exophiala* sp. 15LV1, was previously shown to survive UV-B and C radiation. These two strains survived stratospheric exposure better than *B. subtilis* spores, although ~90% of the viable cells were inactivated. A third yeast strain, *Holtermanniella watticus* 16LV1, isolated from the high-altitude volcanic area of the Atacama Desert, lost most of its viability due to desiccation. When desiccated, *H. watticus* was further weakened by low pressure and temperature and did not survive the stratospheric UV exposure. Additionally, an environmental simulation chamber was used to evaluate effects of desiccation combined with other stressors. Desiccation plus exposure to stratospheric low pressure and temperature had a greater impact on the yeasts than the spores (Pulschen et al., 2018).

Additional microbial related parameters analyzed in the laboratory
Radiation
UV, desiccation and cold conditions exposure in the stratosphere can result in DNA damage including single- and double-stranded breaks. In order to survive damaging UV exposure, terrestrial microbes have been known to use a variety of DNA repair mechanisms (Cuthbertson and Pearce, 2017). Some of the best-known examples of extreme resistance to UV and IR are species in the genus *Deinococcus* (originally *Micrococcus*). These include *Deinococcus radiodurans* and *Deinococcus geothermalis* – nonsporulating microbes known for their efficiency in repairing damaged DNA. These can typically survive acute exposures to ionizing radiation ≥12,000 Gy (with *D. radiodurans* having survived up to 20,000 Gy; Krisko and Radman, 2013), compared to 8,000 Gy of γ-radiation survived by some fungi, while *E. coli* is killed by only 200-800 Gy (Harris et al., 2009). *Deinococcus* species are also resistant to desiccation by maintaining homeostasis using DNA repair mechanisms, ROS detoxification and accumulation of compatible solutes (Ranawat and Rawat, 2017). Recent stratospheric *Deinococcus* isolates include a radioresistant orange-pigmented, desiccation-tolerant, UV- and γ-radiation resistant bacterium *Deinococcus aerius* TR0125 (from 0.8-5.8 km), and *Deinococcus aetherius* ST0316 (from 10-12 km above Japan). They were found to have similar radiation resistance traits as *D. radiodurans* in laboratory studies (Yang et al., 2008; Yang et al., 2009; Yang et al., 2010; Satoh et al., 2018). *D. aerius* was found to encode DNA photolyase involved in UV resistance, as well as, radiation/desiccation response system genes including *pprI*, *pprA*, *recA*, *ddrA*, and *ddrO* also found in *D. radiodurans*, *D. grandis*, and *D. geothermalis*. (Du and Gebicki, 2004; Daly et al., 2007; Yang et al., 2008; Satoh et al., 2018). Some species form aggregates, e.g. *D. aerius* and *D. aetherius* (Kawaguchi et al.,

2013), and *D. radiodurans* is usually found as tetrads (Eltsov and Dubochet, 2005).

Laboratory experiments on UV tolerance, have been largely based on the premise that solar radiation and high vacuum are the main factors affecting the incidence of microorganisms in the stratosphere and mesosphere. Laboratory research on microorganisms isolated from the upper layers of the atmosphere, showed that conidia of *Aspergillus niger* are highly resistant to UV irradiation. Moreover, the conidia of *Penicillium* spp., *Papulaspora anomala*, and *Circinella muscae*, or vegetative cells of *Micrococcus* spp. and *Mycobacterium* spp. are resistant to high vacuum. Their inactivation varied within the range of 2 to 16%, with an exception of *Micrococcus* sp., which was higher, ~40% (Lysenko, 1980). In another study, stratospheric *Bacillus* isolates (*B. aerius*, *B. aerophilus*, *B. stratosphericus* and *B. altitudinis*) were found to be more UV-B-resistant than terrestrial *B. licheniformis* MTCC 429T and *B. pumilus* MTCC 1640T based on CFU, when 100 µl culture samples were spread onto nutrient agar plates and exposed to a UV-B lamp (15Wx4) with the lids open (Shivaji et al., 2006).

In a different experiment, the impact of UV-C radiation on the non-heterocystous cyanobacterium *Microcystis aeruginosa* was examined (Sahu and Šimek, 2013; Phukan et al., 2018). Effects were observed on photo-absorbing pigments such as chlorophyll a, carotenoids, phycocyanin, allophycocyanin and phycoerythrin, proteins involved in photosynthesis such as D1 protein and RuBisCO, and enzymes involved in nitrogen metabolism (Phukan et al., 2018). Similar deleterious effects were also observed, after exposure to UV-C of the nitrogen-fixing heterocystous cyanobacterium, *Nostoc muscorum* Meg1.

Genomes of UV tolerant non-sporulating microbes isolated from the stratosphere often have been found to contain a high percentage of GC. High GC content has been thought to lead to greater tolerance as a result of avoidance of T-T photoproducts (Kennedy et al., 2001; Griffin, 2004) (Table 1).

Another proposed UV shielding strategy is the formation of aggregates of cells, where the outer cells may protect interior cells against radiation. *M. luteus*, for example, forms cell aggregates and is 100 times more resistant to UV than *E. coli* (Wainwright et al., 2003). *Halorubrum lacusprofundi*, which is able to survive trips into the stratosphere, is known to form biofilms and flocculent material, which might also provide cellular shielding (Reid et al., 2006; DasSarma et al., 2017).

Extensive studies have resulted in an increasing understanding of how radiotolerance (tolerance to UV radiation) is expressed in extremophiles and polyextremophiles (DeVeaux et al., 2007; Karan et al., 2014). Many of these microorganisms have developed strategies to tolerate radiation like the production of extremolytes and extremozymes which have potential uses for biotechnology and therapeutic industry (Gabani and Singh, 2013). Derivative strains of *Halobacterium* sp. NRC-1, have been shown to have a high tolerance to high energy ionizing radiation, with an LD_{50} greater than 11 kGy. These derivatives were found to overexpress a single-stranded-binding protein operon (*rfa*3, *rfa*8, *ral*), suggesting a novel mechanism of DNA protection and repair (DeVeaux et al., 2007; Karan et al., 2014). Additionally, it was shown that conidia of the insect pathogen *Metarhizium robertsii* accumulate trehalose and mannitol under nutritive stress conditions leading to an increased UV-B tolerance (Ranawat and Rawat, 2017).

Pigmentation
From the earliest, pigmentation of isolates was noted and thought to be important for survival in the stratosphere. It has long been suggested that microbes use pigments like melanin and carotenoids for UV protection (Imshenetsky et al., 1978; Tong and Lighthart 1997; Singh and Gabani; 2011; Koller et al., 2014). Recent microscopic examination of cloud material showed that 55% of detected bacterial cells were pigmented, as well as up to 41% fungal cells (Vaïtilingom et al., 2012).

Microbes isolated from 48-77 km ASL included pigmented conidia: black from *Aspergillus niger*, green from *Penicillium notatum*, and grey from *Circinella muscae* (Imshenetsky et al., 1979). These strains along with vegetative cells of *Micrococcus albus*, and unpigmented mutants were subjected to UV treatment. The unpigmented mutants were more UV sensitive, with resistance restored by addition of *Aspergillus niger* black pigments (which had a maximum radiation absorption range of 210-370 nm) (Imshenetsky et al., 1979). Further surveys showed that albino conidia of a *Metarhizium robertsii* mutant was less UV-B tolerant than the wild-type green conidia (Braga et al., 2001a, 2001b; Rangel et al., 2006; Dias et al., 2018). Interestingly, survival of an unpigmented entomopathogenic fungus, *Metarhizium acridum*, was noted to be similar to that of the pigmented halophilic *Cladosporium herbarum* fungus isolated from the stratosphere and Chernobyl nuclear reactor (Zhdanova et al., 2000; Butinar et al., 2005; Rangel et al., 2005; Rangel et al., 2006; Rangel et al., 2010; Braga et al., 2015; Della Corte et al., 2014). Therefore, even though conidium pigmentation seems to be related to radiotolerance, it is not the only factor contributing to resistance.

It is also believed that melanin plays the role of a radioprotector, since melanized fungi are able to survive exposure to high doses of γ-radiation, lethal to most non-melanized fungi (Dadachova et al., 2008). For example, the melanized fungi *Cryptococcus neoformans* and *C. antarcticus* are resistant to highly energetic and damaging particulate radiation, e.g. deuterons (Pacelli et al., 2017). Melanin may also be used as an energy transducer, allowing for utilization of ionizing radiation for metabolic processes, and increasing growth rates, compared to non-melanized fungi when exposed to higher than background radiation (Dadachova et al., 2008; Robertson et al., 2012).

Carotenoids are known to be important for protection from oxidative stress and UV irradiation. Not surprisingly, many isolates from the stratosphere, as well as the ones that survive exposure, contain carotenoid pigments. These include halophilic archaea, which contain novel C50 isoprenoids, such as bacterioruberins (DasSarma et al., 2001). In these extremophilic organisms, a number of genes have been shown to be involved in their synthesis, including a cytochrome P450 (Hescox and Carlberg, 1972, Müller et al., 2018). Studies have also been carried out on the fungus *Aschersonia aleyrodis* to show that carotenoids including β-carotene are involved in oxidative stress and UV irradiation protection (van Eijk et al., 1979; Avalos and Limón, 2015; Dias et al., 2018).

Cold temperature
Microbes have developed multiple strategies to protect themselves from the damaging effects of freezing, such as: biofilm-formation (Reid et al., 2006), accumulation of chaotropic metabolites (e.g. fructose and glycerol), increase of polyol levels (e.g. intracellular trehalose), change of membrane lipids and fluidity, secretion of antifreeze proteins, and use of cold-active enzymes (Robinson, 2001; Feller and Gerday, 2003; Chin et al., 2010; Gerday 2013, Martin and McMinn 2018). Many studies have addressed the molecular adaptations that allow growth and survival at cold temperatures on the Earth's surface, but relatively few have addressed this property in the high atmosphere (Karan et al., 2012; Gerday 2013). One important property, for protein function at low temperatures, is greater flexibility and less negative charge at the surface (Laye et al., 2017). Microbial membranes also need to be more fluid, which can be achieved by changing saturation levels of fatty acid modifications and shortening of fatty acid chain length. Synthesis of antifreeze glycoproteins and peptides leads to freezing point depression of water and may improve survival at extremely cold temperatures (Pikuta et al., 2007).

Growth of microbes in the laboratory at subzero temperatures has been observed. The proteomics of *Colwellia psychrerythraea* 34H (Cp34H) at -1 to -10 °C revealed several strategies, including osmolyte regulation and polymer secretion. These appear to be necessary for metabolic activity subzero, while differentially expressed proteins include those involved in DNA repair chemotaxis and sensing (with a drop-in motility-related proteins) (Nunn et al., 2014). The haloarchaeon *Halobacterium lacusprofundi*, isolated from an Antarctic lake, was found to be capable of growth at sub-zero temperatures in high salt brine and also to be more freeze-thaw resistant than mesophilic haloarchaea (Reid et al., 2006; DasSarma et al., 2017). Additional microbes able to withstand the cold, include *Trichococcus patagoniensis*, which was determined to be able to divide at -5 °C under both aerobic and anaerobic conditions (Pikuta et al., 2007).

Studies suggest that polyhydroxyalkanoates (PHA) storage granules may also be involved in cold stress, osmotic shock, and radiation protection (Pavez et al., 2009; Tribelli and López, 2011; Obruca et al., 2016; Obruca et al., 2017). Strains producing PHA have been found to be more UV radiation resistant than mutants that do not produce PHA (Slaninova et al., 2018). This was also observed in *Azospirillum brasilense* when PHA-rich and PHA poor cells were compared (Tal and Okon, 1985). The granules are believed to scatter UV radiation, shielding bound DNA as well as decreasing intracellular ROS levels. This protection was confirmed using genetically modified *E. coli* (Slaninova et al., 2018).

Osmotic stress
The dry conditions, present in large sections of the atmosphere, reduce water availability and induce osmotic stress in cells exposed to them. Osmotic stress often requires multiple stress responses. For example, fungi have been shown to respond by altering ion transport, homeostasis, sodium extrusion, and melanin synthesis. They also adjust their internal solute potentials by accumulating solutes such as glycerol, erythritol, mannitol, and trehalose, modifying the plasma membrane, decreasing fatty acid saturation in membranes, and increasing cell wall thickness to limit osmotic losses (Hallsworth and Magan, 1994; Serrano et al., 1999; Almagro et al., 2001; Turk et al., 2004; Dijksterhuis and de Vries, 2006; Kogej et al., 2007; Rangel et al., 2008; Kralj Kuncic et al., 2010; Rangel, 2011). Further adaptation strategies to osmotic stress have been studied in halophiles, and both salt-in and salt-out strategies have been identified (for more, see review in DasSarma and DasSarma, 2015).

Metabolic activity
Several observations of microbes from the stratosphere show that they are generally dormant, with suppressed metabolic activity, when isolated. This is consistent with the hypothesis that dormancy conditions prevent excessive DNA damage, which mainly occurs in actively dividing cells. As seen on Table 1, the majority of strains isolated from the stratosphere form spores. For example, in a study of microbes isolated from an altitude of 20 km, several spore-forming pigmented fungi and bacteria were found, including *Penicillium* sp. and several bacilli (Griffin, 2004). Other observations have indicated that cells grow more slowly upon return from the stratosphere. Slowed growth was reported for the orange-pigmented, non-motile, non-spore-forming *Deinococcus aerius* TR0125, isolated from 10 km ASL, which was found to grow at a much slower rate than *D. radiodurans*, *D. grandis*, and *D. geothermalis* (Satoh et al., 2018). This is consistent with other findings of slower growth observed for a stratospheric *Bacillus* sp. isolate (Smith et al., 2010).

Halotolerant fungi, such as melanized *Cladosporium* sp., have been isolated from the stratosphere and have been shown to accumulate mycosporines as a response to stress (Della Corte et al., 2014). *Aspergillus penicilliodes* is able to germinate at very low water activity (0.585 aw, approximately 58.5% of relative humidity), which is now considered the lower limit for life. Studies of this fungus should result in a better understanding of life under severe water limiting conditions, such as found in the stratosphere (Stevenson, 2017; Rangel et al., 2018).

Epidemiology

While most microbiological research has been conducted in the troposphere, it has been established that several pathogens may be viable even after exposure to the stratosphere (Chudobova et al., 2015). Molecular-based studies of airborne microbes in the troposphere have determined that, during African dust events, up to 25% present in the Caribbean air are species of bacteria or fungi that are known to be plant pathogens and about 10% were identified as opportunistic human pathogens (Griffin et al., 2001; Kellogg and Griffin, 2006). Potential pathogens isolated from the troposphere include: *Puccinia melanocephala* and *Hemileia vastatrix*, which cause sugar cane and coffee rust respectively; *Puccinia graminis,* a wheat pathogen; *Mycospherella musicola*, which causes banana leaf spot disease; *Bacillus pumilus*, which causes bacterial blotch in peaches; *Bacillus megaterium*, which causes 'wetwood' disease in trees; *Aspergillus sydowii*, which has been implicated in sea-fan disease; *Karenia brevis*, which is a causative agent for algal blooms; and, the causative agent of meningococcal meningitis, *Neisseria meningitis* (Griffin et al., 2001; Griffin et al., 2002). Since both plant and animal (including human) pathogens have been isolated from the troposphere, the atmosphere may indeed be transporting human pathogenic agents (Griffin et al., 2002). However, how many, if any, are present and transported by the stratosphere is yet undetermined. Further studies are needed to establish the potential epidemiology related to the tropospheric and stratospheric transport of pathogens.

Planetary protection

Studies of the stratosphere are relevant for planetary protection, which is the practice of protecting solar system bodies from contamination by Earth life and protecting Earth from possible life forms that may be returned from other solar system bodies (OSMA, 2019). Understanding microbial survival and adaptation to stratosphere conditions allows for better definition of policies and helps in their development and establishment. For example, the 1958 Committee on Space Research (COSPAR) developed the original guidelines to minimize forward- and backward- contamination and was the basis for the 1967 Outer Space Treaty that provided planetary protection policies. This included quarantining both astronauts and material for the Apollo program (1969-72) (Nicholson et al., 2009). In the 1960s, acceptable unmanned spacecraft microbial bioloads were 10^4-10^8 CFU/vehicle. The Viking 1 and 2 lander missions to Mars, in 1976, were sterilized using heat, reducing spore forming CFU to 2 x 10^4/lander. These had the search for life as part of their mission, while all other NASA missions have primarily been geology focused (Fairén et al., 2018). Later missions allowed for higher counts, < 3×10^5 CFU bacterial spores, in order to reduce sterilization costs (Nicholson et al., 2009).

Roughly 85–95% of microbial isolates from Spacecraft Assembly Facilities (SAFs) and spacecraft are indigenous to humans, and the remaining have been found in soil and dust. Spore-forming *Bacillus* spp. isolates from SAFs make up ~10% of total cultured bioloads (Nicholson et al., 2009). When considering how to effectively sterilize outbound materials, we have to consider the destination. For example, Mars has been, and still is naturally "sterilized" against terrestrial microorganisms, with broad-spectrum radiation, extreme cold and dryness, and surface soil chemistry containing highly reactive oxidizing agents and low pressure (Nicholson et al., 2009; Freissinet et al., 2015; Khodadad et al., 2017; Fairén et al., 2017; Fairén et al., 2018).

Planetary Protection constraints exploration of Special Regions of Mars, including the potentially aqueous, briny recurrent slope lineae (RSL), in order not to contaminate these potentially habitable environments. On the other hand, long-term plans include human missions, resulting in transport of large numbers of microbes (Fairén et al., 2017; Rummel and Conley, 2017; Fairén et al., 2018). If a Martian mission leads to the discovery of microbial life, how would one know if this is Martian or hitchhikers from Earth? They could be exobiota from the current climate or from when Mars was warmer, or simply contamination from a launch site or a spacecraft clean-room or assembly facility like the highly UV resistant *Bacillus nealsonii*, or *Bacillus odysseyi* (Venkateswaran et al., 2003; La Duc et al., 2009). The stratosphere is likely to be extremely valuable as a Mars simulation region due to their physical-chemical similarities and also serving to test the efficacy of our planetary protection measures. Understanding how Earth's microbes survive stratospheric stressors is likely to help us understand how life from our planet could survive Martian conditions.

In addition to the links to the astrobiological exploration of Mars, such insights will also prove helpful in the future study of the icy moons of the outer solar system and extra solar planets. Discussions on the planetary protection of the exooceans of these moons is an on-going dialogue between different space agencies (Rettberg et al., 2019), and preparations for their future exploration are gaining increased traction and visibility (e.g. Antunes et al., 2020; Jebbar et al., 2020; Taubner et al., 2020).

Climate change

Burning of fossil fuels since the industrial revolution has resulted in increasing concentrations of greenhouse gases to levels not seen for more than a million years. For example, the most abundant of these, CO_2, ranged from 180-280 ppm for most of the past hundreds of thousands of years, but has increased to >400 ppm in 2016. Widespread climate change is occurring as a result, including global warming, sea level rise, extreme storms and flooding, and droughts, wildfires, and desertification. Greenhouse gases and particulate matter affect radiative transfer, changing atmospheric temperature, density and albedo patterns and weather. The role of the stratosphere in climate change has resulted primarily from effects on ozone. The level of ozone in the stratosphere affects the temperature and the height of the tropopause, with significant consequences for the weather and temperature at the Earth's surface (Trickl et al., 2019).

Certain greenhouse gases, especially refrigerants like halocarbon gases (hydrofluorocarbons and chlorofluorocarbons), are known to deplete ozone in the stratosphere via photoreaction (World Meteorological Organization, 2014). Ozone depletion and the subsequent cooling of the stratosphere results in a rise in the temperature of the tropopause with potential impacts on climate (Hoskins, 2003). *Cumulonimbus* clouds may enter the stratosphere as a consequence, bringing moisture to this layer, but at the same time drying the lower troposphere, resulting in less frequent, but longer-lasting thunderstorms and an effective cooling (Ursem, 2016). A difference in humidity at the tip of the boundary layer in the lower stratosphere can result in nucleation on nanostructured colloidal aerosols and formation of droplets and a visible haze effect – especially in the Arctic, where this haze has become almost permanent, with particle lifetime in the stratosphere predicted to be 1-2 years (compared to <2 weeks in the lower troposphere) (Ursem, 2016).

Increased stratospheric cloud formation may in turn increase the rates of ozone depletion, resulting in a damaging cycle. Since the ozone layer within the stratosphere acts as a radiation shield, protecting the Earth's surface from UV-C damage, increased solar UV radiation reaching the Earth's surface harms both plants and animals. Damage to plants suppresses the net photosynthetic rate, and lowers transpiration rate of crops, reducing the overall CO_2 sink (Lou et al., 2017; Pérez et al., 2017). As a result, efforts have been made over the past few decades to reduce ozone depleting greenhouse gases such as refrigerants, through the Montreal Protocol. While these have been quite successful, challenges to increased usage of hydrofluorocarbons and other ozone-reactive gases continue to be of concern (Bais et al., 2018).

The atmospheric trace gas N_2O has also been implicated in climate change, by ozone-depletion in the stratosphere, and is mainly produced by microbes (~35% from oceanic bacteria and archaea) (Barnes and Upstill-Goddard, 2018). The 100-year global warming potential of N_2O is calculated to be ~300 times stronger than that of CO_2, and its emission has been increasing at a rate of 0.25% per year, with wastewater treatment plants emitting 3.2% of the total anthropogenic N_2O emissions globally (Yan et al., 2017). Methane is another destructive greenhouse gas involved in climate change, although its effects result from ground level, rather than higher atmosphere effects. Natural methane production rates are in turn influenced by climate: as temperature rises, so does methane production in a process known as positive climate feedback (Dean et al., 2018).

Other greenhouse gases are water vapor and molecular hydrogen (H_2) (Meredith et al., 2017). When water vapor enters the lower stratosphere during severe storms, it leads to hygroscopic growth of sulfate aerosols and subsequent ozone loss by up to 17% (Bais et al., 2018). Carbonyl sulfide is the most abundant sulfur compound in the troposphere and can be transported into the stratosphere where it is converted to sulfate by photolysis or reactions with O or OH radicals, resulting in production of sulfate aerosols, which in turn influence the Earth's radiation balance and causes ozone depletion (Ogawa et al., 2017).

There are many natural sources of particulate matter in the stratosphere which decrease heat absorption and result in global cooling, including volcanic eruptions, conical blue jets (from tops of thunderclouds), smoke from large fires, and photophoretic forces. Negatively charged anthropogenic nanoparticles can be transported to the lower stratosphere and accumulate at 18-20 km. These originate from combustion, industry, aircraft, ground transportation, heating with coal and wood and energy production. Proposals for using dust-like substances (like Loess – defined as an eolian sediment, that has been transported and deposited by the wind, and dominated by silt-sized particles of 20–50μm diameter), to cool the environment and limit climate change, have been suggested in a process known as geoengineering (Martínez-Garcia et al., 2011; Lamy et al., 2014).

Clearly, complex dynamics are on-going in the stratosphere. The degree to which microorganisms are affected or are affecting these dynamics is largely unknown. Studies are needed to more fully determine their relevance and impact.

Conclusions

The stratosphere is an extreme environment subject to many simultaneous stressors and is undoubtedly challenging for life. Radiation appears to be the major factor in loss of viability in the stratosphere, but other stressors such as low temperature also limit survival. In addition to being radiation-tolerant and cold-tolerant, cells must also be able to tolerate hypobaric conditions to survive the extremely low pressures (0.1-10 kPa) present in the stratosphere (Rothchild and Mancinelli, 2001; Fletcher et al., 2014). Stratospheric conditions also lead to rapid desiccation, and only the most xerotolerant microbes are able to cope with such extreme conditions. Due to the sparseness of resources in the stratosphere, it is difficult for any living cells to find the building blocks of life needed to metabolize and reproduce. However, attachment to particulate matter (dust) may provide opportunities to circumvent this limitation.

Whether life can exist for any appreciable length of time in the stratosphere, let alone thrive, is still an open question. What is clear though, is that changes in the stratosphere from human activities are disturbing the troposphere-stratosphere boundary and increasing the exchange of materials between the layers. This reflects the anthropomorphic changes and feedback effects from the troposphere, as well as the surface of the Earth and its oceans, rivers, and lakes. Depletion of the ozone layer represents one of the most destructive potential changes in the stratosphere which may adversely affect animal health, plant life and agriculture.

More detailed investigations involving collection and analysis of stratospheric material will provide better insights into many outstanding questions about the higher atmosphere, including the role of the stratosphere in climate change, long-range dispersal of microorganisms, and implications for planetary protection. In addition, utilizing the stratosphere as a proxy for the surface of Mars will be valuable for astrobiology as we consider the potential for life on the red planet.

Acknowledgements
Work in the laboratory of PDS and SDS is supported by NASA grant NNH18ZDA001N.

References
Aguilera, A., De Diego-Castilla, G., Osuna, S., Bardera, R., Mendi, S. S., Blanco, Y., and González-Toril, E. (2018). Microbial Ecology in the Atmosphere: The Last Extreme Environment. In *Extremophilic Microbes and Metabolites-Diversity, Bioprespecting and Biotechnological Applications*. IntechOpen.

Almagro, A., Prista, C., Benito, B., Loureiro-Dias, M.C., and Ramos, J. (2001). Cloning and expression of two genes coding for sodium pumps in the salt-tolerant yeast *Debaryomyces hansenii*. Journal of bacteriology. *183*(10), 3251e3255.

Amato, P. (2012). Clouds Provide Atmospheric Oases for Microbes. Microbe. 7(3), 119-123.

Amato, P., Brisebois, E., Draghi, M., Duchaine, C., Fröhlich-Nowoisky, J., Huffman, J.A., Mainelis, G., Robine, E., and Thibaudon, M. (2017). Main Biological Aerosols, Specificities, Abundance, and Diversity. In: Microbiology of Aerosols. John Wiley & Sons, Inc. pp. 1-21.

Anderson, I.J., DasSarma, P., Lucas, S., Copeland, A., Lapidus, A., Del Rio, T.G., Tice, H., Dalin, E., Bruce, D.C., Goodwin, L. Pitluck, S. (2016). Complete genome sequence of the Antarctic *Halorubrum lacusprofundi* type strain ACAM 34. Standards in Genomic Sciences. *11*(1), 70.

Antunes, A., Olsson-Francis, K., and McGennity, T. (2020). Exploring deep-sea brines as potential terrestrial analogues of oceans in the icy moons of the outer solar system. Current Issues in Molecular Biology (accepted).

Avalos, J. and Limón, C.M. (2015). Biological roles of fungal carotenoids. Current Genetics. *61*(3), 309-324.

Bais, A.F., Lucas, R.M., Bornman, J.F., Williamson, C.E., Sulzberger, B., Austin, A.T., Wilson, S.R., Andrady, A.L., Bernhard, G., McKenzie, R.L., Aucamp, P.J., Madronich, S., Neale, R.E., Yazar, S., Young, A.R., de Gruijl, F.R., Norval, M., Takizawa, Y., Barnes, P.W., Robson, T.M., Robinson, S.A., Ballaré, C.L., Flint, S.D., Neale, P.J., Hylander, S., Rose, K.C., Wängberg, S.-Å., Häder, D.-P., Worrest, R.C., Zepp, R.G., Paul, N.D., Cory, R.M., Solomon, K.R., Longstreth, J., Pandey, K.K., Redhwi, H.H., Torikai, A. and Heikkilä, A.M. (2018). Environmental effects of ozone depletion, UV radiation and interactions with climate change: UNEP Environmental Effects Assessment Panel, update 2017. Photochemical & Photobiological Sciences. *17*(2), 121-258.

Barberán, A., Ladau, J., Leff, J.W., Pollard, K.S., Menninger, H.L., Dunn, R.R. and Fierer, N. (2015). Continental-scale distributions of dust-associated bacteria and fungi. Proceedings of the National Academy of Sciences of the United States of America. *112*(18), 5756-5761.

Barnes, J. and Upstill-Goddard, R.C. (2018). The denitrification paradox: The role of O_2 in sediment N_2O production. Estuarine, Coastal and Shelf Science. *200*, 270-276.

Berera, A. (2017). Space dust collisions as a planetary escape mechanism. Astrobiology. *17*(12), 1274-1282.

Boubriak, I., Ng, W.L., DasSarma, P., DasSarma, S., Crowley, D.J., McCready, S.J. (2008). Transcriptional responses to biologically relevant doses of UV-B radiation in the model archaeon, *Halobacterium* sp. NRC-1. Saline Systems. *4*:13.

Braga, G.U.L., Flint, S.D., Miller, C.D., Anderson, A.J. and Roberts, D.W. (2001a). Both solar UVA and UVB radiation impair conidial culturability and delay germination in the entomopathogenic fungus *Metarhizium anisopliae*. Photochemistry and Photobiology. *74*(5), 734-739.

Braga, G.U.L., Flint, S.D., Miller, C.D., Anderson, A.J., and Roberts, D.W. (2001b). Variability in response to UV-B among species and strains of *Metarhizium anisopliae* isolates from sites at latitudes from 61N to 54S. Journal of Invertebrate Pathology. *78*(2), 98-108.

Braga, G.U.L., Rangel, D.E.N., Fernandes, E.K.K., Flint, S.D. and Roberts, D.W. (2015). Molecular and physiological effects of environmental UV radiation on fungal conidia. Current genetics. *61*(3), 405-425.

Bryan, N.C., Stewart, M., Granger, D., Guzik, T.G. and Christner, B.C. (2014). A method for sampling microbial aerosols using high altitude balloons. Journal of Microbiological Methods. *107*, 161-168.

Bryan, N.C., Christner, B.C., Guzik, T.G., Granger, D.J. and Stewart, M.F. (2019). Abundance and survival of microbial aerosols in the troposphere and stratosphere. International Society for Microbial Ecology Journal. *13*, 2789-2799.

Butinar, L., Sonjak, S., Zalar, P., Plemenitas, A. and Gunde-Cimerman, N. (2005). Melanized halophilic fungi are eukaryotic members of microbial communities in hypersaline waters of solar salterns. Botanica Marina. 48(1), 73-79.

Chen, T.H.B., Feather, G., Keswani, J., and Edgell III, H.D. (2012). U.S. Patent No. 8,205,511. Washington, DC: U.S. Patent and Trademark Office.

Chin, J.P., Megaw, J., Magill, C.L., Nowotarski, K., Williams, J.P., Bhaganna, P., Linton, M., Patterson, M.F., Underwood, G.J.C., Mswaka, A.Y. and Hallsworth, J.E. (2010). Solutes determine the temperature windows for microbial survival and growth. Proceedings of the National Academy of Sciences of the United States of America. *107*(17), 7835-7840.

Christian, E.R. (2012). Cosmic Rays. https://helios.gsfc.nasa.gov/cosmic.html, Accessed on the 12th Jun 2019.

Chudobova, D., Cihalova, K., Jelinkova, P., Zitka, J., Nejdl, L., Guran, R., Klimanek, M., Adam, V. and Kizek, R. (2015). Effects of stratospheric conditions on the viability, metabolism and proteome of prokaryotic cells. Atmosphere. *6*(9), 1290-1306.

Cooper, C.W., Aithinne, K.A., Floyd, E.L., Stevenson, B.S. and Johnson, D L. (2019). A comparison of air sampling methods for *Clostridium difficile* endospore aerosol. Aerobiologia. *35*(3), 411-420.

Creamean, J.M., Suski, K.J., Rosenfeld, D., Cazorla, A., DeMott, P.J., Sullivan, R.C., White, A.B., Ralph, F.M., Minnis, P., Comstock, J.M., Tomlinson, J.M. and Prather, K.A. (2013). Dust and Biological Aerosols from the Sahara and Asia Influence Precipitation in the Western U.S. Science. *6127*(339), 1572-1578.

Crowley, D.J., Boubriak, I., Berquist, B.R., Clark, M., Richard, E., Sullivan, L., DasSarma, S., McCready, S. (2006). The *uvr*A, *uvr*B and *uvr*C genes are required for repair of ultraviolet light induced DNA photoproducts in *Halobacterium* sp. NRC-1. Saline Systems. *2*:11.

Cunningham, D.D. (1873). Microscopic examinations of air (Calcutta: Superintendent of Government Printing).

Cuthbertson, L. and Pearce, D.A. (2017). Aeromicrobiology. In: Psychrophiles: From Biodiversity to Biotechnology by R. Margesin, ed. (Cham, Switzerland: Springer International Publishing), pp. 41-55.

Dadachova E., Bryan R.A., Howell R.C., Schweitzer A.D., Aisen P., Nosanchuk J.D. and Casadevall A. (2008). The radioprotective properties of fungal melanin are a function of its chemical composition, stable radical presence and spatial arrangement. Pigment Cell & Melanoma Research. *21*(2), 192–199.

Daly, M.J., Ouyang, L., Fuchs, P. and Minton, K.W. (1994). *In vivo* damage and recA-dependent repair of plasmid and chromosomal DNA in the radiation-resistant bacterium *Deinococcus radiodurans*. Journal of Bacteriology. *176*(2), 3508–3517.

Daly, M.J., Gaidamakova, E.K., Matrosova, V.Y., Vasilenko, A., Zhai, M., Leapman, R.D., Lai, B., Ravel, B., Li, S.-M.W., Kemner, K.M. and Fredrickson, J.K. (2007). Protein oxidation implicated as the primary determinant of bacterial radioresistance. PLoS Biol. *5*(4), 769–779.

Darwin, C.R. (1846). An account of the fine dust which often falls on vessels in the Atlantic Ocean. Quarterly Journal of the Geological Society of London. *2*(1-2), 26-30.

DasSarma S, DasSarma P. (2015). Halophiles and their enzymes: negativity put to good use. Current Opinion in Microbiology. *25*:120-126.

DasSarma, S., and DasSarma, P. (2018). Survival of microbes in Earth's stratosphere. Current Opinion in Microbiology. *43*, 24-30.

DasSarma, S., Kennedy, S.P., Berquist, B., Ng, W.V., Baliga, N.S., Spudich, J.L., Krebs, M.P., Eisen, J.A., Johnson, C.H. and Hood, L. (2001). Genomic perspective on the photobiology of *Halobacterium* species NRC-1, a phototrophic, phototactic, and UV-tolerant haloarchaeon. Photosynthesis Research. *70*(1), 3–1.

DasSarma, P., Laye, V.J., Harvey, J., Reid, C., Shultz, J., Yarborough, A., Lamb, A., Koske-Phillips, A., Herbst, A., Molina, F., Grah, O., Philips, T. and DasSarma, S. (2017). Survival of halophilic Archaea in Earth's cold stratosphere. International Journal of Astrobiology. *16*(4), 321-327.

DasSarma, S. and Schwieterman, E. (2018). Early evolution of purple retinal pigments on Earth and implications for exoplanet biosignatures. International Journal of Astrobiology. 1-10.

Dean, J.F., Middelburg, J.J., Röckmann, T., Aerts, R., Blauw, L.G., Egger, M., Jetten, M.S.M., de Jong, A.E.E., Meisel, O.H., Rasigraf, O., Slomp, C.P., Zandt, M.H. and Dolman, A.J. (2018). Methane Feedbacks to the Global Climate System in a Warmer World. Reviews of Geophysics. *56*(1), 207-250.

DeLeon-Rodriguez, N., Lathem, T.L., Rodríguez-R., L., Barazesh, J.M., Anderson, B.E., Beyersdorf, A.J., Ziemba, L.D., Bergin, M., Nenes, A. and Konstantinidis, K.T. (2013). Microbiome of the upper troposphere: Species composition and prevalence, effects of tropical storms, and atmospheric implications. Proceedings of the National Academy of Sciences of the United States of America. *110*(7), 2575-2580.

Della Corte, V., Rietmeijer, F.J.M., Rotundi, A. and Ferrari, M. (2014). Introducing a New Stratospheric Dust-Collecting System with Potential Use for Upper Atmospheric Microbiology Investigations. Astrobiology. *14*(8), 694-705.

Delort, A.M., Vaïtilingom, M., Amato, P., Sancelme, M., Parazols, M., Mailhot, G., Laj, P. and Deguillaume, L. (2010). A short overview of the microbial population in clouds: potential roles in atmospheric chemistry and nucleation processes. Atmospheric Research. 98(2-4), 249-260.

DeVeaux, L.C., Müller, J.A., Smith, J., Petrisko, J., Wells, D.P. and DasSarma, S. (2007). Extremely radiation-resistant mutants of a halophilic archaeon with increased single-stranded DNA-binding protein (RPA) gene expression. Radiation Research. 168(4), 507-514.

Dias, L.P., Araújo, C.A.S., Pupin, B., Ferreira, P.C., Braga, G.U.L. and Rangel, D.E.N. (2018). The Xenon Test Chamber Q-SUN® for testing realistic tolerances of fungi exposed to simulated full spectrum solar radiation. Fungal Biology. 122(6), 592-601.

Dijksterhuis, J. and de Vries, R.P. (2006). Compatible solutes and fungal development. Biochemical Journal. 399(2), e3.

Di Liberto, T. (2018). Dust from the Sahara Desert stretches across the tropical Atlantic Ocean in late June/early July 2018. www.climate.gov/news-features/event-tracker/dust-sahara-desert-stretches-across-tropical-atlantic-ocean-late, Accessed on the 12th Jun 2019.

Du, J. and Gebicki, J.M. (2004). Proteins are major initial cell targets of hydroxyl free radicals. The international journal of biochemistry & cell biology. 36(11), 2334–2343.

Dubos, R. (1986). Louis Pasteur: Free Lance of Science. New York: Da Capo Press.

Dyar, H.G. (1894). XII.— On certain bacteria from the air of New York city. Annals of the New York Academy of Sciences. 8(1), 322-380.

Echigo, A., Hino, M., Fukushima, T., Mizuki, T., Kamekura, M. and Usami, R. (2005). Endospores of halophilic bacteria of the family *Bacillaceae* isolated from non-saline Japanese soil may be transported by Kosa event (Asian dust storm). Saline Systems. 1(1), 8.

Eltsov, M. and Dubochet, J. (2005). Fine structure of the *Deinococcus radiodurans* nucleoid revealed by cryoelectron microscopy of vitreous sections. Journal of Bacteriology, 187(23), 8047-8054.

Fairén A.G., Parro V., Schulze-Makuch D. and Whyte L. (2017). Searching for life on Mars before it is too late. Astrobiology. 17(10), 962–970.

Fairén, A.G., Parro, V., Schulze-Makuch, D. and Whyte, L. (2018). Is Searching for Martian Life a Priority for the Mars Community? Astrobiology. 18(2), 101-107.

Feller, G. and Gerday, C. (2003). Psychrophilic enzymes: hot topics in cold adaptation. Nature reviews microbiology. 1(3), 200–208.

Fletcher, C.V., Staskus, K., Wietgrefe, S.W., Rothenberger, M., Reilly, C., Chipman, J.G., Beilman, G.J., Khoruts, A., Thorkelson, A., Schmidt, T.E., Anderson, J., Perkey, K., Stevenson, M., Perelson, A.S., Douek, D.C., Haase, A.T. and Schacker, T.W. (2014). Persistent HIV-1 replication is associated with lower antiretroviral drug concentrations in lymphatic tissues. Proceedings of the National Academy of Sciences of the United States of America. 111(6), 2307-2312.

Freissinet, C.; Glavin, D.P.; Mahaffy, P.R.; Miller, K.E.; Eigenbrode, J.L.; Summons, R.E.; Brunner, A.E., Buch, A., Szopa, C., Archer Jr, P.D., Franz, H.B., Atreya, S.K., Brinckerhoff, W.B., Cabane, M., Coll, P.,

Conrad, P.G., Des Marais, D.J., Dworkin, J.P., Fairén, A.G., François, P., Grotzinger, J.P., Kashyap, S., ten Kate, I.L., Leshin, L.A., Malespin, C.A., Martin, M.G., Martin-Torres, F.J., McAdam, A.C., Ming, D.W., Navarro-González, R., Pavlov, A.A., Prats, B.D., Squyres, S.W., Steele, A., Stern, J.C., Sumner, D.Y., Sutter, B., Zorzano, M.-P. and the MSL Science Team. (2015). Organic Molecules in the Sheepbed Mudstone, Gale Crater, Mars. Journal of Geophysical Research: Planets. *120*, 495–514.

Fröhlich-Nowoisky J., Pickersgill D.A., Despres V.R. and Pöschl U. (2009). High diversity of fungi in air particulate matter. Proceedings of the National Academy of Sciences of the United States of America. *106*(31), 12814-12819.

Fröhlich-Nowoisky, J., Ruzene Nespoli, C., Pickersgrill, D.A., Galand, P.E., Müller-Germann, I., Nunes, T., Gomes Cardoso, J., Almeida, S.A., Pio, C., Andreae, M.O., Conrad, R., Poschl, U. and Desorés, V.R. (2014). Diversity and seasonal dynamics of airborne archaea. Biogeosciences, European Geosciences Union. *11*(21), 6067-6079.

Gabani, P. and Singh, O.V. (2013). Radiation-resistant extremophiles and their potential in biotechnology and therapeutics. Applied Microbiology and Biotechnology. *97*(3), 993-1004.

Gerday, C. (2013). Psychrophily and catalysis. Biology (Basel). *2*(2), 719-741.

Gettelman, A., Hoor, P., Pan, L.L., Randel, W., Hegglin, M.I. and Birner, T. (2011). The extratropical upper troposphere and lower stratosphere. Reviews of Geophysics. *49*(3).

Griffin, D.W. (2004). Terrestrial microorganisms at an altitude of 20,000 m in Earth's atmosphere. Aerobiologia. *20*(2), 135-140.

Griffin, D.W. (2008). Non-spore forming Eubacteria isolated at an altitude of 20,000 m in Earth's atmosphere: extended incubation periods needed for culture-based assays. Aerobiologia. *24*(1), 19-25.

Griffin, D.W., Garrison, V.H., Herman, J.R. and Shinn, E.A. (2001). African desert dust in the Caribbean atmosphere: Microbiology and public health. Aerobiologia. *17*(3), 203-213.

Griffin, D.W., Gonzalez-Martin, C., Hoose, C. and Smith, D.J. (2018). Global-Scale Atmospheric Dispersion of Microorganisms. In: Microbiology of Aerosols, First Edition. Delort, A.M. and Amato, P. John Wiley & Sons, Inc. Hoboken, NJ, USA. pp. 155-194.

Griffin, D.W., Gonzalez, C., Teigell, N., Petrosky, T., Northup, D.E. and Lyles, M. (2011). Observations on the use of membrane filtration and liquid impingement to collect airborne microorganisms in various atmospheric environments. Aerobiologia. *27*(1), 25–35.

Griffin, D.W., Kellogg, C.A., Garrison, V.H., Lisle, J.T., Borden, T.C. and Shinn, E.A. (2003). Atmospheric microbiology in the northern Caribbean during African dust events. Aerobiologia. *19*(3-4), 143-157.

Griffin, D.W., Kellogg, C.A., Garrison, V.H., and Shinn, E.A. (2002). The Global Transport of Dust: an intercontinental river of dust, microorganisms and toxic chemicals flows through the Earth's atmosphere. American Scientist. *90*(3), 228-235.

Hallsworth, J.E. and Magan, N. (1994). Effects of KCl concentration on accumulation of acyclic sugar alcohols and trehalose in conidia of three entomopathogenic fungi. Letters in Applied Microbiology. 18(1), 8-11.

Harris, D.R., Pollock, S.V., Wood, E.A., Goiffon, R.J., Klingele, A.J., Cabot, E.L., Schackwitz, W., Martin, J., Eggington, J., Durfee, T.J., Middle, C.M., Norton, J.E., Popelars, M.C., Li, H., Klugman, S.A., Hamilton, L.L., Bane, L.B., Pennacchio, L.A., Albert, T.J., Perna, N.T., Cox, M.M. and Battista, J.R. (2009). Directed evolution of ionizing radiation resistance in *Escherichia coli*. Journal of Bacteriology. 191(16), 5240-5252.

Harris, M.J., Wickramasinghe, N.C., Lloyd, D., Narlikar, J.V., Rajaratnam, P., Turner, M.P., Al-Mufti, S., Wallis, M.K., Ramadurai, S. and Hoyle, F. (2002). The detection of living cells in stratospheric samples. In: Instruments, Methods, and Missions for Astrobiology IV (Vol. 4495, pp. 192-198). Paper presented at: International Symposium on Optical Science and Technology (San Diego, CA, United States).

Henderson, T.J. and Salem, H. (2016). The Atmosphere: Its Developmental History and Contributions to Microbial Evolution and Habitat. In Aerobiology: The Toxicology of Airborne Pathogens and Toxins. Salem, H., Katz, S.A. (Eds.). Royal Society of Chemistry. pp. 1-41.

Hescox, M.A. and Carlberg, D.M. (1972). Photoreactivation in *Halobacterium cutirubrum*. Canadian Journal of Microbiology. 18(7), 981–985.

Ho, H.-M., Rao, C.Y., Hsu, H.-H., Chiu, Y.-H., Liu, C.-M. and Chao, H.J. (2005). Characteristics and determinants of ambient fungal spores in Hualien, Taiwan. Atmospheric Environment. 39(32), 5839-5850.

Hoskins, B.J. (2003). Climate change at cruising altitude? Science. 301(5632), 469–470.

Huard, M., Briens, C., Berruti, F. and Gauthier, T. A. (2010). A review of rapid gas-solid separation techniques. International Journal of Chemical Reactor Engineering. 8(1).

Hutchinson, F. (1985). Chemical changes induced in DNA by ionizing radiation. Progress in nucleic acid research and molecular biology. 32, 115-154.

Imlay, J.A. (2006). Iron-sulphur clusters and the problem with oxygen. Molecular Microbiology. 59(4), 1073-1082.

Imshenetsky, A.A., Lysenko, S.V., Kasakov, G.A. and Ramkova, N.V. (1977). Resistance of Stratospheric and Mesospheric Micro-organisms to Extreme Factors. Life sciences and space research. 15, 37-39. Paper presented at: COSPAR Life Sciences and Space Research (Philadelphia, Pennsylvania, USA: Pergamon Press, A. Wheaton & Co., Exeter).

Imshenetsky, A.A., Lysenko, S.V. and Kazakov, G.A. (1978). Upper Boundary of the Biosphere. Applied and environmental microbiology. 35(1), 1-5.

Imshenetsky, A.A., Lysenko, S.V., Kazakov, G.A. and Ramkova, N.V. (1976). On Micro-organisms of the Stratosphere. Life sciences and space research. 14, 359-362.

Imshenetsky, A.A., Lysenko, S.V. and Lach, S.P. (1979). Microorganisms of the Upper Layer of the Atmosphere and the Protective Role of their Cell Pigments. In: Life sciences and space research, XVII: Proceedings of the Open Meeting of the W. pp. 105-110.

Jang, G.I., Hwang, C.Y. and Cho, B.C. (2018). Effects of heavy rainfall on the composition of airborne bacterial communities. Frontiers of Environmental Science & Engineering. *12*(2), 12.

Jebbar, M., Cavalazzi, B., Hickman-Lewis, K., Taubner, R.-S., Rittmann, S.K.-M.R. and Antunes, A. (2020). Microbial Diversity and Biosignatures. Space Science Reviews **(accepted)**.

Johnson, S., Nguyen, V. and Coder D. (2013). Assessment of Cell Viability. Current Protocols in Cytometry. *64*(1), 9.2.1-9.2.26.

Joung, Y.S., Ge, Z. and Buie, C.R. (2017). Bioaerosol generation by raindrops on soil. Nature communications. *8*, 14668.

Karan, R., Capes, M.D. and DasSarma, S. (2012). Function and biotechnology of extremophilic enzymes in low water activity. Aquatic Biosystems. *8*(1), 4.

Karan, R., DasSarma, P., Balcer-Kubiczek, E., Weng, R.R., Liao, C.-C., Goodlett, D.R., Ng, W.V. and DasSarma, S. (2014). Bioengineering radioresistance by overproduction of RPA, a mammalian-type single-stranded DNA-binding protein, in a halophilic archaeon. Applied Microbiology and Biotechnology. *98*(4), 1737-1747.

Kawaguchi, Y., Yang, Y., Kawashiri, N., Shiraishi, K., Takasu, M., Narumi, I., Satoh, K., Hashimoto, H., Nakagawa, K., Tanigawa, Y., and Momoki, Y. H. (2013). The possible interplanetary transfer of microbes: assessing the viability of *Deinococcus* spp. under the ISS environmental conditions for performing exposure experiments of microbes in the Tanpopo mission. Origins of Life and Evolution of Biospheres. *43*(4-5), 411-428.

Kellogg, C.A. and Griffin, D.W. (2006). Aerobiology and the global transport of desert dust. Trends in Ecology and Evolution. *21*(11), 638-644.

Kennedy, S.P., Ng, W.V., Salzberg, S.L., Hood, L. and DasSarma, S. (2001). Understanding the Adaptation of *Halobacterium* Species NRC-1 to Its Extreme Environment through Computational Analysis of Its Genome Sequence. Genome Research. *11*(10), 1641–1650.

Kennedy, G.P. (2018). The two Explorer stratosphere balloon flights. Generated on the 25th Feb. Accessed on the 10th August 2019, http://stratocat.com.ar/artics/explorer-e.htm

Khodadad, C.L., Wong, G.M., James, L.M., Thakrar, P.J., Lane, M.A., Catechis, J.A. and Smith, D.J. (2017). Stratosphere conditions inactivate bacterial endospores from a Mars spacecraft assembly facility. Astrobiology. *17*(4), 337-350.

Klein, A. M., Bohannan, B. J., Jaffe, D. A., Levin, D. A., & Green, J. L. (2016). Molecular evidence for metabolically active bacteria in the atmosphere. Frontiers in Microbiology. *7*, 772.

Kogej, T., Stein, M., Volkmann, M., Gorbushina, A.A., Galinski, E.A. and Gunde-Cimerman, N. (2007). Osmotic adaptation of the halophilic fungus

Hortaea werneckii: role of osmolytes and melanization. Microbiology. *153*(12), 4261-4273.

Koller, M., Muhr, A. and Braunegg, G. (2014). Microalgae as versatile cellular factories for valued products. Algal Research. *6*, 52–63.

Kralj Kuncic, M., Kogej, T., Drobne, D. and Gunde-Cimerman, N. (2010). Morphological response of the halophilic fungal genus *Wallemia* to high salinity. Applied and Environmental Microbiology. *76*(1), 329-337.

Krisko, A., and Radman, M. (2013). Biology of extreme radiation resistance: the way of *Deinococcus radiodurans*. Cold Spring Harbor perspectives in biology. *5*(7), a012765.

Kulmala, M., Vehkamäki, H., Petäjä, T., Dal Maso, M., Lauri, A., Kerminen, V.M., Birmili, W. and McMurry, P.H. (2004). Formation and growth rates of ultrafine atmospheric particles: a review of observations. Journal of Aerosol Science. *35*(2), 143-176.

La Duc, M.T., Osman, S., Vaishampayan, P., Piceno, Y., Andersen, G., Spry, J.A. and Venkateswaran, K. (2009). Comprehensive Census of Bacteria in Clean Rooms by Using DNA Microarray and Cloning Methods. Applied and Environmental Microbiology. *75*(20), 6559–6567.

Lamy, F., Gersonde, R., Winckler, G., Esper, O., Jaeschke, A., Kuhn, G., Ullermann, J., Martinez-Garcia, A.5, Lambert, F. and Kilian, R. (2014). Increased Dust Deposition in the Pacific Southern Ocean During Glacial Periods. Science. *343*(6169), 403-407.

Laye, V.J., Karan, R., Kim, J.-M., Pecher, W.T., DasSarma, P. and DasSarma, S. (2017). Key amino acid residues conferring enhanced enzyme activity at cold temperatures in an Antarctic polyextremophilic β-galactosidase. Proceedings of the National Academy of Sciences of the United States of America. *114*(47), 12530-12535.

Leary, K. (2018). In 2017, the US Led the World in Successful Orbital Launches. https://futurism.com/2017-us-led-world-successful-orbital-launches

Lederberg, J. (1960). Exobiology: Approaches to Life beyond the Earth. Science. *132*(3424), 393-400.

Lou, Y., Gu, X. and Zhou, W. (2017). Effect of Elevated UV-B Radiation on Microbial Biomass and Soil Respiration in Different Barley Cultivars Under Field Conditions. Water, Air, and Soil Pollution. *228*(3), 96.

Lutgens, F.K. and Tarbuck, E.J. (1995). The Atmosphere: an introduction to meteorology. Prentice Hall, 6th edition. pp. 14-17.

Lysenko, S.V. (1980). Resistance of microorganisms of upper layers of the atmosphere to ultraviolet radiation and a high vacuum. Mikrobiologiia. *49*(1), 175-177.

Maki, T., Ishikawa, A., Kobayashi, F., Kakikawa, M., Aoki, K., Mastunaga, T., Hasegawa, H. and Iwasaka, Y. (2011). Effects of Asian dust (KOSA) deposition event on bacterial and microalgal communities in the Pacific Ocean. Asian Journal of Atmospheric Environment. *5*(3), 157-163.

Martin, A. and McMinn, A. (2018). Sea ice, extremophiles and life on extra-terrestrial ocean worlds. International Journal of Astrobiology. *17*(1), 1-6.

Martínez-Garcia, A., Rosell-Melé, A., Jaccard, S.L., Geibert, W., Sigman, D.M. and Haug, G.H. (2011). Southern Ocean dust–climate coupling over the past four million years. Nature. *476*(7360), 312–315.

Meier, F.C. (1936). *Paper, US Army Air Corps Stratosphere Flight of 1935 in Balloon Explorer II*. National Geographic Society; Washington DC, pp. 152-153.

Meier, F.C. and Lindbergh, C.A. (1935). Collecting Micro-Organisms from the Arctic Atmosphere: With Field Notes and Material. Scientific Monthly. *40*, 5-20.

Meredith, L.K., Commane, R., Keenan, T.F., Klosterman, S.T., Munger, J.W., Templer, P.H., Tang, J., Wofsy, S.C. and Prinn, R.G. (2017). Ecosystem fluxes of hydrogen in a mid-latitude forest driven by soil microorganisms and plants. Global Change Biology. *23*(2), 906–919.

Moeller, R. and Horneck, G. (2004). *Bacillus* Endospores-an ideal exobiological Tool. In 35th COSPAR Scientific Assembly *35*. pp. 2596.

Mohanakumar, K. (2008). Structure and composition of the lower and middle atmosphere. Stratosphere Troposphere Interactions: An Introduction. pp. 1-53.

Müller, W.J., Smit, M.S., van Heerden, E. Capes, M.D. and DasSarma, S. (2018). Complex Effects of Cytochrome P450 Monooxygenase on Purple Membrane and Bacterioruberin Production in an Extremely Halophilic Archaeon: Genetic, Phenotypic, and Transcriptomic Analyses. Frontiers in Microbiology. *9*, 2563.

Nicholson, W.L., Schuerger, A.C. and Race, M.S. (2009). Migrating microbes and planetary protection. Trends in Microbiology. *17*(9), 389-392.

Nunn, B.L., Slattery, K.V., Cameron, K.A., Timmins–Schiffman, E. and Junge, J. (2014). Proteomics of Colwellia psychrerythraea at subzero temperatures – a life with limited movement, flexible membranes and vital DNA repair. Environmental Microbiology. *17*(7), 2319-2335.

Obruca, S., Sedlacek, P., Krzyzanek, V., Mravec, F., Hrubanova, K., Samek, O., Kucera, D., Benesova, P. and Marova, I. (2016). Accumulation of Poly(3-hydroxybutyrate) Helps Bacterial Cells to Survive Freezing. PLoS One. *11*(6), e0157778.

Obruca, S., Sedlacek, P., Mravec, F., Krzyzanek, V., Nebesarova, J., Samek, O., Kucera, D., Benesova, P., Hrubanova, K., Milerova, M. and Marova, I. (2017). The presence of PHB granules in cytoplasm protects non-halophilic bacterial cells against the harmful impact of hypertonic environments. New Biotechnology. *39*, 68–80.

Ogawa, T., Hattori, S., Kamezaki, K., Kato, H., Yoshida, N. and Katayama, Y. (2017). Isotopic Fractionation of Sulfur in Carbonyl Sulfide by Carbonyl Sulfide Hydrolase of *Thiobacillus thioparus* THI115. Microbes and Environments. *ME17130*.

Oren, A. (2014). Halophilic archaea on Earth and in space: growth and survival under extreme conditions. Philosophical Transactions of the Royal Society A: Mathematical, Physical and Engineering Sciences. *372*(2030), 20140194.

Orr Jr, C. and Keng, E.Y. (1964). Photophoretic effects in the stratosphere. Journal of the Atmospheric Sciences. *21*(5), 475-478.

Office of Safety and Mission Assurance - OSMA. (2019). Planetary Protection. NASA. https://sma.nasa.gov/sma-disciplines/planetary-protection (last accessed 10th Nov 2019).

Pacelli C., Selbmann L., Moeller R., Zucconi L., Fujimori A. and Onofri, S. (2017). Cryptoendolithic Antarctic Black Fungus *Cryomyces antarcticus* Irradiated with Accelerated Helium Ions: Survival and Metabolic Activity, DNA and Ultrastructural Damage. Frontiers in Microbiology. *8*, 2002.

Pasko, V.P., Stanley, M.A., Mathews, J.D., Inan, U.S. and Wood, T.G. (2002). Electrical discharge from a thundercloud top to the lower ionosphere. Nature. *416*(6877), 152-14.

Pasteur, M.L. (1860). Nouvelles experiences relatives aux generations dites spontanees (Paris: Mallet-Bachelier, Imprimeur-Libraire).

Pavez, P, Castillo, J.L., González, C. and Martínez, M. (2009). Poly-β-hydroxyalkanoate exert a protective effect against carbon starvation and frozen conditions in *Sphingopyxis chilensis*. Current Microbiology. *59*(6), 636-640.

Pérez, V., Hengst, M., Kurte, L., Dorador, C., Jeffrey, W.H., Wattiez, R., Molina, V. and Matallana-Surget, S. (2017). Bacterial survival under extreme UV radiation: a comparative proteomics study of *Rhodobacter* sp., isolated from high altitude wetlands in Chile. Frontiers in Microbiology. *8*, 1173.

Pérez-Díaz, J.L., Ivanov, O., Peshev, Z., Álvarez-Valenzuela, M.A., Valiente-Blanco, I., Evgenieva, T., Dreischuh, T., Gueorguiev, O., Todorov, P.V. and Vaseashta, A. (2017). Fogs: Physical Basis, Characteristic Properties, and Impacts on the Environment and Human Health. Water. *9*(10), 807.

Phukan, T., Rai, A.N. and Syiem, M.B. (2018). Dose dependent variance in UV-C radiation induced effects on carbon and nitrogen metabolism in the cyanobacterium *Nostoc muscorum* Meg1. Ecotoxicology and Environmental Safety. *155*, 171-179.

Pikuta, E.V., Hoover, R.B. and Tang, J. (2007). Microbial Extremophiles at the Limits of Life. Critical Reviews in Microbiology. *33*, 183-209.

Proctor, B. E. (1934). The microbiology of the upper air. I. In *Proceedings of the American Academy of Arts and Sciences* (Vol. 69, No. 8, pp. 315-340). American Academy of Arts & Sciences.

Pulschen, A.A., de Araujo, G.G., de Carvalho, A.C.S.R., Cerini, M.F., de Mendonça Fonseca,L., Galante, D. and Rodrigues, F. (2018). Survival of extremophilic yeasts to the stratospheric environment on balloon flights and laboratory simulations. Applied and Environmental Microbiology. *84*(23), e01942-18.

Radosevich, J.L., Wilson, W.J., Shinn, J.H., DeSantis, T.Z. and Andersen, G.L. (2002). Development of a high-volume aerosol collection system for the identification of air-borne micro-organisms. Letters in Applied Microbiology. *34*, 162–167.

Ranawat, P. and Rawat, S. (2017). Radiation resistance in thermophiles: mechanisms and applications. World Journal of Microbiology & Biotechnology. *33*, 111-113.

Rangel, D.E.N. (2011). Stress induced cross-protection against environmental challenges on prokaryotic and eukaryotic microbes. World Journal of Microbiology & Biotechnology. *27*(6), 1281-1296.

Rangel, D.E.N., Alston, D.G. and Roberts, D.W. (2008). Effects of physical and nutritional stress conditions during mycelial growth on conidial germination speed, adhesion to host cuticle, and virulence of *Metarhizium anisopliae*, an entomopathogenic fungus. Mycological Research. *112* (11), 1355-1361.

Rangel, D.E.N., Braga, G.U.L., Anderson, A.J. and Roberts, D.W. (2005). Influence of growth environment on tolerance to UV-B radiation, germination speed, and morphology of *Metarhizium anisopliae* var. acridum conidia. Journal of Invertebrate Pathology. *90*(1), 55-58.

Rangel, D.E.N., Butler, M.J., Torabinejad, J., Anderson, A.J., Braga, G.U.L., Day, A.W. and Roberts, D.W. (2006). Mutants and isolates of *Metarhizium anisopliae* are diverse in their relationships between conidial pigmentation and stress tolerance. Journal of Invertebrate Pathology. *93*(3), 170-182.

Rangel, D.E.N., Fernandes, E.K.K., Dettenmaier, S.J. and Roberts, D.W. (2010). Thermotolerance of germlings and mycelium of the insect-pathogenic fungus *Metarhizium* spp. and mycelial recovery after heat stress. Journal of Basic Microbiology. *50*(4), 344-350.

Rangel, D.E.N., Finlay, R.D., Hallsworth, J.E., Dadachova, E. and Gadd, G.M. (2018). Fungal strategies for dealing with environment- and agriculture-induced stresses. Fungal Biology. *122*(6), 602-612.

Rastelli, E., Corinaldesi, C., Dell'Anno, A., Lo Martire, M., Greco, S., Facchini, M.C., Rinaldi, M., O'Dowd, C., Ceburnis, D. and Danovaro, R. (2017). Transfer of labile organic matter and microbes from the ocean surface to the marine aerosol: an experimental approach. Sci Rep. *7*(1), 11475.

Reche, I., D'Orta, G., Mladenov, N., Winget, D.M. and Suttle, C.A. (2018). Deposition rates of viruses and bacteria above the atmospheric boundary layer. The ISME Journal. *12*, 1154–1162.

Reid, I., Sparks, W., Lubow, S., McGrath, M., Livio, M., Valenti, J., Sowers, K., Shukla, H., MacAuley, S. and Miller, T. (2006). Terrestrial models for extraterrestrial life: methanogens and halophiles at Martian temperatures. International Journal of Astrobiology. *5*, 89-97.

Rettberg, P., Antunes, A., Brucato, J., Cabezas, P., Collins, G., Haddaji, A., Kminek, G., Leuko, S., McKenna-Lawlor, S., Moissl-Eichinger, C. and Fellous, J.L. (2019). Biological Contamination Prevention for Outer Solar System Moons of Astrobiological Interest: What Do We Need to Know? Astrobiology. *19*, 951-974.

Robertson, K.L., Mostaghim, A., Cuomo, C.A., Soto, C.M., Lebedev, N., Bailey, R.F. and Wang, Z. (2012). Adaptation of the Black Yeast

Wangiella dermatitidis to Ionizing Radiation: Molecular and Cellular Mechanisms. PLoS ONE. *7*(11), e48674.

Robinson, C.H. (2001). Cold adaptation in Arctic and Antarctic fungi. New Phytologist. *151*, 341–353.

Rogers, L.A. and Meier, F.C. (1936). National Geographic Society, Technical Papers. 146-151.

Rohatschek, H. (1984). The Role of Gravitophotophoresis for Stratospheric and Mesospheric Particulates. Journal of Atmospheric Chemistry. *1*, 377-389.

Rohatschek, H. (1996). Levitation of Stratospheric and Mesospheric Aerosols by Gravito-Photophoresis. Journal of Aerosol Science. *27*(3), 467-475.

Rothschild, L. and Mancinelli, R.L. (2001). Life in extreme environments. Nature (London). *409*, 1092-1101.

Rummel, J.D. and Conley, C.A. (2017). Four Fallacies and an Oversight: Searching for Martian Life. Astrobiology. *17*, 10.

Sahu, J.K. and Šimek, M. (2013). Effect of UV-C on thylakoid arrangement, pigment content and nitrogenase activity in the cyanobacterium *Microchaete* sp. Indian Journal of Experimental Biology. *51*, 388-392.

Satoh, K., Arai, H., Sanzen, T., Kawaguchi, Y., Hayashi, H., Yokobori, S.-i., Yamagishi, A., Oono, Y. and Narumic, I. (2018). Draft Genome Sequence of the Radioresistant Bacterium *Deinococcus aerius* TR0125, Isolated from the High Atmosphere above Japan. Genome announcements. *6*.

Saunders, D.H., Arato, E.G., and Davies, O.M. (2003). U.S. Patent No. 6,531,066. Washington, DC: U.S. Patent and Trademark Office.

Shivaji, S., Chaturvedi, P., Suresh, K., Reddy, G.S.N., Dutt, C.B.S., Wainwright, M., Narlikar, J.V. and Bhargava, P.M. (2006). *Bacillus aerius* sp. nov., *Bacillus aerophilus* sp. nov., *Bacillus stratosphericus* sp. nov. and *Bacillus altitudinis* sp. nov., isolated from cryogenic tubes used for collecting air samples from high altitudes. International Journal of Systematic and Evolutionary Microbiology. *56*, 1465-1473.

Serrano, R., Mulet, J.M., Rios, G., Marquez, J.A., de Larrinoa, I.F., Leube, M.P., Mendizabal, I., Pascual-Ahuir, A., Proft, M., Ros, R. and Montesinos, C. (1999). A glimpse of the mechanisms of ion homeostasis during salt stress. Journal of Experimental Botany. *50*, 1023-1036.

Singh, O.V. and Gabani, P. (2011). Extremophiles: radiation resistance microbial reserves and therapeutic implications. Journal of Applied Microbiology. *110*(4), 851–861.

Slaninova, E., Sedlacek, P., Mravec, F., Mullerova, L., Samek, O., Koller, M., Hesko, O., Kucera, D., Marova, I. and Obruca, S. (2018). Light scattering on PHA granules protects bacterial cells against the harmful effects of UV radiation. Applied Microbiology and Biotechnology. *102*(4), 1923–1931.

Smets, W., Leff, J.W., Bradford, M.A., McCulley, R.L., Lebeer, S. and Fierer, N. (2016). A method for simultaneous measurement of soil bacterial abundances and community composition via 16S rRNA gene sequencing. Soil Biology & Biochemistry. *96*, 145-151.

Smith, D.J., Griffin, D.W. and Schuerger, A.C. (2010). Stratospheric microbiology at 20 km over the Pacific Ocean. Aerobiologia. *26*, 35–46.

Smith, D.J., Griffin, D.W., McPeters, R.D., Ward, P.D. and Schuerger, A.C. (2011). Microbial survival in the stratosphere and implications for global dispersal. Aerobiologia. *27*(4), 319-332.

Smith, D.J., Jaffe, D.A., Birmele, M.N., Griffin, D.W., Schuerger, A.C., Hee, J. and Roberts, M.S. (2012). Free Tropospheric Transport of Microorganisms from Asia to North America. Environmental Microbiology. *64*, 973-985.

Smith, D.J. and Sowa, M.B. (2017). Ballooning for Biologists: Mission Essentials for Flying Life Science Experiments to Near Space on NASA Large Scientific Balloons. Gravitational and Space Research. *5*, 52-73.

Smith, D.J., Timonen, H.J., Jaffe, D.A., Griffin, D.W., Birmele, M.N., Perry, K.D., Ward, P.D. and Roberts, M.S. (2013). Intercontinental Dispersal of Bacteria and Archaea by Transpacific Winds. Applied and Environmental Microbiology. *79*, 1134-1139.

Spring, A.M., Docherty, K.M., Domingue, K.D., Kerber, T.V., Mooney, M.M., AND Lemmer, K.M. (2018). A Method for Collecting Atmospheric Microbial Samples From Set Altitudes for Use With Next-Generation Sequencing Techniques to Characterize Communities. Air, Soil and Water Research. *11*, 1178622118788871.

Soffen, G.A. (1965). Atmospheric Collection at 130,000 Feet. Proceedings of the Atmospheric Biology Conference, 1965, Minneapolis, Minnesota, USA 213-219.

Solomon, P.A., Moyers, J.L. and Fletcher, R.A. (1983). High volume dichotomous virtual impactor for the fractionation and collection of particles according to aerodynamic size. Aerosol Science and Technology. *2*, 455-464.

Stadtman, E.R. and Levine, R.L. (2003). Free radical-mediated oxidation of free amino acids and amino acid residues in proteins. Amino Acids. *25*, 207-218.

Stevenson, D.S. (2017). Ultimately, Can Life Survive? In: The Nature of Life and Its Potential to Survive, Astronomer's Universe. Springer International Publishing. 341-384.

Tal, S. and Okon, Y. (1985). Production of the reserve material poly-β-hydroxybutyrate and its function in *Azospirillum brasilense* Cd. Canadian journal of microbiology. *31*(7), 608–613.

Taubner, R.-S., Olsson-Francis, K., Vance, S., Ramkissoon, N.R., Postberg, F., de Vera, J.-P., Antunes, A., Casas, E.C., Sekine, Y., Noack, L., Barge, L., Goodman, J., Jebbar, M., Journaux, B., Karatekin, Ö., Klenner, F., Rabbow, E., Rettberg, P., Rückriemen-Bez, T., Saur, J., Shibuya, T., and Sonderlund, K. (2020). Experimental and Simulation Efforts in the Astrobiological Exploration of Exooceans. Space Science Reviews **(accepted)**.

Tong, Y. and Lighthart, B. (1997). Solar Radiation Is Shown to Select for Pigmented Bacteria in the Ambient Outdoor Atmosphere. Photochemistry and Photobiology. *65*(1), 103-106.

Tribelli, P.M. and López, N.I. (2011). Poly(3-hydroxybutyrate) influences biofilm formation and motility in the novel Antarctic species *Pseudomonas extremaustralis* under cold conditions. Extremophiles. *15*(5), 541–547.

Trickl, T., Vogelmann, H., Ries, L. and Sprenger, M. (2019). Very high stratospheric influence observed in the free troposphere over the Northern Alps – just a local phenomenon? Atmospheric Chemistry and Physics Discussions. https://doi.org/10.5194/acp-2019-588, in review.

Turk, M., Mejanelle, L., Sentjurc, M., Grimalt, J.O., Gunde-Cimerman, N. and Plemenitas, A. (2004). Salt-induced changes in lipid composition and membrane fluidity of halophilic yeast-like melanized fungi. Extremophiles. *8*(1), 53-61.

UNSCEAR "Sources and Effects of Ionizing Radiation" page 339 retrieved 2011-6-29 www.unscear.org/docs/reports/2008/09-86753_Report_2008_Annex_B.pdf

Ursem, B. (2016). Climate Shifts and the Role of Nano Structured Particles in the Atmosphere. Atmospheric and Climate Sciences. *6*(1), 51-76.

Vaïtilingom, M., Attard, E., Gaiani, N., Sancelme, M., Deguillaume, L., Flossmann, A.I., Amato, P. and Delort, A.-M. (2012). Long-term features of cloud microbiology at the puy de Dôme (France). Atmospheric Environment. *56*, 88-100.

Vaïtilingom, M., Deguillaume, L., Vinatier, V., Sancelme, M., Amato, P., Chaumerliac, N. and Delorta, A.-M. (2013). Potential impact of microbial activity on the oxidant capacity and organic carbon budget in clouds. Proceedings Academy of Science USA. *110*(2), 559–564.

van Eijk, G.W., Mummery, R.S., Roeymans, H.J. and Valadon, L.R. (1979). A comparative study of carotenoids of *Aschersonia aleyroides* and *Aspergillus giganteus*. Antonie Van Leeuwenhoek. *45*(3), 417-422.

von Sonntag, C. (1987). The chemical basis of radiation biology. London: Taylor & Francis.

Vargin, P.N., Volodin, E.M., Karpechko, A.Y. and Pogoreltsev, A.I. (2015). Stratosphere–Troposphere Interactions. Herald of the Russian Academy of Sciences/Vestnik Rossiiskoi Akademii Nauk. *85*, 56–63.

Venkateswaran, K., Kempf, M., Chen, F., Satomi, M., Nicholson, W. and Kern, R. (2003). *Bacillus nealsonii* sp. nov., isolated from a spacecraft-assembly facility, whose spores are gamma-radiation resistant. International Journal of Systematic and Evolutionary Microbiology. *53*(1), 165-172.

Wainwright, M., Wickramasinghe, N.C., Narlikar, J.V. and Rajaratnam, P. (2003). Microorganisms cultured from stratospheric air samples obtained at 41 km. FEMS Microbiology Letters. *218*(1), 161-165.

Wells, W.F. and Fair, G.M. (1935). Viability of *B. coli* exposed to ultra-violet radiation in air. Science. *82*(2125), 280-281.

Whisler, B. (1940). The efficacy of ultra-violet light sources in killing bacteria suspended in air. Iowa State College Journal of Science. *14*, 215-231.

World Meteorological Organization, Scientific Assessment of Ozone Depletion. (2014). World Meteorological Organization, Global Ozone

Research and Monitoring Project-Report No. 55, Geneva, Switzerland, 2014. pp. 416

Wilson, L., Sparks, R.S.J., Huang, T.C. and Watkins, N.D. (1978). The control of volcanic column heights by eruption energetics and dynamics. Journal of Geophysical Research: Solid Earth. *83*(B4), 1829-1836.

Wu, P.-C., Tsai, J.-C., Li, F.-C., Lung, S.-C. and Su, H.-J. (2004). Increased levels of ambient fungal spores in Taiwan are associated with dust events from China. Atmospheric Environment. *38*(29), 4879-4886.

Xu, C., Wei, M., Chen, J., Sui, X., Zhua, C., Li, J., Zheng, L., Sui, G., Li, W., Wang, W., Zhang, Q. and Mellouki, A. (2017). Investigation of diverse bacteria in cloud water at Mt. Tai, China. Science of the Total Environment. *580*, 258-265.

Yan, X., Zheng, J., Han, Y., Liu, J. and Sun, J. (2017). Effect of influent C/N ratio on N_2O emissions from anaerobic/anoxic/oxic biological nitrogen removal processes. Environmental Science and Pollution Research. *24*(30), 23714-23724.

Yang, Y., Itahashi, S., Yokobori, S. and Yamagishi, A. (2008). UV-resistant bacteria isolated from upper troposphere and lower stratosphere. Biological Sciences in Space. *22*(1), 18-25.

Yang, Y., Itoh, T., Yokobori, S., Shimada, H., Itahashi, S., Satoh, K., Ohba, H., Narumi, I. and Yamagishi, A. (2010). *Deinococcus aetherius* sp. nov., isolated from the stratosphere. International Journal of Systematic and Evolutionary Microbiology. *60*(4), 776-779.

Yang, Y., Itoh, T., Yokobori, S., Itahashi, S., Shimada, H., Satoh, K., Ohba, H., Narumi, I. and Yamagishi, A. (2009). *Deinococcus aerius* sp. nov., isolated from the high atmosphere. International Journal of Systematic and Evolutionary Microbiology. *59*(8), 1862–1866.

Zhdanova, N.N., Zakharchenko, V.A., Vember, V.V. and Nakonechnaya, L.T. (2000). Fungi from Chernobyl: mycobiota of the inner regions of the containment structures of the damaged nuclear reactor. Mycological Research. *104*(12), 1421–1426.

Lightning Source UK Ltd.
Milton Keynes UK
UKHW051114060223
416527UK00003B/22